Disclaimer and Copyright

Text Copyright © Siim Land 2018

All rights reserved. No part of this guide may be reproduced in any form without permission in writing from the publisher except in the case of brief quotations embodied in critical articles or reviews.

Legal & Disclaimer

The information contained in this book is not designed to replace or take the place of any form of medicine or professional medical advice. The information in this book has been provided for educational and entertainment purposes only.

The information contained in this book has been compiled from sources deemed reliable, and it is accurate to the best of the Author's knowledge; however, the Author cannot guarantee its accuracy and validity and cannot be held liable for any errors or omissions. Changes are periodically made to this book. You must consult your doctor or get professional medical advice before using any of the suggested remedies, techniques, or information in this book.

Upon using the information contained in this book, you agree to hold harmless the Author from and against any damages, costs, and expenses, including any legal fees potentially resulting from the application of any of the information provided by this guide. This disclaimer applies to any damages or injury caused by the use and application, whether directly or indirectly, of any advice or information presented, whether for breach of contract, tort, negligence, personal injury, criminal intent, or under any other cause of action.

You agree to accept all risks of using the information presented inside this book. You need to consult a professional medical practitioner in order to ensure you are both able and healthy enough to participate in this program.

All rights reserved. No part of this publication may be reproduced, distributed, or transmitted in any form or by any means, including photocopying, recording, or other electronic or mechanical methods, without the prior written permission of the publisher, except in the case of brief quotations embodied in critical reviews and certain other noncommercial uses permitted by copyright law. For permission requests, contact the publisher, at the address below.

http://www.siimland.com.

Cover design by Siim Land.

Table of Contents

Contents

Disclaimer and Copyright 1

Table of Contents 3

Introduction In Search of the Elixir of Life 7

Chapter I The Code of Longevity 13

    The Hallmarks of Aging 15

    Mitochondrial Theory of Aging 16

    The Insulin IGF-1 Pathway 21

    Insulin/IGF-1 Signaling in *Caenorhabditis Elegans* 23

    Caloric Restriction and Longevity 27

    Stress Adaptation and Longevity 35

    Telomeres and Longevity 38

    Hormesis and Longevity 41

    The Price of Longevity 44

Chapter II The Hedonic Treadmill 45

    Hedonic Adaptation 47

    The Physiology of Fasting 56

Chapter III Why Intermittent Fasting 64

    Fasting and the Mitochondria 68

    Fasting and the Brain 71

    Fasting and the Immune System 75

    Fasting and Gut Health 79

    Fasting and Fat Loss 83

Chapter IV What We Know About Autophagy So Far 88

    The Negative Side Effects of Autophagy 96

    Balancing Autophagy and mTOR 102

Chapter V Squaring the Curve 106

    Muscle and Longevity 107

    What Kind of Muscle Benefits Longevity 119

    Endurance and Cardio 122

Chapter VI HyperTORphyc Growth 128
Enter TOR 129
Benefits of mTOR 136
Is IGF-1 Good or Bad 141
Chapter VII Starting With Strength 147
Training Principles 150
How Should You Train 154
Chapter VIII Anabolic Autophagy 179
Can You Build Muscle and Lose Fat at the Same Time 180
Losing Muscle While Fasting 188
Can Fasting Make You Build Muscle 192
Chapter IX Protein Absorption and Anabolism 195
How Much Protein Does Your Body Need 197
Should You Consume Protein Before Working Out 204
Post-Workout Nutrition While Fasting 208
Chapter X Food Fallacy 213
Enter the Lipid Hypothesis 213
The Standard American Diet 222
Saturated Fat 230
Meat Kills? 234
Chapter XI The Case Against Sugar (and Fat) 240
Insulin Resistance 247
The Case Against Fat 255
Chapter XII WTF Should I Eat 261
What Humans Evolved to Eat 262
The „Natural Diet" Fallacy 265
Essential Nutrients 269
Nutrient Density 273
Eat Less Move More? 279
Chapter XIII The Keto-Adaptation Process 285
Why Bother With Keto? 293
How to Increase Metabolic Flexibility 300
Chapter XIV The Anabolic/Catabolic Score of Food 305

- High mTOR (HiTOR) Foods ... 306
- Moderate mTOR (ModTOR) Foods ... 308
- Low mTOR (LowTOR) Foods ... 310
- mTOR Neutral (nTOR) Foods ... 312
- Low Autophagy (LowATG) Foods ... 313
- High Autophagy (HiATG) Foods ... 315

Chapter XV Metabolic Autophagy Foods ... 318
- Protein and Amino Acids ... 318
- Carbohydrates and Vegetables ... 324
- Fats and Lipids ... 336
- Drinks and Beverages ... 346
- What NOT to Eat ... 349

Chapter XVI Supplementation ... 355
- The 3 Main Supplements ... 356
- Essential Nutrients ... 357
- Longevity Supplements ... 361
- Anabolic Supplements ... 367
- The Extra Edge ... 370
- When to Take Supplements ... 373

Chapter XVII Metabolic Autophagy in Practice ... 377
- Targeted Intermittent Fasting ... 379
- Metabolic Autophagy Cycle ... 385
- Principles of Metabolic Autophagy ... 390

Chapter XVIII What Breaks a Fast ... 394
- How to Break a Fast ... 405

Chapter XIX How to Fast for Days and Days ... 409

Chapter XX When Not to Fast ... 421
- Should You Fast with Low Thyroid ... 422
- Should You Do Intermittent Fasting When Trying to Bulk ... 426
- Should Pregnant Women Fast ... 427
- Should Children Do Intermittent Fasting ... 428
- Should Old People Do Intermittent Fasting ... 429
- Should You Fast Every Day ... 430

Chapter XXI Circadian Rhythms and Autophagy 433
    Circadian Rhythm Basics 435
    Food Intake and Circadian Rhythms 440
Chapter XXII Sleep Optimization 448
Bonus Chapter How to Drink Coffee Like a Strategic MotherF#%ka! 457
Conclusion 465
Extras 469
    Glycemic Index 469
    Insulin Index 474
About the Author 477
More Books From the Author 478
References 483

Introduction
Wisdom of the Body

"The reasonable man adapts himself to the world:
the unreasonable one persists in trying to adapt the world to himself.
Therefore, all progress depends on the unreasonable man."

George Bernard Shaw

Living organisms have developed complex systems of behavior and physiological processes to adapt to their environment.

There are countless stimuli we get exposed to on a daily basis and all of them send a certain signal to the body. This, in turn, triggers a chain reaction of events that determine how your metabolism, nervous system, and psychology respond. In 1878, Claude Bernard wrote:

> *The constancy of the 'milieu interieur' is the condition of a free and independent existence.*[1]

He used it to describe how the first organisms left the oceans by developing the ability to balance their internal liquids and "carry the ocean with them" in the form of kidneys. This internal milieu describes the phenomenon of *homeostasis* also called equilibrium, inner balance, and 'wisdom of the body'[2]. The term itself was picked up in 1939 by behaviorist Walter Cannon[3]. It also describes the necessity for maintaining equanimity within oneself to truly experience freedom and independence in the world. Being stressed out and fatigued literally blindfolds you to the joys of life and prevents you from enjoying it fully.

This sort of equilibrium is something all living things and systems strive for. We want to be able to control the world around us as to avoid discomfort, chaos, disorder, pain, and ultimately death.

Entropy is the tendency of complex systems, including living organisms, to progressively move towards chaos, disorder, death, and deterioration. It's based on the Second Law of Thermodynamics, which states that the total entropy of an isolated system cannot decrease with time. Ideally, it can remain in a constant equilibrium or undergo reversible processes.

Aging is a phenomenon that describes a gradual process of entropy in the physical body. Birth or conception is the beginning point for a journey of growing and proliferation, which ultimately reaches deterioration and death. This is called senescence and it's caused by the phenomenon of entropy.

Basically, aging and death are the entropy of life – the slow waning away of the body that happens because of many things.

Immortality and eternal youth have been ideas of great curiosity throughout the entire history of the human species. Thousands of texts and practices have been invented in attempts to fight off entropy or at least postpone it.

Ancient aboriginal shamans, Ayurvedic yogis, Medieval alchemists, pharmacists, and contemporary biohackers all have many things in common. They seek to hone their bodies and sharpen their minds in some shape or form. The difference is only in degree – to what extent are they capable of altering the human physiology and how. With modern technology and science, we now have access to more resources and information that will not only make our lives better and more vigorous but can also lengthen it.

This book is not about trying to live forever or bestowing you with the water of eternal youth. It's more like an attempt to optimize the human body for both longevity and performance. There's not going to be a specific blueprint or a one-size-fits-all template for all to follow. Instead, I'm going to give you the tools and principles you can use to improve your healthspan, get more things done, and potentially increase life expectancy based on the research we currently have.

This homeostatic equilibrium and entropy manifest themselves inside the physiological processes of our bodies as well.

The word 'metabolism' comes from Greek, meaning 'change', and it describes the collection of all the life-sustaining chemical reactions inside the organism. These actions enable us to produce energy, maintain our physical framework, and eliminate the waste we get exposed to. It's the biological circuitry of life.

The metabolism has two sub-categories or sub-processes called anabolism and catabolism. They're polar opposites of each other that undergo a constant dance that directs the body to conduct certain metabolic functions and lead it to a certain direction of development.

- **Anabolism, meaning 'upward' in Greek, describes the synthesis of biological molecules to build up new physical matter in the body.** These reactions require energy and nutrients. Being anabolic means you're growing and building the framework that will decrease the accumulation of entropy.
- **Catabolism, meaning 'downward' in Greek, describes the breaking down of biological molecules to release energy.** This can apply to the breakdown of bodily tissue as well as the digestion of food that then gets assimilated into the body through anabolic processes. Being catabolic means you're fragmenting larger structures into smaller ones and using it to produce new energy.

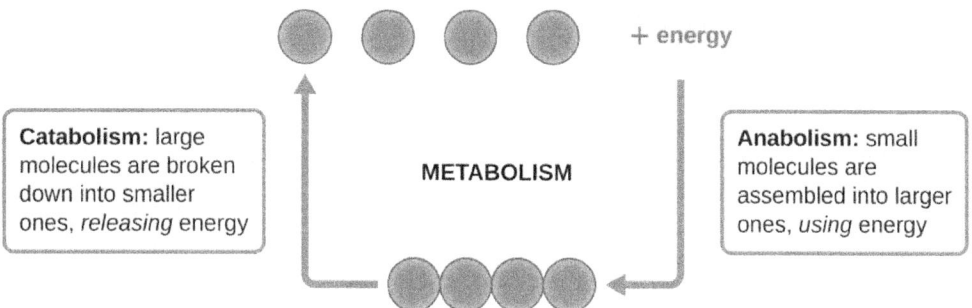

Figure 1 The Cycles of Anabolism and Catabolism

These oppositions can be seen everywhere around you. There are day and night, winter and summer, the sun and the moon, man and woman, black and white, wakefulness and sleep. It's how the world balances itself out as to maintain homeostasis and prevent excess.

Too much of either anabolism or catabolism will be unfavourable because it leads you in only a single direction and out of balance. You'd think that for health and longevity it's better to always stay anabolic as to prevent any loss of bodily tissue. However, being catabolic has many beneficial and even essential qualities that actually make you live longer.

This book is a collection of guidelines about the principles of the anabolic-catabolic cycles in regards to nutrition and exercise. It's definitely not a panacea – a solution or remedy for all conditions and circumstances. Instead, it's a very specific protocol that's not supposed to apply for all situations. Truth be told that everyone could do it. It's just that most won't because of fear, lack of understanding, or simply because of poor decisions.

The quote at the beginning of this chapter by George Bernard Shaw is perfect for describing this book: *"The reasonable man adapts himself to the world: the unreasonable one persists in trying to adapt the world to himself. Therefore, all progress depends on the unreasonable man."* It means that rather than subjecting yourself to the inevitable side of entropy, you take arms up against it. In so doing, you not only increase your healthspan but will also make the process more enjoyable.

In addition to '*Metabolic*', you can also find another word in the title - '*Autophagy*', which translates from Ancient Greek into 'self-devouring' or 'eating of self'. This is central to the main practice of this book. By maintaining a balance between anabolism and catabolism, you can effectively extend your lifespan.

The process of autophagy entails your healthy cells devouring the old, worn-out, weak ones and converting them back into energy. It's literally your body eating itself and using that to maintain homeostasis. There are many longevity-boosting benefits to this as illustrated in virtually all other species. What and how it all actually works will be covered in the upcoming chapters.

Here's How the Book Is Structured:

- In the first chapter, I'll reveal the code of longevity by talking about what causes aging, how it happens in other species and what metabolic pathways are associated with increased lifespan in humans.
- Chapter Two looks into why humanity is standing on a slippery slope of disease and metabolic disorders. More importantly, what to do about it.
- Chapter Three is about the most well-documented and effective ways of improving your health and longevity, which is caloric restriction and intermittent fasting.
- In Chapter Four, we're going to delve into the topic of autophagy and what we know about it so far.
- Chapters Five-Nine give you arms to combat entropy with and prevent the deterioration of your body.

- Chapters Ten to Thirteen are about food and what you should eat. I'll uncover some of the misconceptions and fallacies you might have heard about healthy eating and bring more context into the story.
- In Chapters Fourteen until Sixteen, I'm going to outline the main dietary components of Metabolic Autophagy. It's something you don't see any other nutritional book really talk about, namely, how do certain foods affect the anabolic-catabolic cycles of your body.
- Chapter Seventeen introduces the central practice of this protocol – *targeted intermittent fasting*. It's a way of crossing the chasm between the positive and negative side effects of anabolism as well as catabolism. The following chapters Eighteen to Twenty will complement the entire concept and give additional guidelines.
- Chapters Twenty One and Two are about another critical component to equilibrium. In addition to nutrition, you have to take into consideration things like light, sleep, and circadian rhythms.

Even if you think all of this seems overwhelming and confusing then fear not. As you start reading the book, you'll begin to have several revelations about the main message and how everything falls into place. What's more, you can choose to make your own adjustments based on your requirements and lifestyle factors.

Another quote by Bernard Shaw: *„Men are wise in proportion not to their experience but to their capacity for experience."* Essentially, it means that wisdom results from not only practical experience but also from realizing how much there is yet to learn. So it is with this book – although you need to make it a part of your everyday life, you still have to honor the complexity and nuance to what goes on in the body. For full mastery, you have to always consider the context of the situation because it's going to determine the final outcome.

The principles and guidelines of Metabolic Autophagy are supposed to become a permanent part of your daily routines because they require constant practice. If you want to be healthier and live longer, then you have to be working at it all the time, starting today. You can't place your bets on some externalities like a miracle longevity supplement, genetic modifications, biohacking, or artificial intelligence – all the bets are on you and what you do throughout the entire day.

Before embarking on this quest, let me introduce myself as well. My name is Siim Land and I'm an author, content creator, public speaker, entrepreneur, biohacker, and a high-performance coach. I like to think of myself as a self-actualizing Renaissance man who seeks to enhance human performance and life overall. In my previous books, I've talked about ketogenic dieting, intermittent fasting, exercise, mindset, and other biohacks. This one will be focused more on balancing the aspect of performing at your peak without sacrificing longevity or health.

Metabolic Autophagy is as much a book about nutrition as it is about personal transformation. In order to grow and take yourself to the next level, you have to leave behind the previous version of yourself. You literally have to die and be reborn from the ashes like the Phoenix – to bite your own tail and eat yourself like the mythological Ouroboros.

Death and destruction are only the beginning and the anabolic-catabolic cycles of life will subsist despite what path you end up following. This book will shed some light into the inner-workings of your metabolism and can show in what direction you're heading towards. Let's begin.

Chapter I
The Code of Longevity

"As for man, his days are numbered, whatever he might do, it is but wind."

Andrew George, The Epic of Gilgamesh

In nature, you don't see a lot of animals with exceptionally long lifespans. Part of it has to do with predation, high death rates, harsh living conditions, habitat destruction, and food scarcity. However, we know that some organisms have the capacity to live extremely long. When put under ideal conditions they could out-live us all. Here are a few examples:

- **Greenland Shark** – These sharks can live up to 200 with the oldest one found at 400 years of age.
- **Bowhead Whale** – The second largest mammal after the blue whale has an average lifespan of 200 years.
- **Galapagos Giant Tortoise** – The longest living turtles that live about 100-150 years.
- **Ocean Quahog** – Marine clams that are found to be more than 400-500 years old.
- **Red Sea Urchin** – Small spiny creatures that live in the Pacific Ocean and are known to live over 200 years.
- **Antarctic Glass Sponges** are estimated to live up to 15 000 years but it's not sure how old they are right now because no one was there to check them 15 000 years ago.
- **Glass Sponges in the East China Sea** have been found to live up to 10 000 years.
- **Pando the Trembling Giant** – This colony of quaking aspen in Fishlake National Park Utah is considered the oldest single living thing on the planet. Its root system has been alive for over 80 000 years although single trees are at some 4000+ years of age.
- **The Immortal Jellyfish** – A jellyfish that can revert back to its premature state when exposed to stress or injury thus being almost immortal. In reality, most jellyfish still get eaten or die to too many injuries.

Lifespan length also differs greatly between individuals of the same species as there are some people who die very young whereas there are others who stay youthful all the way up until elderhood.

Average lifespans in humans range from 55-100 years old. The oldest recorded human alive was Jeanne Calment from France who lived to the age of 122 years and 164 days (1875-1997). There may be some genetics and environmental conditions at play but by far the most important variable is lifestyle.

If a species' mortality rate doesn't increase after maturity, then it won't age and is considered biologically immortal. The word 'immortal' describes cells that aren't subject to the Hayflick Limit, which is the point where cells can't divide anymore because of DNA damage or shortened telomeres.

- Hydra species seem to show no signs of aging when observed in laboratory conditions. They may have escaped senescence and are biologically immortal[4].
- Hydrozoan species *Turritopsis dohrnii* (The Immortal Jellyfish) can revert from an adult stage back into an immature polyp stage and back again several times, which suggests that there isn't a natural limit to its lifespan.
- The larvae of skin beetles can also go back into an earlier stage of maturity if they're starved. This cycle can be repeated many times[5].

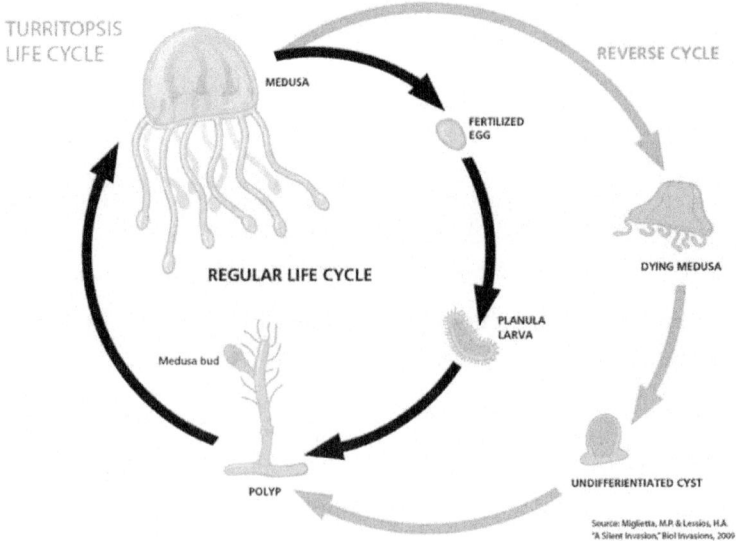

Figure 2 The Life Cycle of Turritopsis (The Immortal Jellyfish)

Some species and bacteria can also be revived after thousands of years of hibernation. But there's a difference between aging and immortality. Both of them are relevant for determining an organism's lifespan.

The Hallmarks of Aging

In 2013, Lopez-Otin and colleagues published a review of the hallmarks of aging in different species including mammals[6]. They are:

- **Genomic Instability** – genetic damage and mutations throughout life
- **Telomere Attrition** – shortening of protective telomere caps on top of chromosomes that occurs during DNA repair
- **Epigenetic Alterations** – alterations in methylation patterns, post-translational modification of histones, and chromatin remodeling
- **Loss of Proteostasis** – dysfunctional protein folding, proteolysis, and proteotoxicity. Basically, not being able to put muscle back together correctly.
- **Deregulated Nutrient Sensing** – inadequate growth hormone production, related to the Insulin/IGF-1 signaling pathway
- **Mitochondrial Dysfunction** – old worn out mitochondria begin to produce more reactive oxygen species and oxidative stress, which damages all other cells
- **Cellular Senescence** – the accumulation of dead cells and cancer proliferation
- **Stem Cell Exhaustion** – decline in regenerative potential of tissues and lack of swapping old cells with new cells
- **Altered Intercellular Communication** – miscommunication in endocrine, neuroendocrine, or neuronal systems that cause inflammation and other problems

There are many things that determine the lifespan of an organism, such as its genetics, phylogeny, mutations, and life history but the biggest role probably has to do with the ecological niche. What kind of an environment both internal and external the animal is living in and how it's going to respond to it.

Biological organisms develop certain adaptations based on the conditions they get exposed to in their environment. That's why some animals have completely different metabolic profiles as well as physical traits than humans. They also live differently partly because of how they've adapted to their surroundings over the course of eons. That's why most of these hallmarks of aging are controllable and epigenetic. You can influence your rate of aging and longevity by simply understanding these mechanisms and changing your lifestyle.

Mitochondrial Theory of Aging

What causes death and aging has been a topic of inquiry amongst humans for eons. It's often explained away by spiritual, religious, or purely physical reasons. But why does it happen?

In 1956, Denham Harman was the first to propose the Free Radical Theory of Aging (FRTA)[7] and furthered the idea in 1970 to describe mitochondrial production of reactive oxygen species (ROS)[8].

The free radical theory of aging states that organisms die because of the accumulation of free radical damage on the cells over time.

A free radical is an atom or molecule with a single unpaired electron in its outer layer. Most free radicals are very reactive and they cause oxidative damage. That oxidation can be lowered by antioxidants and other reducing agents.

Mitochondria are the cells' energy manufacturers that generate adenosine triphosphate (ATP). It's the energetic currency needed for life. This process occurs by reacting hydrocarbons from calories or sunlight with oxygen.

The classical free radical theory of aging proposes that energy generation by the mitochondria damages mitochondrial macromolecules, including mitochondrial DNA (mtDNA), which promotes aging[9]. After a certain threshold, this produces too many reactive oxygen species (ROS), which cause cell death and degradation.

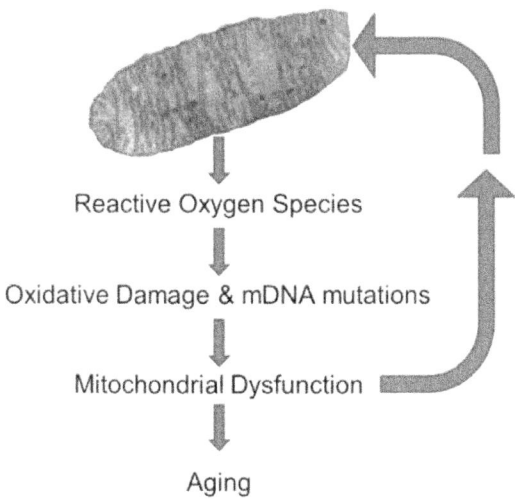

Figure 3 Reactive Oxygen Species and Mitochondrial Aging

Mitochondrial Free Radical Theory of Aging was introduced in 1980, which implicate the mitochondria as the main targets of ROS damage[10]. This happens when electrons get out of the electron transport chain and react with water to create ROS, such as superoxide radical. These radicals damage DNA and other proteins.

Age-related impairments in the mitochondrial respiratory chain decrease ATP synthesis, damage DNA, and make the cells more susceptible to oxidative stress.

However, it's been now shown that mutations in mtDNA can result in premature aging without increasing ROS production by mutating the polymerase Pol-γ that's responsible for mitochondrial DNA synthesis[11].

There are many factors that affect lifespan. Excessive generation of ROS and mutations in mtDNA both are central to the mitochondrial theory of aging. However, it's suggested that ROS aren't the primary or initial cause of it.

Reactive Oxygen Species and Aging

The ability to cope with oxidative stress and other stressors is compromised in aging thus making you more vulnerable to free radicals as you get older. Mutant mtDNA increases with age, especially in tissues with higher energy demands like the heart, brain, liver, kidneys, etc. These notions support the theory of mitochondrial aging[12].

In some species like yeast and fruit flies, reducing oxidative stress can extend lifespan[13]. However, blocking the antioxidant system in mice doesn't shorten lifespan in most cases[14]. Likewise, in roundworms, inhibiting the natural antioxidant superoxide dismutase has been shown to actually increase lifespan[15].

Taking a lot of antioxidants and lowering oxidative stress with supplements have failed to be effective in fighting disease and in fact may promote the chances of getting sick. Treatment with high doses of anti-oxidants like beta-carotene, vitamin A, and vitamin E may actually increase mortality[16]. Consuming more fruit and vegetables doesn't seem to have a significant effect on reducing cancer risk[17].

Free radicals and ROS are involved in most human diseases and cancers but their degree of influence is still uncertain. Increasing your body's own endogenous antioxidant levels may be a better option for disease prevention[18].

Oxidative stress and free radicals can increase life expectancy in nematodes by inducing a bi-phasic response to the stress. This phenomenon is called *mitohormesis* or mitochondrial hormesis.

Hormesis is a dose-specific response to a toxin or a stressor that makes the organism stronger than it was before (See Figure 4). The idea is that you experience a small shock that makes the body want to deal with it better in the future thus becoming more resilient. Sublethal mitochondrial stress with a minute increase in ROS may cause a lot of the beneficial effects found in caloric restriction, intermittent fasting, exercise, and dietary phytonutrients[19].

- If you experience no stress and zero exposure to free radicals, then your body is by default weaker because of having no fighting reference from the past.

- If you experience too much stress and excessive accumulation of ROS, then you promote disease and sickness because of not having enough time to recover.

- If you experience just the right dose of stress, then you'll be able to deal with it, recover from the shock, and thus augment your cells against future stressors.

- If you block all mitochondrial stress and eliminate free radicals, then your body won't have the time nor the means to promote mitohormesis. That's why antioxidants all the time won't have a positive effect.

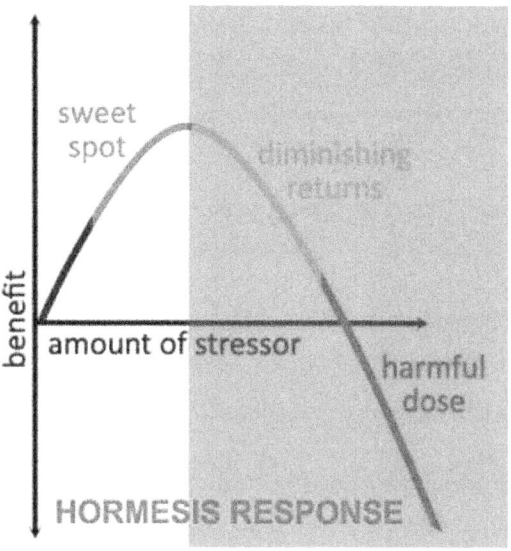

Figure 4 The Dose Matters for Hormesis

Mitochondrial decay is central to the mitochondrial theory of aging and it's one of the major causes of aging.

Here are some of the factors that have been shown to produce oxidative stress and induce mitochondrial aging.

- **Insulin and high glucose environments generate free radicals and promote oxidative stress**[20]. The insulin signaling pathway is one of the main mechanisms of accelerated aging. However, this is dose specific and some is beneficial for ROS production[21].

- **Chronic stress accelerates aging and disease.** Over-production of free radicals because of too many stress hormones decreases mitochondrial functioning and makes you more prone to disease because of a weakened immune system, high insulin, and damage to cells.

- **Sleep deprivation and circadian mismatches promote all disease.** If your body's biological clocks are misaligned with its circadian rhythms, then you'll cause more cellular stress and predispose yourself to all types of dysfunctions.

- **Avoid environmental toxins and pollution.** Polluted air, water, heavy metal exposure, mercury in food, pesticides, glyphosate, GMO crops, toxic personal care products, house cleaning chemicals – all of them will create more reactive oxygen species and oxidative stress. The amount of these stressors is beyond our body's natural ability to cope with them, thus they don't have a beneficial effect in the long run.

- **Avoid inflammation like wildfire.** Inflammation is correlated with most diseases, as it directly decreases the body's immune system. Processing food and over-cooking it increases the number of free radicals and carcinogens in it.

There's a difference between beneficial stressors and too many free radicals. Of course, excessive oxidative stress is still bad for you and accelerates aging. The key is to differentiate it from the positive hormetic stress.

The mitochondria are one of the most important organelles in your body as they govern everything related to energy metabolism and cellular homeostasis. Dysfunctional mitochondria will not only speed up aging but also make you feel more tired, exhausted, lethargic, weak, and experience atrophy.

All of the ideas related to hormesis, mitohormesis, avoiding oxidative stress and getting the right amount of reactive oxygen species will be talked about in the upcoming chapters. Having covered this, I'm going to move on with the other related pathways and mechanisms that are shown to affect aging and longevity.

Longevity Pathways in Humans

Like said before, different individuals of the same species may exhibit drastically different life spans and rates of aging. This is so because aging is regulated by many genetic pathways and biological processes.

In humans, there are several longevity pathways recognized to control the aging process and its constituent mechanisms. I'm going to outline the ones that are currently most recognized to regulate longevity and lifespan. Then I'll go through them one by one in closer detail.

- **The Growth Hormone/Insulin and Insulin-Like Growth Factor-1 Signaling Pathway**, which regulates cell replication, nutrient partitioning, and storage.
- **The FOXO/Sirtuin Pathway,** which includes proteins and transcription factors responsible for energy homeostasis. They manage homeostasis under harsh conditions and stress.
- **Hormesis and General Stress Adaptation** mediated by FOXO proteins and mitochondrial functioning. This phenomenon makes the organism more resilient against environmental stressors.
- **The mTOR/AMPK Pathway,** which governs homeostasis between anabolism and catabolism. Basically, it's the body going to grow or eat itself.

All of these pathways interact with each other and they're affecting longevity in different ways. These are the mechanisms that affect aging and lifespan. Let's now walk through them in closer detail one by one.

The Insulin IGF-1 Pathway

One of the most understood pathways of longevity is the Insulin/IGF-1 Signaling Pathway (IIS).

- **Insulin is the main storage hormone** that directs nutrient partitioning and glycogen replenishment. It basically helps to unlock the cells so they could store glucose into liver and muscle glycogen.
- **Insulin-Like Growth Factor (IGF-1)** or somatomedin C is an IGF-1 encoded human gene. It's also been referred to as the *'sulfation factor'*. IGF-1s role is to promote tissue growth and development.

The effects of IGF-1 are mediated through the IGF-1 receptor (IGF-1R), which is similar to the receptor of the storage hormone insulin.

IGF-1 gets produced in the liver by the stimulation of Human Growth Hormone (HGH). IGFBP is a binding protein that carries IGF-1 around the body and it's regulated by insulin.

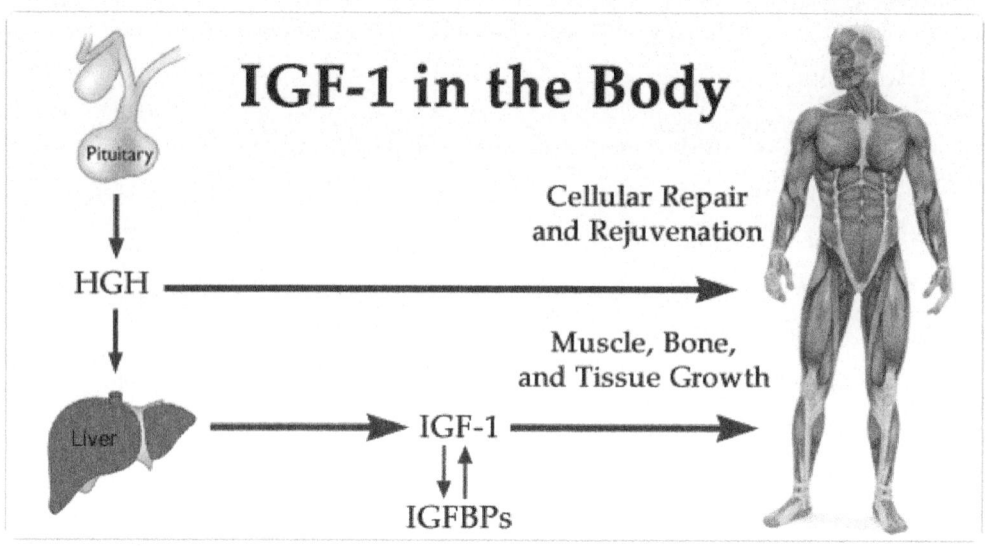

Figure 5 The Effects of IGF-1 in the Body

Reduced insulin signaling has been found to increase the lifespan of fruit flies, nematodes, and rodents[22]. Here are some of the studies in other living organisms:

- **In 1993, it was discovered that mutating an insulin-like receptor called DAF-2 by suppressing it in nematodes doubled their lifespan**[23]. This extension required the presence of another gene DAF-16, which encodes together with a FOXO transcription factor[24]. DAF-2 activates the signaling pathway of PI-3 kinase, which shortens lifespan by activating insulin/IGF-1 signaling and phosphorylating DAF-16[25]. Suppressing DAF-2 does the opposite. In humans, DAF-2 is the equivalent of IGF-1 and insulin signaling.
- **Reducing glucose and carbohydrate intake increases FOXO activity by suppressing insulin/IGF-1 signaling (IIS)**. One study done by Lee *et al.* found that adding just 2% of glucose into the diets of roundworms shortened their lifespan by 20% because of inhibiting the activity of DAF-16/FoxO and heat-shock factor HSF-1[26]. *Not a good trade-off.*
- **Insulin/IGF-1 receptor mutations can increase the lifespan of fruit flies by up to 85%**[27]**!** Additionally, mutated suppression of an insulin receptor substrate (IRS)-like signaling protein called CHICO increases the lifespan of fruit flies by 48%[28]. This seems to be dependent on FOXO proteins, which are transcription

factors of longevity. Drosophila dFOXO signaling controls lifespan and regulates insulin signaling in the brain and fat tissue[29].

- **Knocking out the IGF-1 receptor in mice makes them live 26-33% longer[30].** Mice who lack the insulin receptor in fat tissue live 18% longer[31].
- **In yeast, mutating the insulin-dependent AKT ortholog SCH9 triples their mean lifespan[32].** Additionally, overexpression of SIR2, which is part of the insulin/IGF-1 pathway in C. Elegans extends the lifespan of worms and yeast by up to 50%[33].

All these findings show that the insulin/IGF-1 system is a critical regulator of the organisms longevity by controlling many downstream pathways. They vary between species but their orthologs are found in humans as well.

Insulin/IGF-1 Signaling in *Caenorhabditis Elegans*

The Insulin/IGF-1 Signaling Pathway has been found to play a crucial role in the aging and development of lower organisms such as Caenorhabditis Elegans, nematodes, roundworms, and larva[34].

Nematodes go through several cycles of development during their average 3-week lifespan. One of them is the Dauer stage that gets activated during periods of higher environmental stress[35] (Figure 6 Notice how entering into the Dauer stage can extend the larva's life by several months. Figure 6).

Dauer larvae are morphologically specialized roundworms that adapt to harsh conditions such as starvation, temperature stress, and oxygen deprivation. Normally, they would exit the period of conservation according to the influx of energy from their environment, but this cycle can be prolonged in laboratories. Dauer larvae can survive up to 8 times longer under laboratory conditions[36]. That's a huge difference based on a simple elongation of a certain life stage and environmental input.

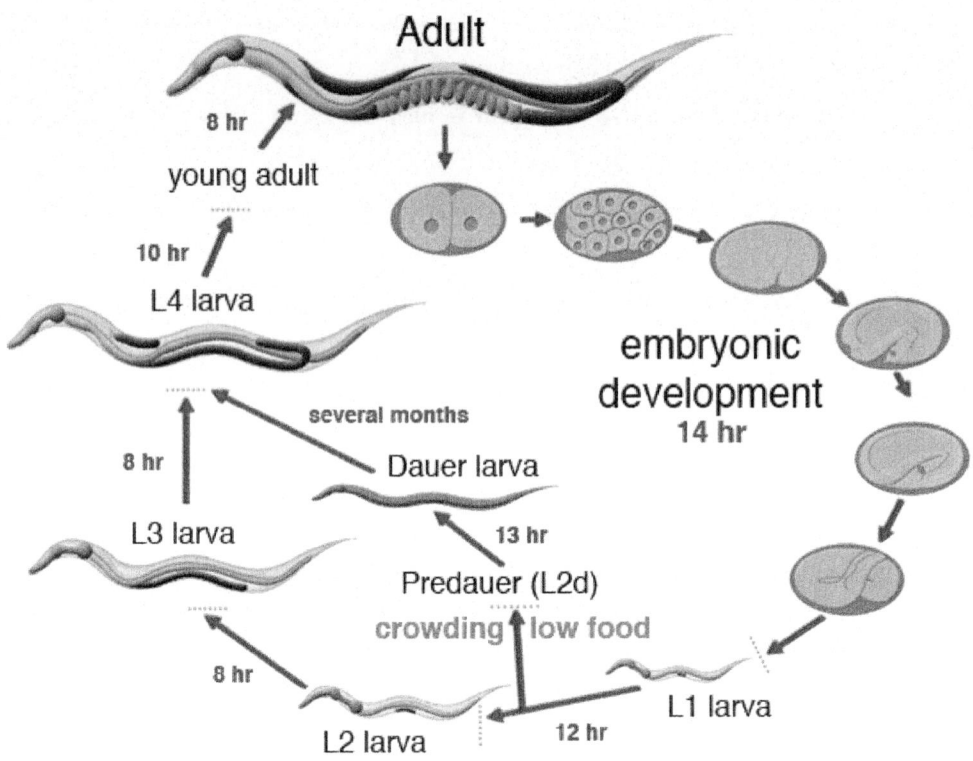

Figure 6 Notice how entering into the Dauer stage can extend the larva's life by several months.

In 1999, it was found that Sir2 overexpression in yeast extended their lifespan by 50%[37]. Sirtuin genes have been found to have similar anti-aging functions in other species, such as yeast, fruit flies, and roundworms[38].

Sirtuins are a family of proteins that act as metabolic sensors. They deacetylase a coenzyme NAD+ into free nicotinamide. Basically, they break down acetyl from proteins to maintain their functioning for longer. The ratios of NAD+ and NADH determine the nutritional status of the cell and sirtuins are there to respond.

NAD+ is an essential currency for energy metabolism and DNA repair. Sirtuins are proteins that evolved to respond to the availability of NAD+ in the body.

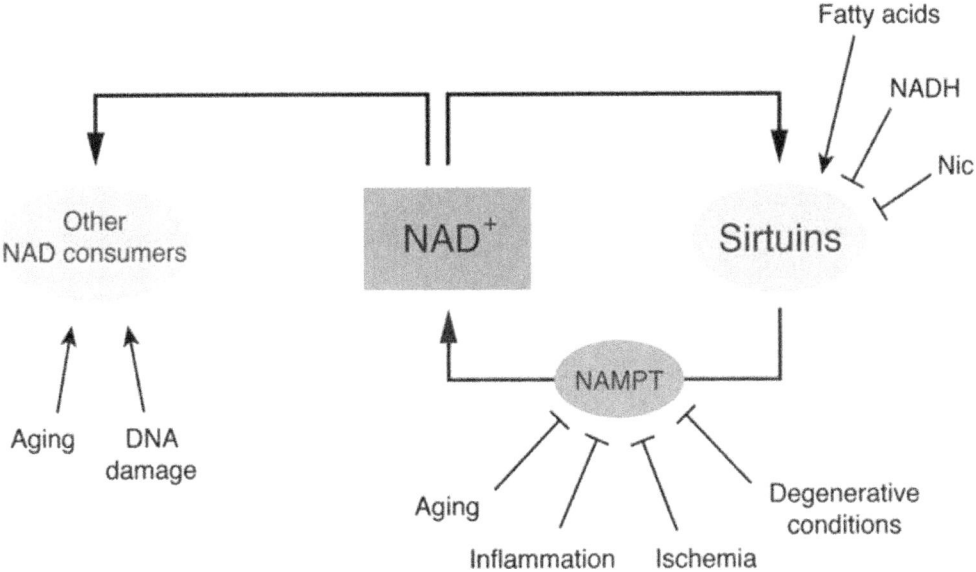

Figure 7 The Mechanisms of NAD+ and Sirtuins

SIRT6 overexpression has been found to lengthen the lifespan of male mice by as much as 15,8%[39]. SIRT6 deficiencies in mice accelerate their aging[40]. This may be due to the anti-cancer effects of SIRT6, not the anti-aging itself.

Cellular deterioration and senescence are thought to be caused mostly by the accumulation of unrepairable DNA damage[41]. SIRT1 plays an important role in activating DNA repair proteins[42]. It's specifically involved with repairing the double helix of DNA[43].

SIRT1 can also induce cellular autophagy by directly deacetylating AuTophaGy (ATG) proteins such as Atg5, Atg7, and Atg8[44]. This then promotes *mitophagy* or mitochondrial autophagy and helps to eliminate old worn out cells.

How to Increase Sirtuins for Longevity

There's a lot of evidence pointing to the longevity benefits of increased sirtuin activity in humans as well. If not in over-expression, then increasing sirtuins can still be good for your health in most cases.

- **Glucose restriction extends the lifespan of human fibroblasts because of increased NAD+ and sirtuin activity**[45]. Inhibiting insulin shuttles SIRT1 out of the cell's nucleus into the cytoplasm.

- **Caloric restriction and fasting increase SIRT3 and deacetylate many mitochondrial proteins**[46]. Reduction of calorie intake without causing malnutrition is the only known intervention that increases the lifespan of many species including primates[47][48]. It's thought that these effects in longevity require SIRT1[49].

- **Activating AMPK elevates NAD+ levels, leading to increased SIRT1 activity**[50]. AMPK is the fuel sensor that mobilizes the body's energy stores such as fat and it promotes autophagy as well.

- **Ketosis and ketone bodies like beta-hydroxybutyrate (BHB) promote sirtuin activity.** SIRT3 deficient mice can't produce normal levels of ketones while fasting[51]. However, a ketogenic diet increases brown fat and improves mitochondrial function but lowers SIRT3 in mice[52]. This may be due to the signaling of mTOR by a nutrient-dense diet high in too much fat. Natural ways of caloric restriction and fasting are still the best ways of signaling energy deprivation which promotes longevity. In an everyday context, a low carb diet is still pro-sirtuin to a certain extent because of the low levels of insulin and glucose. More about optimizing the macronutrient ratios in future chapters.

- **Exercise has anti-inflammatory effects and it increases SIRT1**[53]. The long-term benefits of exercise are even thought to be regulated by SIRT1.

- **Cyclic-AMP (cAMP) pathway activates SIRT1 very rapidly to promote fatty acid oxidation independent of NAD+**[54]. cAMP is linked with AMPK which gets activated under high energy demands while being energy deprived.

- **Heat exposure and saunas increase NAD+ levels which promote SIRT1** as well[55]. Sweating, cardio, yoga, or infrared saunas will probably have a similar effect on activating heat shock proteins.

- **Chronic oxidative stress and DNA damage deplete NAD+ levels and decrease sirtuin activity.** This will then disrupt DNA repair and impair mitochondrial functioning. That's why you want to keep stressors acute and followed by recovery.

- **Melatonin can activate sirtuins and has anti-aging effects**[56]. It's also the main sleep hormone and a powerful antioxidant that helps the brain get more recovery from deeper stages of sleep. Most of the repair processes and growth happen when you're sleeping and melatonin has an important role.

- **Sirtuins also affect the circadian clocks so keeping a consistent circadian rhythm is incredibly important for longevity**. NAD+ is under circadian control and when you're misaligned you'll have less energy and lower SIRT1 and SIRT3 activity[57]. Circadian rhythm mismatches are linked to many metabolic disorders, glucose intolerance, and brain degeneration. That's the opposite of what you want for longevity.

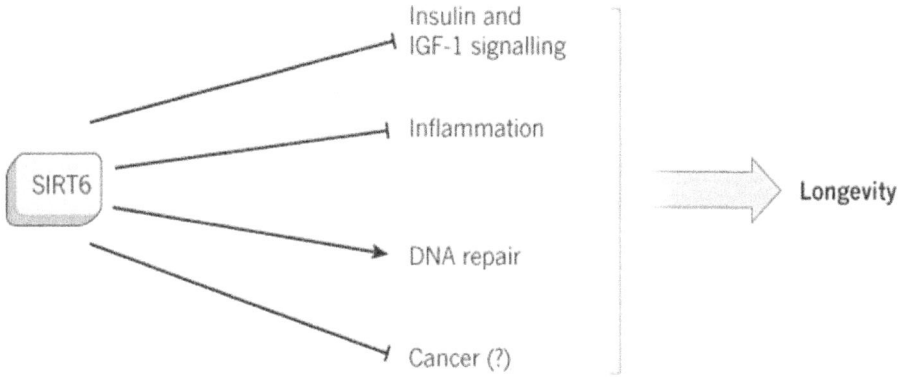

Figure 8 How SIRT6 Affects Longevity

These are the main ways of increasing sirtuin activity but they're not only exclusive to the SIRT family. In fact, most of them also interact with the other longevity-promoting pathways we mentioned beforehand. So, let's carry on.

Caloric Restriction and Longevity

One of the few known ways of actually extending lifespan in most species is caloric restriction.

Caloric restriction and energy deprivation lower mTOR signaling, which in turn upregulate other pathways of energy homeostasis, such as AMPK and autophagy.

Autophagy and cellular turnover are essential for the life extension effects in DAF-2 mutants[58]. Suppressing autophagy in other species has been found to negate the longevity benefits of caloric restriction as well. For instance, if you block genetically modified mices' autophagy genes, then they won't live longer even under caloric restriction, whereas normal mice who have autophagy activated do[59].

Therefore, the life-extension benefits of caloric restriction and fasting are mostly induced by autophagy and increased sirtuin activity that promote cellular turnover and recycling of old cells. That's an important point because it means you can side-step some of the negative side-effects of prolonged caloric restriction by knowing what you're doing and elevating autophagy with other means. In Chapter IV, I'll talk about this in closer detail.

Nematode Worms and Anti-Aging

In 2017, a study published in the journal Cell Metabolism showed that aging and age-related diseases are associated with a decrease in the cells' ability to process energy efficiently[60]. The scientists used nematode worms who live for only 2 weeks to carry out an experiment on their mitochondria.

They found that restricting the worms' calories and manipulating AMPK promoted longevity by maintaining mitochondrial networks and increasing fatty acid oxidation. This happened in communication with other organelles called peroxisomes that regulate fat metabolism (See Figure 9). Essentially, more fatty acid oxidation from their own energy stores led up to living longer because they were put under caloric restriction.

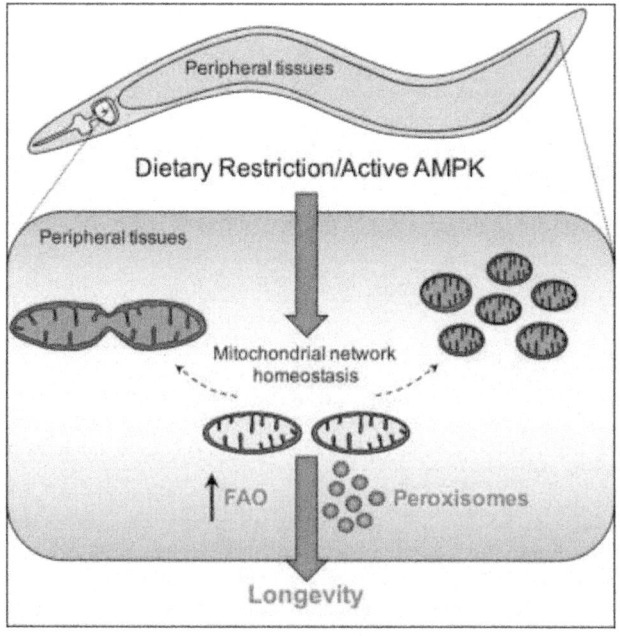

Figure 9 Dietary Restriction Increases Longevity Through Mitochondrial Network Homeostasis

The researchers proposed that fission and fusion amongst the mitochondria's network and fatty acid oxidation are required for the longevity benefits. For intermittent fasting-mediated lifespan increase, you need the dynamic remodeling of mitochondrial networks, which happen in response to various physiological and pathological stimuli.

The life cycles of mitochondria are characterized by fission and fusion events (See Figure *10*).

- Fusion states happen when several mitochondria mix and organize themselves into a network. They basically merge together into a single much larger mitochondrion.

- Fission states happen when the fused mitochondria get split into 2 out of which the one with a higher membrane potential will return to the fission-fusion-cycle and the one with a more depolarized membrane will stay solitary until its membrane potential recovers. If its membrane potential remains depolarized it'll lose its ability to fuse and eventually will be eliminated by mitophagy.

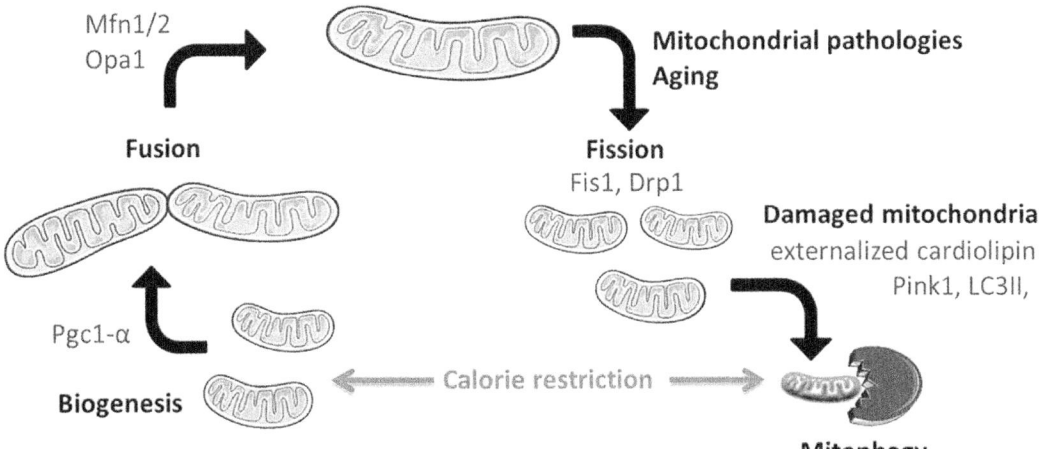

Figure 10 Mitochondrial Fission-Fusion Cycles and Longevity

Changes in nutrient and energy availability can make the mitochondria stay in either one of these states for longer.

- Post-Fusion State is called Elongation, which is characteristic to states of energy efficiency, such as starvation, acute stress, caloric restriction, and senescence.

- Post-Fission State is called Fragmentation, which shortens the mitochondria and keeps them separate. This is typical to bioenergetic inefficiency that's caused by high energy supply and extended exposure to excess nutrients.

Basically, caloric restriction promotes energy efficiency because the organism is required to sustain itself with fewer calories. Having access to an abundance of energy, however, leads to inefficient mitochondrial function because every single mitochondrion has to expend less effort to carry its weight so to say. That can lead to the accumulation of dysfunctional components.

These mechanisms show that the mitochondria evolved to adapt to drastic changes in nutrient availability in the form of fasting and feasting. Fasting and caloric restriction promote mitochondrial efficiency by fusing together several mitochondria. Nutrient excess in the eating phase fragments the mitochondria and decreases their ability to produce energy. Chapter XII will also talk about nutrient/energy density but for now, let's carry on with caloric restriction.

Caloric Restriction of Rats and Monkeys

Caloric restriction and intermittent fasting have been linked to longevity previously already. Here are some of the main findings in other species.

- **In 1946, a study on rats found that fasting 1 day out of 3 increased lifespan in males by 20% and in females by 15%**[61]. They didn't experience any retardation of growth but what did happen was the death of tumors increased in proportion to the amount of fasting. Other studies on rodents have noted reduced inflammation and other age-related health issues[62].

- **Fasting has been shown to increase the lifespan of bacteria and yeast by 10-fold**[63]. Yeast have a very short lifespan of just a few days and weeks, but a 10x boost is still phenomenal.

- **Caloric restriction shows increased lifespan of brain neurons in both humans and monkeys**[64]. Maintaining cerebral health is a critical factor for longevity because you wouldn't be able to enjoy your life.

In 2009, a group of scientists from the University of Wisconsin reported improved biomarker and longevity benefits in rhesus monkeys who ate less[65]. However, in 2012, a follow-up study done by the National Institute of Aging noted there to be no improvements in survival, but they did find a trend toward better health. After working through the conflicting outcomes, it's thought that the different results were caused by several things[66].

- Caloric restriction is more beneficial in adults and older monkeys but not as so in younger animals. That's because growing organisms need more nutrition for proper development. Similar effects may be true in humans as well.
- How much less food was eaten also affected the differences in survival rates. A severe caloric restriction isn't sustainable and not beneficial. In humans, malnourishment and nutrient deficiencies aren't healthy in the long term either.
- The monkeys in the National Institute of Aging ate naturally sourced foods whereas the ones in Wisconsin ate processed food with higher sugar content, which made them substantially fatter. That's probably due to the increased inflammation and insulin/IGF-1 pathway stimulation.
- There were also sex differences, where females seemed to have less adverse effects of obesity than males. This makes sense, as women are more prone to carrying extra fat for their offspring and thus not be that affected by it as much.

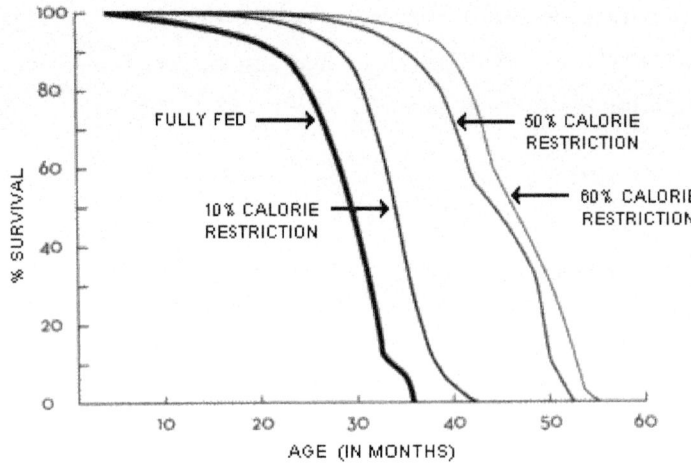

Figure 11 Graph of increased lifespan from higher caloric restriction in monkeys

What about humans? Do caloric restriction and intermittent fasting have a similar effect on longevity in humans? We do share 93% of the genes with rhesus monkeys. Other species like Chimpanzees are also practically genetically identical to us.

One human study on 3 weeks of alternate day fasting discovered an increase in SIRT1, which is associated with longevity[67]. Why this happens is still unclear, but it's suggested that caloric restriction induces cellular respiration, which increases NAD+ and reduces NADH levels. NADH inhibits Sir2 and SIRT1.

SIRT1 has been shown to also activate PGC-1α, which triggers the growth of new mitochondria[68]. SIRT3, SIRT4, and SIRT5 improve mitochondrial function as well[69]. This may be evidence that if not increased lifespan, then caloric restriction and fasting will improve the longevity of the cells nonetheless.

When your body faces a shortage of energy whether through caloric restriction, fasting, starvation, or anything the like, then you're going to promote the fusion of mitochondria. This lowers your energetic demands because the organelles in your cells are better connected. It'll also make you recycle old worn out cell components and convert them back into energy through the process of autophagy. Mitophagy is a layer deeper and happens inside the mitochondrial fission-fusion cycle.

Energy restriction also upregulates the other genes that increase energy efficiency by improving insulin sensitivity and fat oxidation. Remember – the increased lifespan of the nematode worms happened because of peroxisome-mediated fat metabolism. During states of fasting or depletion of exogenous calories, your mitochondria rev up their functioning and boost endogenous energy production from internal sources.

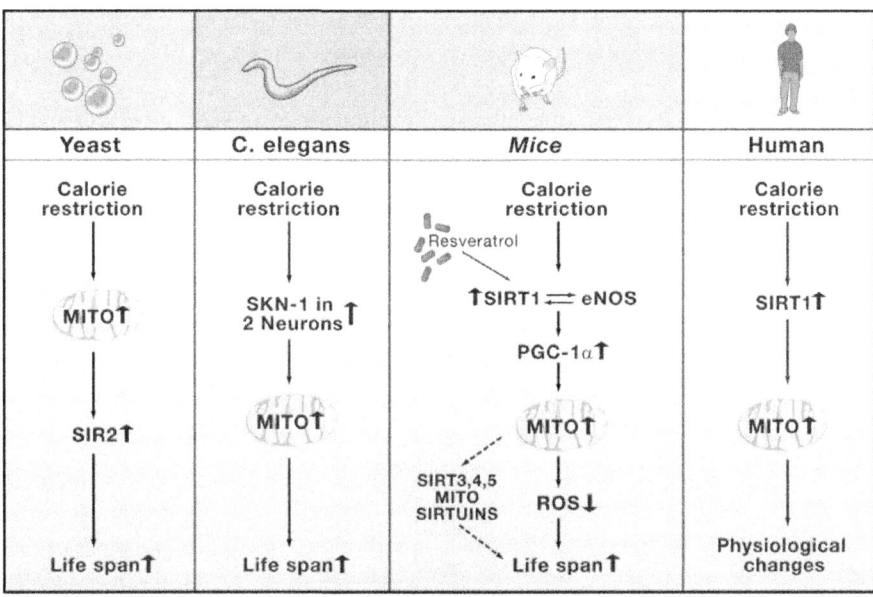

Figure 12 Increased lifespan in many species is linked to improved mitochondrial functioning and increased sirtuin activity

Ketosis is another vital component to the survival of the mitochondria as it allows them to become more energy efficient. This shift of starting to burn ketones preserves muscle tissue, gives adequate energy to the brain, and keeps you satiated by downregulating some of the hunger signaling.

From an evolutionary perspective, it makes perfect sense, as the mechanisms of longevity came from organisms trying to survive periods of nutrient deprivation and avoid age-related damage.

The mitochondrial fission-fusion cycles are also dependent on autophagy modulating pathways such as AMPK and mTOR[70]. We'll be getting to know these three very well in the upcoming chapters but to give you some basic overview:

- mTOR or mammalian target of rapamycin is responsible for cell growth, protein synthesis, and anabolism. It will make the body build new tissue.

- AMPK or AMP-activated protein kinase is a fuel sensor that is involved in balancing energy deprived states.

- Autophagy is the process of self-eating and cellular turnover in which the body recycles its old worn out components back into energy.

mTOR inhibits autophagy because it makes your body grow, which requires expending energy and upregulating the metabolism, whereas AMPK supports autophagy due to the energy-deprived state.

One of the subunits of AMPK, AAK-2, is required for DAF-2 mutations to promote longevity in C. Elegans[71]. The mechanism by which this happens is unknown but overexpression of AAK-2 increases lifespan in worms by 13%.

Nutrient starvation allows unneeded proteins to be broken down and recycled into amino acids that are essential for survival. That keeps the organism alive longer because of increased mitochondrial efficiency. Therefore, the key to longevity and increased lifespan still gets traced back to decreased energy intake and improved energy usage within the body itself.

Mice who lack the insulin receptor in adipose tissue live longer because of increased leanness[72]. mTOR interacts with the insulin pathway to regulate the lifespan and development of C. Elegans and fruit flies[73].

There's some evidence to show how excessive mTOR and insulin signaling are related to accelerated aging and disease but it's not that black and white. In some situations, they can actually be beneficial and even pro-longevity. The same applies to caloric restriction and intermittent fasting. One of the core principles of this book is that you want to use these anabolic hormones only at the right time under the right circumstances. That's why it's called Metabolic Autophagy. Let's carry on with more of the pathways related to longevity.

Stress Adaptation and Longevity

Adaptation to harsh environmental conditions has been mentioned a few times already. Heat stress seems to have many health benefits and some of it has to do with the insulin/IGF-1 signaling.

In C. Elegans, activation of heat shock transcription factor 1 (HSF-1) is also required for the DAF-2 mutations to extend lifespan[74]. This is thought to be because HSF-1 and DAF-16 activate specific genes that turn on small heat-shock proteins and promote longevity.

Essentially, exposure to stress whether that be the cold, caloric deprivation, or the heat makes the organism live longer because of forcing hormetic adaptation. Hormesis through heat stress increases the lifespan of flies and worms[75].

Interestingly, increased insulin/IGF-1 signaling mutations prevent the localization of DAF-16 by heat shock, which raises the possibility that the increased lifespan due to stress adaptation occurs because of lower insulin[76].

Stress resistance has been found to be a major contributing factor to increased longevity in animals with mutated insulin/IGF-1 signaling. Activation of the stress-response JNK pathway in fruit flies increases their lifespan by up to 80%[77]!

The adaptation to stress and harsh conditions is mediated through certain transcript factors that regulate energy homeostasis and longevity. They're called FOXO proteins.

Increase FOXO Factors for Longevity

'FOX' stands for 'Forkhead box' and it represents a class of proteins and transcript factors that have many functions in the human body.

FOXO proteins are transcript factors that regulate longevity through the insulin and insulin-like growth factor signaling[78].

Invertebrates have a single FOXO gene, whereas mammals have four: FOXO1, FOXO3, FOXO4, and FOXO6. In mammals, FOXO proteins regulate stress resistance, cellular turnover, apoptosis, glucose and lipid metabolism, and inflammation[79,80].

FOX represents the class of proteins, the letter 'O' is the subclass, and the number represents the member of that group. There are over 100 subclasses of FOX proteins in humans, such as FOXA, FOXR, FOXE, etc. and they have many functions. FOX proteins with the class 'O' are regulated by the insulin/Akt/mTOR signaling pathway.

Theoretically, upregulated FOXO pathway activities increase lifespan in many species because of promoting stress adaptation in harsh environments. The FOXO pathway is an evolutionarily viable mechanism for adapting to low levels of insulin and energy deprivation.

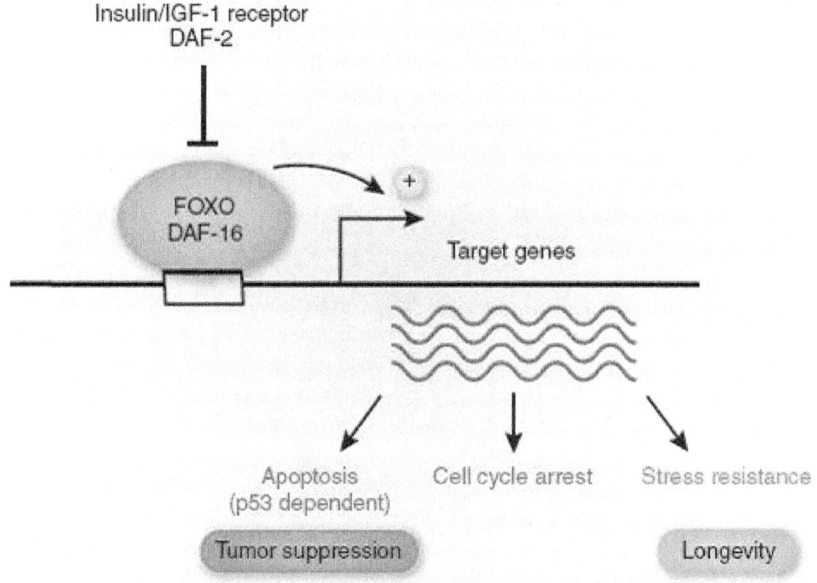

Figure 13 Blocking the Insulin/IGF-1 Receptor promotes FOXO proteins, which leads to stress resistance, tumor suppression, and longevity

Anabolic mechanisms such as insulin, mTOR, and IGF-1 tell the body to grow and replicate but this may come at the expense of longevity and accelerated aging. Which is why you'd want to know how to balance it with the catabolic processes of autophagy, AMPK, and FOXO factors.

In fruit flies, an unhealthy high sugar diet in early adulthood curtails survival later in life despite subsequent dietary improvements[81]. Meaning that having a bad diet in your earlier youth may cause irreversible damage to your metabolic health and longevity. One of the reasons for that is the suppression of FOXO transcription factors and insulin resistance.

SIRT1 increases FOXO DNA binding by deacetylating FOXO in response to oxidative stress. FOXO proteins get increased in response to cellular stress and increased energy depletion.

- **Calorie restriction increases sirtuins as well as FOXO factors**[82][83]. Something we already know.
- **Fasting for 48 hours elevates FOXO1,3, and 4 by 1.5 fold and refeeding drops it back to baseline**[84]. FOXO1 is also critical for adapting to fasting by activating gluconeogenesis in the liver[85]. This makes the liver produce its endogenous glucose whether from amino acids or fatty acids[86].
- **Even just acute exercise increases FOXO1 phosphorylation, improves insulin sensitivity and promotes mitochondrial biogenesis**[87]. However, chronic exercise may decrease this exercise-induced FOXO expression[88]. FOXO factors are important for regulating muscle energy homeostasis and adapting to the stimulus[89].
- **In response to heat stress, Drosophila dFOXO contributes to increased heat shock protein levels, which will protect DNA damage and maintains cellular resistance**[90]. Taking a sauna, exercising and sweating can promote FOXO activation.
- **Exposure to cold stress before heat stress lowers FOXO translocation in fruit flies but it doesn't compromise longevity and resistance to the heat**[91]. Cold exposure actually can boost longevity and lifespan.
- **FOXO3a is activated in response to hypoxic stress and inhibits apoptosis**[92]. Hypoxia can be trained during exercise or some breathing methods.

The general trend for increasing FOXO follows the same pattern as the other longevity pathways such as AMPK and Sirtuins. Energy deprivation and adaptation to stress make the organism more resilient and longer lived. It forces the body to continue producing energy and survive in situations of low nutrients thus becoming really efficient at its own metabolic processes.

However, there's a dichotomy between experiencing too much stress and getting just enough. Being chronically taxed out and under harsh conditions will inevitably lead to an accelerated deterioration just because of the accumulated damage. The cells themselves react to this by looking at the functioning of telomeres.

Telomeres and Longevity

The human body consists of trillions of cells and genes that compose the superorganism you are.

- DNA molecules carry the genetic code that's necessary for cellular growth, replication, and functioning of all living organisms.

- Chromosomes are DNA molecules that contain the genetic material or genome of an organism.

- **Telomeres are nucleotide sequences at the ends of each chromosome that function as protective caps against deterioration.**

The sequence of telomeres is TTAGGG, which gets replicated over 2500 times in humans. In Greek, telomeres are derived from '*telos*' (end) and '*meros*' (part). And they're the tip of your genome.

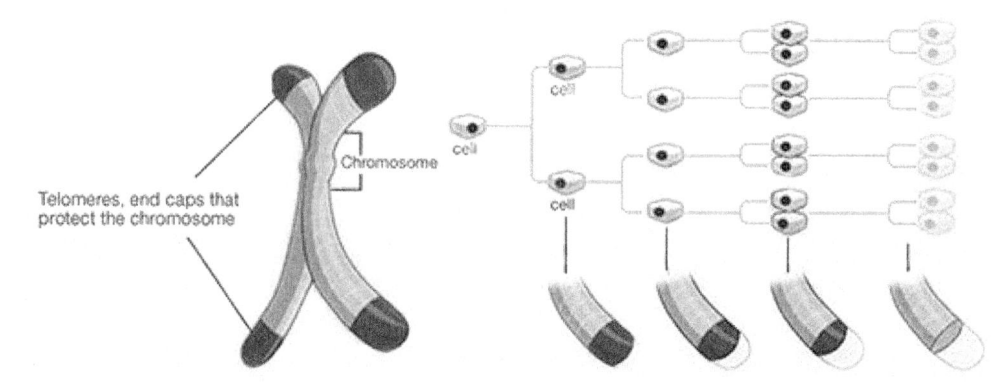

Figure 14 Telomeres are protective caps at the ends of chromosomes

Telomeres were discovered in 1933 by an American cytogeneticist Barbara McClintock who saw that chromosomes without ending points became sticky and hypothesized there are some caps that help to maintain chromosome stability[93].

In 2009, Elizabeth Blackburn, Carol Greider, and Jack Szostak were given the Nobel Prize in Physiology or Medicine for the discovery of how chromosomes are protected by telomeres and the enzyme telomerase[94].

In order to keep the organism alive, chromosomes are continuously replicating themselves and repairing DNA. Every bout of chromosome duplication causes a shortening in telomeres because the enzymes involved in duplicating DNA can't continue their duplication all the way to the end of a chromosome. So, telomeres are vital protective caps that are supposed to protect the genes from damage during this process. They're expendable, basically...

Telomere shortening prevents cells from replicating themselves by limiting the number of cell divisions. Shortened telomeres also weaken the immune system, increasing risk of cancer[95]. Many age-related diseases are linked to shortened telomeres[96]. Longer telomeres are linked to longer lifespans than shorter telomeres[97]. Whether or not short telomeres are just a sign of cellular aging or the main contributors to the aging process is still unknown. Probably both.

On average, human telomeres shorten from about 11 kilobases as a newborn[98] to less than 4 kilobases in elderhood[99]. You lose more than half of your telomeres once you become older. However, your biological age can differ from your chronological age as your telomeres can either shorten or lengthen based on lifestyle factors.

Telomere length can be replenished by an enzyme called *telomerase reverse transcriptase* (TERT). TERT is a subunit of *Telomerase* which adds TTAGGG sequences to the ends of chromosomes[100].

Telomerase activity can prevent the shortening of telomeres that occurs with aging. Telomerase is responsible for the self-renewal properties of stem cells by elongating the telomeres of stem cells, which prevent telomere shortening and increase the lifespan of stem cells[101].

Telomerase is more active in rapidly dividing cells such as embryonic stem cells and adult stem cells and they're quite low in neuronal, skin, pancreas, adrenal, cortex, kidney and mesenchymal stem cells.

- In embryonic stem cells, telomerase is constantly active and thus maintaining telomere length and making the cell almost immortal.

- However, most stem cells in the body don't have enough telomerase to prevent telomere shortening and thus they're still subject to aging albeit at a slower rate than in regular somatic cells[102].

Telomerase activity determines how many times a cell can divide before it dies off completely. It takes about 30-50 cycles of replication until the cell becomes senescent and dead.

Although telomere shortening is linked to aging, over-expression of TERT can promote cancer and tumor formation[103]. Telomerase activity can immortalize cancer cells[104] and about 90% of cancers are characterized by increased telomere activity[105]. This is probably due to the all-encompassing anabolic effects of cellular growth caused by the main growth pathway in the body mTOR that will grow healthy cells but unfortunately sustains cancerous cells as well if they're already present.

Telomeres are highly susceptible to oxidative stress and stress-mediated DNA damage is a huge contributing factor to telomere shortening[106].

Too much stress shortens telomeres but it also damages the mitochondria. That's why a healthy lifestyle should include active stress management. Here's how to increase telomere length and prevent telomere shortening:

- **Meditation helps to maintain telomere length and protect them from getting damaged[107].** It increases telomere activity and reduces oxidative stress[108]. Meditating lowers psychological as well as physiological stress and boosts the immune system[109] – it creates emotional resiliency against all types of stressors thus promoting lifespan.
- **Telomeres and fasting are also quite tied with each other.** Intermittent fasting can promote telomere activity and function through several mechanisms. It lowers oxidative stress, removes senescent cells through autophagy, and boosts stem cell production.
- **Resistance training and muscle building slow down aging.** As you age you decrease the number of satellite cells that are precursors to skeletal muscle cells and you lose muscle mass. Shortening of satellite cell telomeres prevents satellite cell replication and contributes to age-related sarcopenia[110]. Chapter VII is dedicated to longevity-oriented exercise.

Frequent exercise may lower oxidative stress and protect telomeres from reactive oxygen species (ROS)[111]. ROS are detrimental in large amounts but they can actually make your body stronger in the right dose by the principle of hormesis. Chronic stress leads to oxidative stress and reduced longevity.

Hormesis and Longevity

From an evolutionary perspective, all of these pathways evolved the way they did as to promote the survival of the organism under harsh conditions. The vast majority of history on Earth has been governed by caloric deprivation, environmental challenges, and energy restriction.

It's actually the „norm" to be deprived and in a conserved state as to survive the fray so to say. Such adaptation would've allowed species to enter into metabolic states that promote a longer lifespan.

The biological phenomenon to this is called '*HORMESIS*', which comes from the Greek word *hórmēsis* and it means 'rapid motion, eagerness' or 'to set in motion.'

Basically, hormesis is a biphasic response to a toxin or a stressor. (1) The initial contact causes injury to the body. (2) The following reaction leads to adaptation, leaving the body in a better condition than it was before.

Hormesis was first described by a German pharmacologist named Hugo Schulz in the year 1888. He discovered that a very small dose of lethal poison didn't kill off the yeast he was experimenting with but actually made them grow. The term 'hormesis' itself was coined and first used in a scientific paper by Chester M. Southam and J. Ehrlich in 1943 in the journal Phytopathology (Volume 33, pp. 517-541).

In 2012, Mark Mattson explained that cells respond to bioenergetic stressors by increasing DNA repair proteins, antioxidant enzymes and the production of neurotrophic factors (such as BDNF)[112][113][114]. It's also believed that this is the reason eating vegetables, tea or coffee can improve brain health.

- Plants contain 'noxious' chemicals that are supposed to protect them from being eaten by insects and other organisms.

- However, thanks to the constant evolutionary arms race between us and animal kingdom, we've developed counter-adaptations in the form of hormesis and we trigger a beneficial response when eating these foods.

Friedrich Nietzsche said: *"That which does not kill us makes us stronger."* He was a philosopher, not a biologist, but his quote fits hormesis perfectly.

Examples of Hormesis

- **Physical Exercise is a clear example of hormesis**. People with low levels of physical activity are said to be more prone to oxidative stress – not working out makes you more susceptible to a variety of diseases. How crazy is that? Exercise itself causes oxidative stress, especially at high intensities, like lifting weights or HIIT cardio. However, it triggers adaptations that increase mitochondrial density and biogenesis through mitochondrial hormesis.

- **Alcohol is another hypothetical form of hormesis** as it's believed to prevent heart disease and stroke. However, there's evidence to show that these benefits are exaggerated. I would say that maybe a shot of vodka or a glass of wine once a week can be good for you but when you start drinking anything beyond that then you're probably just lying to yourself – „*Ahhh, got to get my hormesis in*" and you end up drinking every day. Beer, ciders, long drinks, cocktails they're all anti-hormetic with zero advantages, like zero. You're only killing your brain cells and getting fatty liver disease. That's why I've chosen to go no alcohol because there are no significant benefits to it.

- **Red wine contains resveratrol**, which is one of those protective plant compounds and it's greatly associated with anti-aging effects. Maybe 1-2 glasses of red wine a week can be beneficial but, again, I wouldn't suggest drinking every day because alcohol is still a neurotoxin that directly damages nerve cells.

- **Exposure to Sunlight** at low or moderate doses has a lot of health benefits. It's one of the most effective ways of synthesizing Vitamin D in the body but it also supports most metabolic processes.

- **Cold Exposure** triggers AMPK, which causes your mitochondria to grow and improve their efficiency. It's also a positive adaptation to lower temperatures with many other health benefits, such as reduced inflammation, stronger immune system, and greater tolerance to pain.

- **Heat Exposure** activates these so-called 'heat shock proteins', which allow the cells to resist the damaging effect of heat. High temperatures can also stimulate the lymphatic system, which works like an inner pump for moving liquids and toxins in the body.

- **Caloric Restriction and Intermittent Fasting** cause mild oxidative stress that trigger protective sirtuin proteins. Short-term fasting and starvation can also fight cancer and make cancer patients more resilient against chemotherapy.

- **Mental Stress** is another example of hormesis where you're forced to flex your neural muscles. Learning new things, gaining skills, being in unpredictable high-stress situations, novel environments and challenges all trigger neuroplasticity and neurogenesis that make you grow new brain cells and create new synaptic connections. This actually bolsters your brain for future stressors and teaches you how to deal with them better mentally.

Whenever you're in difficult situations or whenever you're going through tough times in your life, then reminding yourself that it's okay – that this pain will make me stronger in the future – then you'll not only find more strength to deal with the present moment but you'll also augment your mindset for anything else that's to come.

There's the quote: *"Pain is inevitable. Suffering is optional."* Which means that you can't escape the physical stimuli – the pain, the discomfort, the emotions – but you can change your perception of it. Your perception of the things that are happening to you quite literally alter the way you experience and feel them. You're using your mind to cause a different response, which is just powerful.

The Price of Longevity

There are also reasons to think that increased longevity will have tradeoffs in reproduction or other performance-related functioning. However, the tradeoff theory has been found to be inconsistent in research:

- Guppies who live in nature grow and reproduce faster because of predatory pressures than guppies without natural predators. When predators are removed, these same guppies will still go and have more offspring and live longer[115].
- Fruit fly mutants that live longer have even more progeny than normally[116].
- Removing germline precursors in C. Elegans extends their lifespan[117]. This is not due to sterility but because of hormonal signaling of the DAF-16/FOXO pathway that localizes DAF-16 in the adipose tissue.

There's also the idea that stimulating growth with mTOR and IGF-1 will accelerate aging because of the growth mechanisms. However, there are as many reasons to think that inadequate levels of these pathways are as detrimental for longevity.

- mTOR helps to build muscle and prevents muscle wasting, which is increasingly more important as you get older**[118]**.
- IGF-1 and mTOR increase bone density and joint strength, especially in cartilage and tendons.
- Both high and low levels of IGF-1 are associated with increased mortality in a U-shape curve manner[119].
- IGF-1 fights autoimmune disorders by increasing T-cells[120] and supports proper growth.
- IGF-1 improves blood sugar regulation. Lower IGF-1 is associated with metabolic syndrome and insulin resistance[121]. This may be partly due to low insulin sensitivity that can be improved by building more muscle and stimulating mTOR.

Therefore, it's not all black and white with longevity and these pathways. This book seeks to optimize all of these mechanisms we've been talking about.

Chapter II
The Hedonic Treadmill

"A nation is born stoic and dies epicurean"

Will Durant

Modern society is completely bizarre compared to what wild animals experience. Humans have undergone a massive change in our surroundings and habitat. If you're reading this, then you're probably quite well off. At least in comparison to what our species faced in the past. The world in which we live isn't perfect and devoid of all suffering but, in my opinion, it's still better than ever before. Of course, our dangers may be much more threatening, such as nuclear explosions and global epidemics, but the majority of people are quite fortunate. However, this affluence has its cost, especially in the wealthier countries.

Paradoxically, the richer the country the more health problems it tends to face. Diseases of the Civilization as they're called are more predominant amongst upper-middle to high-income populations.

The issue isn't just about being obese or diabetic. The problem is that in the rise of civilization's comfort and abundance, we're teaching ourselves to lose some of the positive qualities of human nature. Too many suffer from poor self-control, bad dietary habits, horrible biomarkers, and the general hedonic attitude towards life.

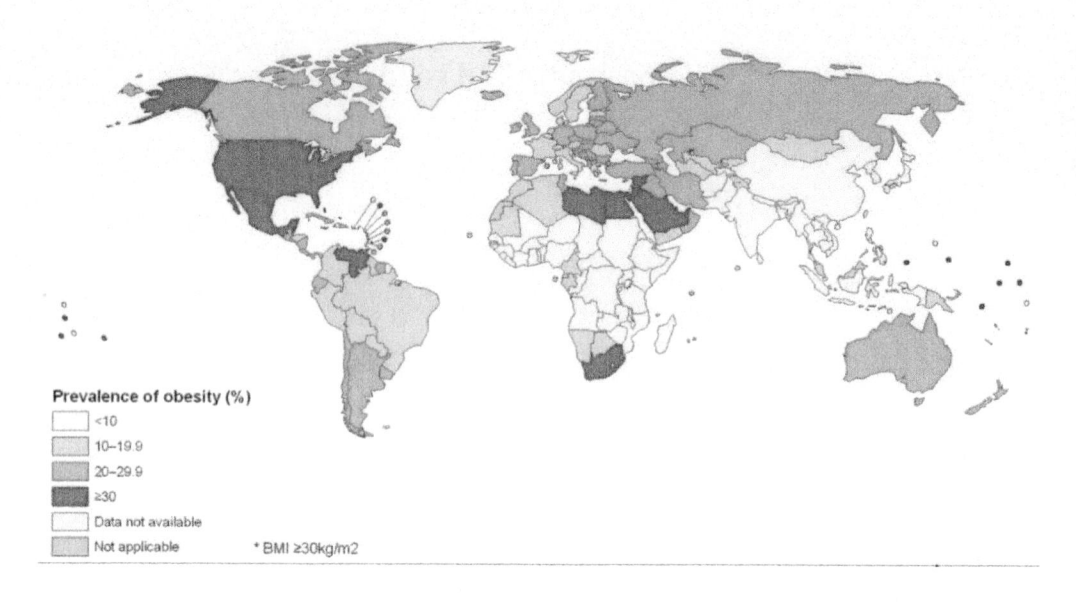

Figure 15 Rates of Obesity in the World

In an environment where food is more scarce and harder to come by, the situation would be slightly different. People either have to expend a lot more energy to get their calories or they simply wouldn't be able to consume them excessively. That's definitely not the only contributing factor to why richer societies tend to suffer from diseases but the correlation is easy to see.

Will Durant, a famous 20th-century historian said: *"A nation is born Stoic and dies Epicurean"*. Stoicism is a branch of Hellenistic philosophy that emphasizes personal ethics, logic, and virtuous living whereas Epicureanism places pleasure as the greatest good. When you look at history, then it's so true. Great civilizations of the past like Ancient Greece, Rome, Mesopotamia, the French Monarchy of Louis XIV all fell into the trap of excess glamour and comfort. As the people became wealthier, they became softer and thus more vulnerable to foreign invaders or upheavals. They became the victims of their own hedonic downfall.

As we know, the lessons of history tend to repeat themselves and human nature is very stubborn to change. That's why it's so important to create these situations of deliberate discomfort and voluntary challenges that elicit a hormetic response. Otherwise, you'd predispose yourself to not only metabolic disorders but you'd be less capable to face unpredictable events in the future. In fact, it can be said that facing adversity, overcoming obstacles, and experiencing suffering is an essential component to having a meaningful human experience.

I'm not saying we should have unnecessary pain in our lives. Instead, those uncomfortable and adverse situations help us to put things into perspective. They also contribute to our longevity through hormesis. Think of it as the Dauer state in larva who exponentially extend their lifespan just because of being put under environmental pressure. I know the resemblance isn't easy to accept but that's the general principle of stress adaptation.

Hedonic Adaptation

It's funny to be talking about such things in the first chapters of a nutrition book but I think it's very important. The more exposure you get to any stimuli, the more resistant you become towards it. Let's take insulin resistance as an example.

- Your blood sugar gets elevated → the pancreas releases insulin to bring it back down.

- Normally, insulin would unlock the cells to shuttle that glucose into glycogen stores but after a while, the pancreas gets taxed.

- Then the pancreas can't keep up with pumping out more and more insulin and your blood sugar levels remain elevated for too long.

- You become insulin resistant - not being able to produce enough insulin to metabolize carbohydrates properly.

Insulin resistance is one of the main driving factors of obesity and metabolic disorders in the Western diet. We'll be talking about it a lot more in Chapter XI.

In the example of caffeine, you start off by being satisfied with a single cup, then you build up your tolerance and you need more to feel the same effect.

This is called hedonic adaptation also known as the hedonic treadmill. It points out how we adapt to a stable level of happiness despite the ups and downs of positive gains or negative losses in our life. Despite the fortunate or disastrous events that we encounter, eventually, we'll reach the homeostatic level of happiness we were at previously.

- You win a lottery and you're on cloud 9 - new house, nice cars, fancy clothes, traveling the world and all that fun stuff. Then, after a few months, you adapt to this new lifestyle and it becomes normal. You're back at baseline.

- You break up with someone and feel heartbroken. After a while, you get over it and you're back in the game.

- You lost your job and had to cut down on your expenses. At first, it's uncomfortable and displeasing but soon you'll get used to it. Instead of eating $200 dinners every night you're satisfied with eating less glamorously.

- You start smoking only one cigar a day but over the course of a year you've built up to an entire pack to get the same effect.

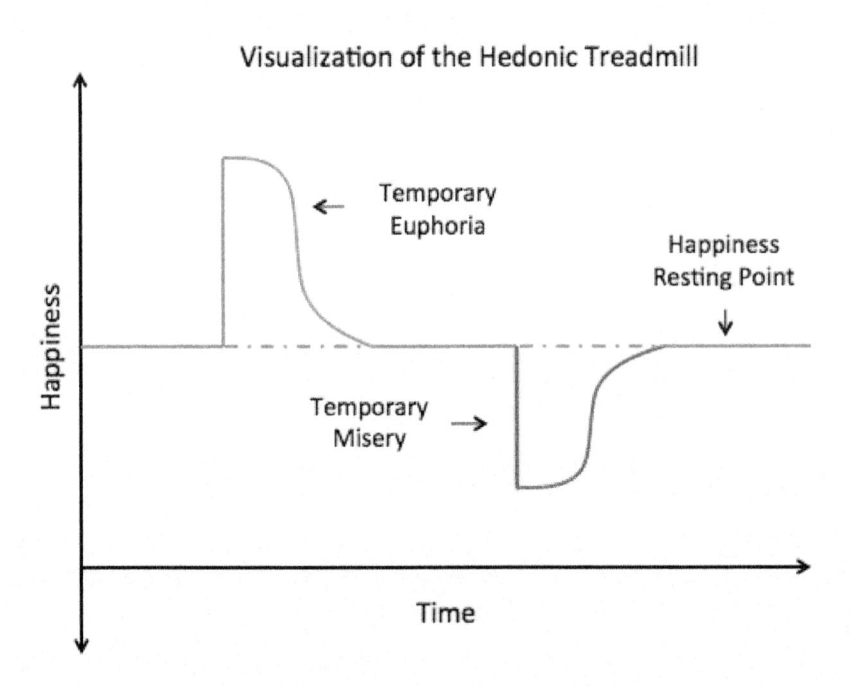

Figure 16 The ups and downs of the Hedonic Treadmill

Why is this so anyway? Why do people eat food past satiety until indulgence and get addicted to these simple pleasures that aren't even meaningful in the grand scheme of things?

The reason for that has to do with our primal origins again. You see, the change in our environment from scarcity to abundance has happened too quickly. We may live in the modern world, but our body thinks it's still in the ancestral landscape.

Because of this *"evolutionary time-lag",* our brain is always trying to motivate us to consume the most valuable sources of calories – salt, sugar, and fat. These 3 have the highest calorie density and can be easily stored for the dark times to come. Unfortunately, those times are happening less and less often.

The deadly combination of salt, sugar, and fat is like a drug, as it stimulates our taste buds in an addictive way and lights up the reward mechanisms in the brain. Michael Moss' book *Salt Sugar Fat: How the Food Giants Hooked Us* talks about how fast food companies have deciphered this secret code that makes us crave more food. It's called the bliss point – the specific amount of those 3 ingredients, which optimizes palatability. There's not too much nor too little, but just enough. By themselves, they're not inherently bad but when together they cause conflicting metabolic and hormonal effects within the body that leads to diabetes and obesity.

The reason why some people can't get enough enjoyment from healthy food is that their bliss point is too high. Refined carbohydrates, sweets, pastries and pizzas have overstimulated their taste buds. They simply can't even feel the taste of anything less than that. To keep up with their primal urges they want to increase their sensations even further. Instead of being satisfied, they keep craving for more and more.

Another thing that causes binge eating is leptin resistance. It's the *satiety hormone*, that regulates the feeling of hunger. Leptin's role is to signal the brain that there's dire need for calories. Once we get full it sends another message, saying that we've had enough.

However, if we're leptin resistant then the lines of communication will be cut short and our mind will never get the information that we've received enough food. In this case, the body is satisfied but the brain is still starving and keeps on craving for more stuff.

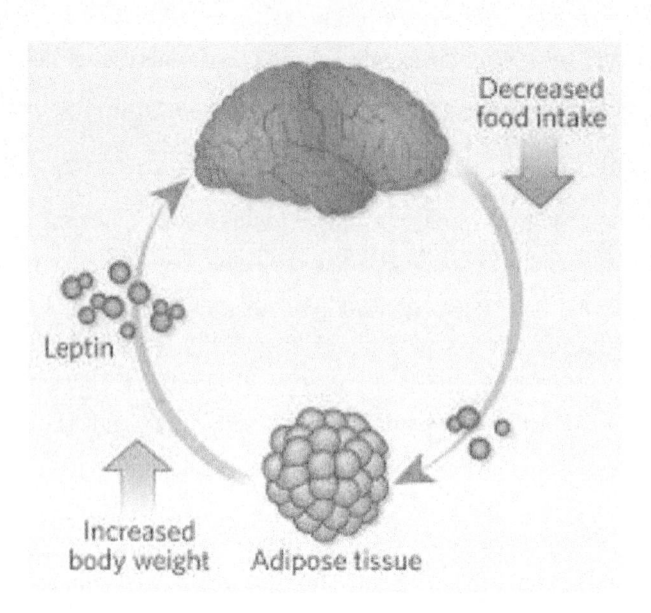

Figure 17 Leptin regulates appetite and body weight in response to food intake

Leptin resistance is caused mostly by emotional binge eating. Usually, it goes hand in hand with insulin resistance, as it's created by the consumption of simple carbohydrates and sugar with a lot of fat at the same time. These combinations of foods affect the mental processes of people and are the most common cause of obesity and diabetes.

Living organisms are hard-wired towards preserving energy to guarantee survival and avoid pain, which gets regulated by the homeostatic balance of the body. Core temperature, blood pressure, daily caloric expenditure, and hormetic conditioning are all linked to this.

Basically, hedonic adaptation is about you getting comfortable with a particular stimulus and it becomes your default state. You reach a new homeostasis. That can be part of the reason why richer countries tend to over-indulge – they've simply gotten used to the pleasurable and comforting aspects of food.

As a conscious human being who seeks to optimize their health and overall life, you should want to pay close attention to where your hedonic homeostasis is. We're always moving up and down the hedonic ladder with some activities making us less comfortable whereas others less so. The level we dwell on the most becomes our default state – the place where we feel the best and satisfied.

If your baseline for feeling pleased and joyous about food is way up in the clouds – needing 6-course gourmet dinners or even highly stimulating fast food that hijacks your taste perception – then you need to be eating more of those things to feel satisfied. On the flip side, if you were to habituate yourself with healthier but slightly blander meals, you'd get the same satisfaction and blissfulness. The difference is that you're not over-indulging on empty calories or teaching yourself to like only very rich foods. Subjectively, the feelings are the same but in terms of the final outcome and calories, it's a completely different story.

Obesity and metabolic diseases are primarily the outcomes of physiological ailments in the body as well as psychological hedonic adaptation to the dopamine rush of highly stimulating foods that make the person follow certain bad lifestyle practices. A rational person wouldn't consciously harm itself with a diet they know is bad for them in the long run. It's just that many people are very much controlled by their emotions, feelings, sensations, and thoughts, which unfortunately can be easily manipulated.

Now, the point of all this isn't that all pleasure and comfort should be avoided. Of course, life is to be enjoyed but what hedonic adaptation shows us is that the feeling of joy and happiness depend on our subjective homeostasis - just our perceptions and conditioning. The question you'd want to ask yourself is what emotions, pleasures, discomforts, foods, stimuli are you exposed to most often?

All addictions are not as much caused by any particular stimulus but more so by our attachment to it. You've habituated yourself to doing something, like smoking a pack a day, having a drink in the evening or putting extra cheese on your food. Your happiness isn't linked to how high your homeostasis is in particular but in your adaptation to the stability and comfort you experience. Which means you can be as happy as you are right now if you were to smoke only a single cigar a day or eating just plain broccoli. You just have bad habits.

The key to overcoming any addiction is to detach yourself from the thing that you're addicted to - you have to reduce your exposure to the stimulus. What I advise you to do is to abstain from it completely - just start fasting.

- Intermittent Fasting resets your taste buds and makes healthy food taste amazing. Junk food will actually become too stimulating.

- Avoiding caffeine for a certain period of time will lower your tolerance to it. You'll get more energy from less coffee.

- Not consuming social media and entertainment for a while will give your brain a break from being constantly stimulated and triggered. This will help you to become more mindful and focused in life.

- Sleeping on the floor or outside every once in a while reminds you how fortunate you really are for having even just a roof over your head. It can also condition you to hold a better posture.

If you understand this concept, then you can see that it's a massive hack for happiness. You'll become happier with less.

However, we shouldn't go extreme with it - i.e. full monk mode in a cave with no clothes or material possessions. We can still experience the pleasures of life and have a greater meaning at the same time. The key is to not make the mistake of scaling up our homeostasis and never coming down from it. We should experience the highs and the lows so that we could appreciate the things we already have. Buy yourself nice things, but get accustomed to being happy without them.

This is what's taught in Stoicism as well. One of the more renown writers of this philosophy, Seneca had an exercise where he voluntarily practiced poverty. He said:

> *Set aside a certain number of days, during which you shall be content with the scantiest and cheapest fare, with coarse and rough dress, saying to yourself the while: 'Is this the condition that I feared?'*

Seneca was the richest banker in Rome, yet he deliberately put himself through difficult situations, discomfort, and pain. This not only made him resistant against turmoil but also allowed him to calibrate his hedonic homeostasis as he wished. His happiness wasn't dependent on anything external because he found it within himself.

Hedonic adaptation can be experienced in pursuit of longevity and performance as well. The idea that you're going to die is both a catalyst for living out your true potential as a human being and a source of existential discomfort at the same time. If you were to live forever, then you'll come across the dichotomy of having all the time in the world to do anything you want, thus getting nowhere, and having so much time that you'll be able to do everything. It's a paradoxical situation the outcome of which depends on your hedonic homeostasis and mindset. Whatever the case may be, the process of living, fulfilling your desires, pursuing your goals, and dying itself are wherein you'll find the most happiness. You just have to make sure that it does give you some sort of a greater sense of meaning. Otherwise, you'll end up with immortal suffering and apathy.

This is still a book about diet and nutrition but I want you to realize how big of a role your mindset and psychology plays in all of this. How healthy you are and how long you'll live are very much dependent upon your daily habits – the small decisions and activities you do all the time. They might not seem that significant in the short term but they're actually the pillar stones of your longevity in the long run.

Think about it…If you're the kind of person who is falling off the rails with their diet or skipping workouts on a regular basis, then it's going to add up. After a while, you may gain 10-20 pounds of unnecessary extra weight just out of habit. You don't have to even notice that you're doing something wrong. It's just the culmination of certain things that you simply didn't pay enough attention to. That extra serving of cheese, sporadic snacking, taking the elevator more often, and dropping the ball many times will scale up the hedonic ladder.

Lao Tzu, one of the most renown philosophers of Ancient China and the founder of Taoist philosophy has a quote: *"Deal with the big while it is still small."* The health of your body is THE BIGGEST THING in your life because it literally anchors your psychosomatic experience of the world. That's why you shouldn't take it for granted. Furthermore, your life expectancy and longevity are another thing that you can't expect to focus on only when you're about to die. It'll be too late by then. Hell, it was probably too late several decades ago…

If you're planning on living a healthier life that's long, full of vigor, bliss, and happiness, then you have to be actively working at it all the time. You may be already eating right, exercising, and sleeping properly. However, by picking up this book, you've now embarked on a much more thorough and conscientious routine of health optimization that will not only improve longevity but increases performance in everything you do as well.

In the context of diet and nutrition, then one of the most efficient ways of avoiding metabolic disease as well as increase life expectancy is intermittent fasting. That's the main practice of the Metabolic Autophagy Protocol. That's what we'll talk about next.

Why Intermittent Fasting

Although currently living in a modern world, our bodies still think we're in the ancestral savannah with lions and stuff. This reflects in the way we metabolize food, experience stress, adapt to physical conditions and also how our psychology functions.

You see, the eating patterns of our hunter-gatherer ancestors were highly unpredictable. Quite frankly, completely random. Sometimes they had a whole lot and at others nothing at all. Whatever the case, they were always in between fasting and feasting. They did both intermittently. This cycle wasn't deliberate but created by the scarcity of food in their environment.

Intermittent fasting (IF) not only has incredible health benefits that are linked to increased longevity but also has psychological effects that help to escape the hedonic treadmill. In a world of unlimited empty calories and too much stimulation, the easier thing to do is to just say 'NO.' This is such a huge life hack that once you've tried it you wish you started sooner. I've been practicing IF since high school practically every day and it's one of the best decisions I've made in regards to nutrition. You might ask: *"Who would be crazy enough to do this? It doesn't make any sense."* This is what decades worth of brainwashing and misleading diet advice may do to unwary minds.

It's quite paradoxical that the majority of people in the world go to bed while starving every single night, but at the same time, almost everyone in Western society is obsessively obese.

The Pareto principle applies here perfectly. He was an Italian economist and in his 1896 paper showed that, in most cases, about 80% of the effects come from 20% of the causes. 80% of the wealth belongs to 20% of the people, 80% of car accidents happen to 20% of the drivers, 80% of the food is consumed and stored as fat in 20% of the world's population etc. Mainly it's used in economics, but this 80-20 rule, or the law of the vital few, is evident in the distribution of calories and obesity as well.

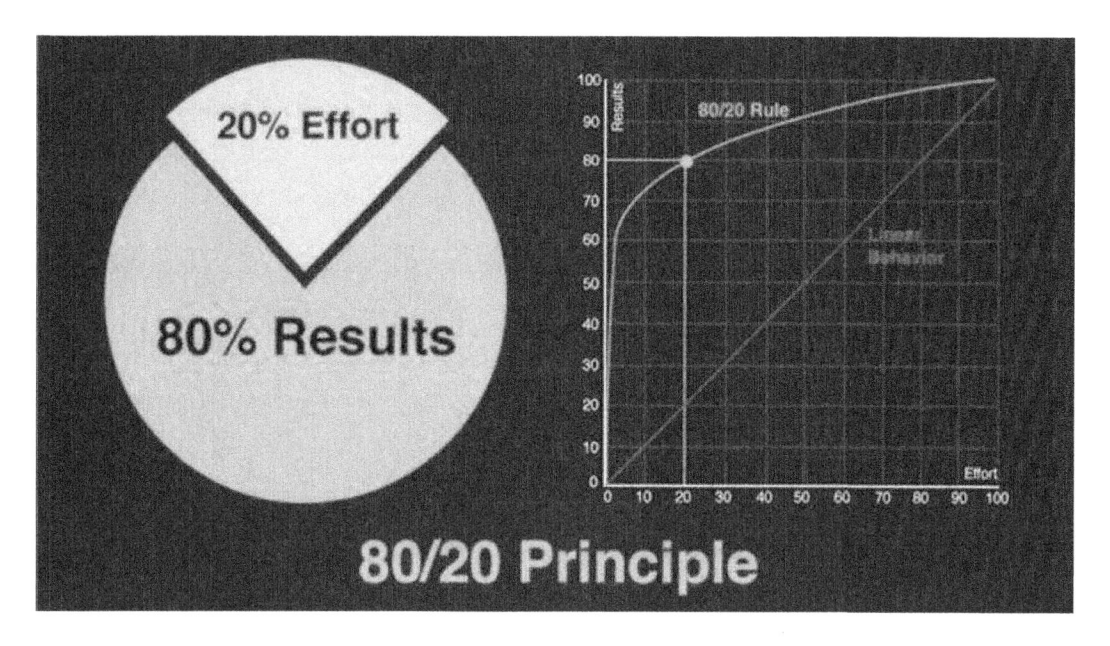

Figure 18 The Pareto Principle - 80% of results from 20% of the effort

Our civilization has reached a point where we don't have to worry about our most primary needs as much and can now spend more time on other activities that develop us further as a species. There's nothing wrong with that. In a perfect world, no animal would have to kill another one and everyone would always be fed and satisfied. However, we don't live in such a place yet, at least for the time being.

Get your head around fasting. In contemporary nutrition advice, it's the F word - the forbidden fruit. *Don't skip a meal or else...* Or else what? IF is such a natural way of eating and it's even more suitable for the modern environment where most people could use some positive restriction. So, how does it work?

The Physiology of Fasting

Intermittent fasting (IF) is a way of eating (or not eating), where the food consumed is restricted to a certain time frame. This means that no calories whatsoever get put into the body in any shape or form. In a way, it's simply timing when you eat..

The two governing states of metabolism are fed and fasted – anabolism and catabolism. The former is when we're using the macronutrients eaten, that have been digested and are now circulating the bloodstream. The latter happens when all of that fuel has run out and our gas tank is empty, so to say. It happens after several hours (7-8) of not eating.

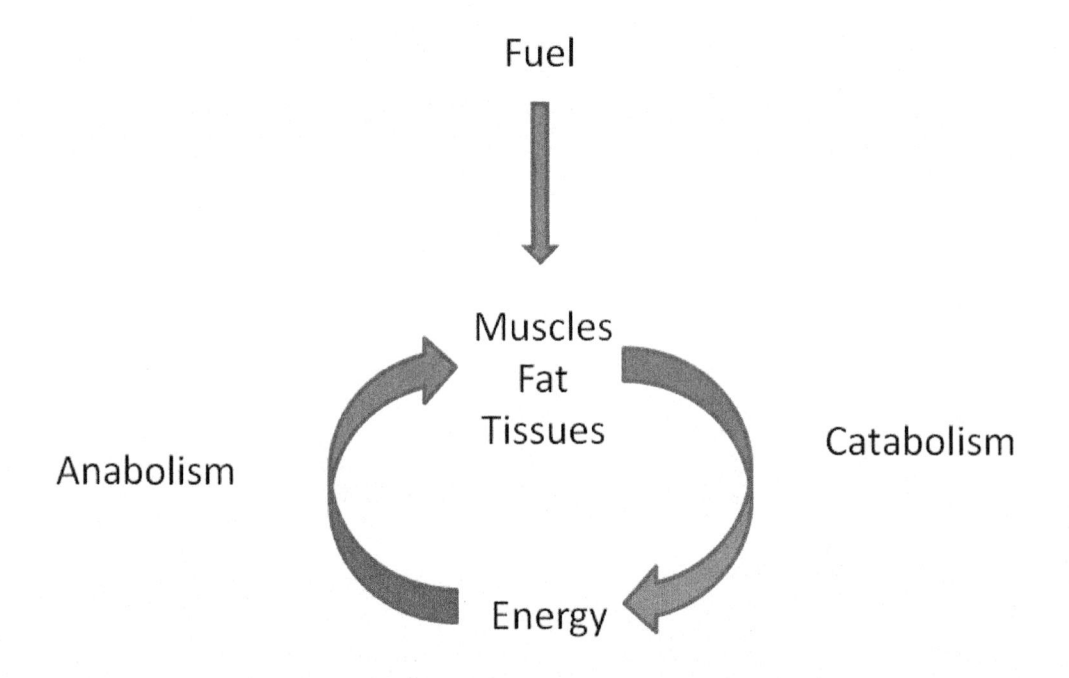

Figure 19 Anabolism and Catabolism in Action

Anabolism also refers to building up whereas catabolism entails breaking down. Both of them can contribute to each other's processes – you need to catabolically dissolve the food you've eaten to become anabolic etc. They're the different sides of the same coin.

How Does the Body Produce Energy While Fasting?

The body's default fuel source is glucose, which exogenously (externally) comes in the form of sugar and carbohydrates and is stored endogenously (internally) as glycogen. The liver can deposit 100-150 grams of glycogen and muscles about 300-500 grams. They're used for back-up.

Liver glycogen stores will be depleted already within the first 18 to 24 hours of not eating - almost overnight. This decreases blood sugar and insulin levels significantly, as there are no exogenous nutrients to be found.

Insulin is a hormone released by the pancreas in response to rising blood sugar, which happens after the consumption of food. Its role is to unlock the receptors in our cells to shuttle the incoming nutrients into our muscles, or when they're full into our adipose tissue (body fat).

The counterpart to insulin is glucagon and also gets produced by the pancreas. It gets released when the concentration of glucose in the bloodstream gets too low. The liver then starts to convert stored glycogen into glucose.

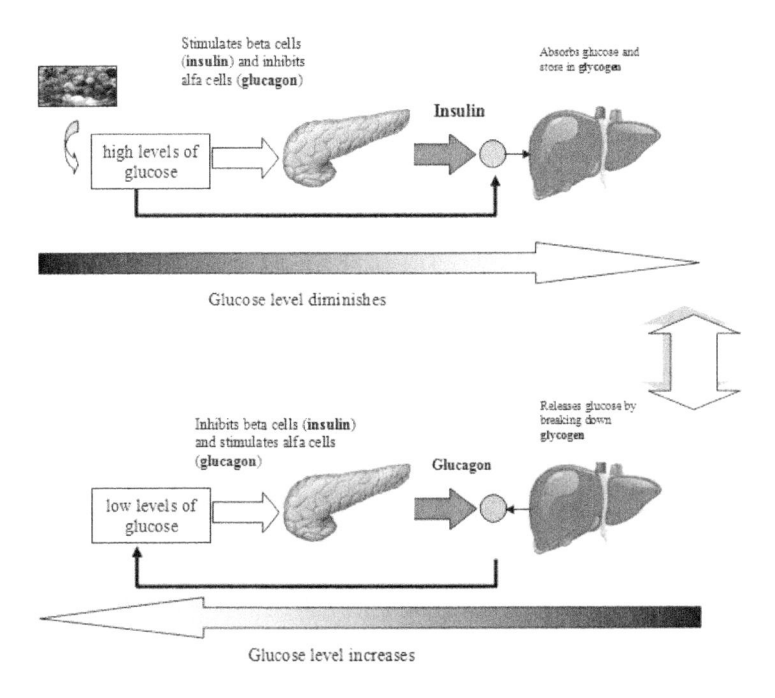

Figure 20 Insulin and Glucagon are Constantly Balancing Blood Sugar

Fasting and Ketosis

As fasting continues, the liver starts to produce ketone bodies which are derived from our own fat cells. Lipolysis (breakdown of stored triglycerides in the adipose tissue) and *ketogenesis* increase significantly due to fatty acid mobilization and oxidation.

Ketosis can occur already after 2-3 days of fasting. Triglycerides (molecules of stored body fat) are broken down into glycerol, which is used for gluconeogenesis (creation of new glucose) and three fatty acid chains. Fatty acids can be used for energy by most of the tissue in the body, but not the brain. They need to be converted into ketone bodies first.

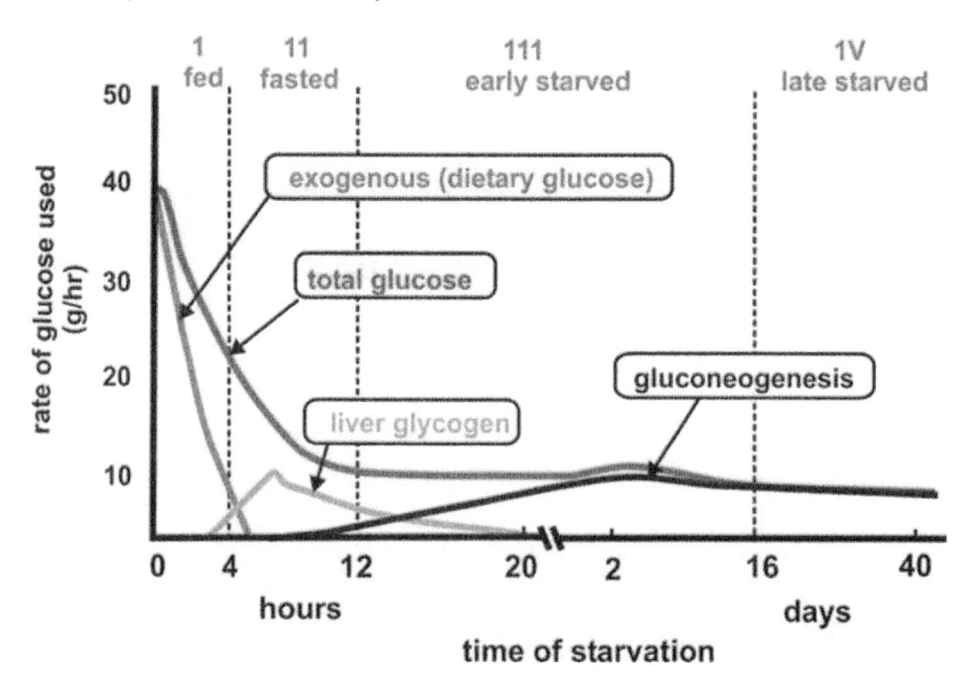

Figure 21 The stages of energy usage during fasting

Ketosis is a metabolic state in which fat is the primary fuel source, instead of glucose, and can be achieved either through fasting or by following a ketogenic diet. Fasting induces ketosis very rapidly and puts the body into its more efficient metabolic state. The more keto-adapted you become the more ketones you'll successfully utilize. At first, the brain and muscles are quite glucose dependent. But eventually, they start to prefer fat for fuel.

Ketone bodies may rise up to 70-fold during prolonged fasts[122]. After several days of fasting, approximately 75% of the energy used by the brain is provided by ketones. This also allows other species, such as king penguins, to survive for 5 months without any food[123].

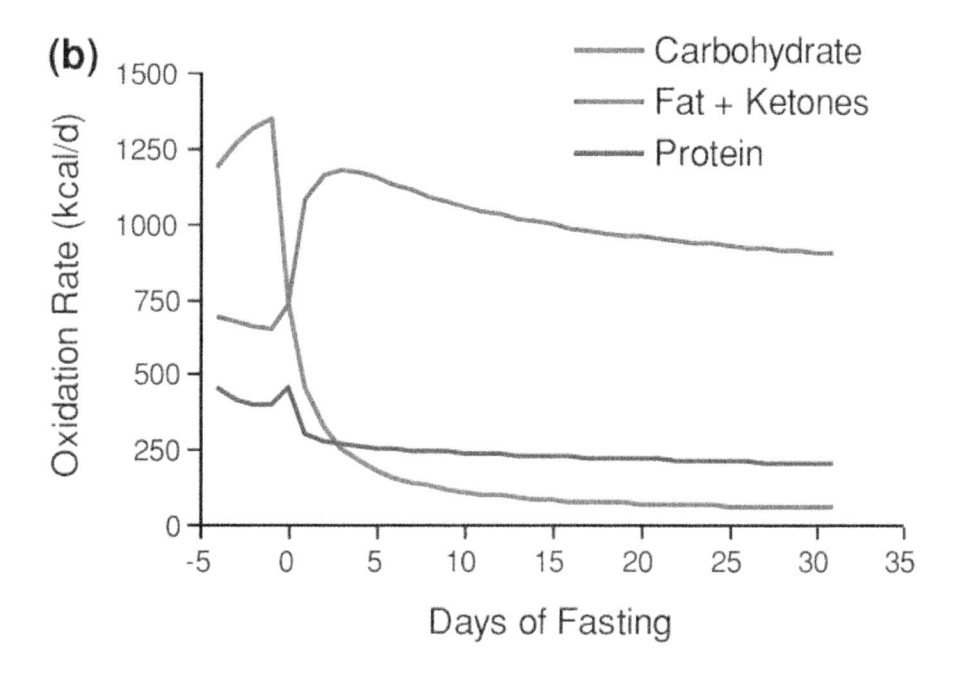

Figure 22 While fasting, glucose drops and ketones soar

Protein catabolism decreases significantly, as fat stores are mobilized and the use of ketones increases. Muscle glycogen gets used even less and the majority of our energy demands will be derived from the adipose tissue. This can be accomplished by following a well-formulated ketogenic diet as well, which actually mimics the physiology of fasting almost entirely.

The Krebs cycle is a sequence of reactions taking place in our mitochondria that generate energy during aerobic respiration (a fancy way of saying breathing normally). When glucose enters this metabolic furnace it goes through glycolysis, which creates the molecule pyruvate. In the case of fatty acids, the outcome is a ketone body called acetoacetate, which then gets converted further into beta-hydroxybutyrate and acetone.

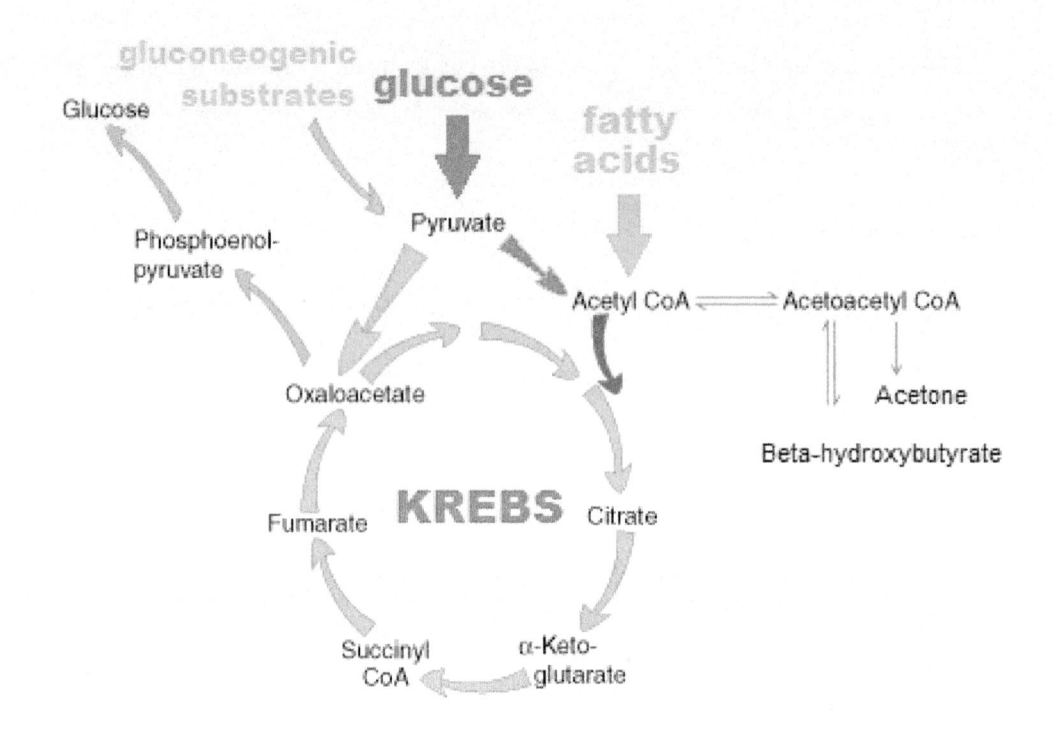

Figure 23 The Krebs Cycle That Works Like the Furnace of Your Metabolism

The difference between pyruvate and ketone bodies is that the latter can create 25% more energy. On top of that, the by-products of glycolysis are advanced glycation end-products (AGEs), which promote inflammation and oxidative stress[124][125], by binding a protein or lipid molecule with sugar. They speed up aging[126] and can cause diabetes.

Fasting VS Caloric Restriction VS Starvation

An important notion is to distinguish fasting from starvation. One is *voluntary* and *deliberate,* the other is *involuntary* and *forced upon.* It's like the difference between suicide and dying of old age. Abstention from food is the art of manipulating our metabolic system and can be done for many reasons. Malpractice might look like the person is starving, but if done correctly it's very healthy and good for you. You just have to know how to balance it with sufficient anabolism and get enough essential nutrients.

- **Starvation is a severe deficiency in energy intake.** The body doesn't have access to essential nutrients and is slowly wasting away by cannibalizing its vital organs. It's a gradual process of degradation that's often characterized by the skinny-fat look or the bloated stomach called *kwashiorkor* which is caused by insufficient protein even in the presence of sufficient caloric intake.

- **Caloric Restriction reduces calorie intake without causing malnutrition or starvation**[127]. You're simply consuming fewer calories needed to maintain your body's current energy demands. This will make you burn your stored fat and also lowers the body's overall metabolic rate, down-regulated reproductive hormones, thyroid functioning and promotes gluconeogenesis. The difference between caloric restriction and starvation is that when calorically restricted, your body still gets access to the energy it needs to maintain its daily energy demands. It's just that those energy demands have adapted to be lower and more efficient in terms of energy gained per calorie.

- **Fasting is a state of metabolic suspension in which you're not consuming any calories.** Despite that, your body is still nourished and gets the energy it needs. This happens by shifting into ketosis, in which you'll be burning your body fat almost exclusively.

Fasting doesn't equal starvation because your body is in a distinct metabolic state. Being fasted and fed is quite binary – even small amounts of food will shift you into a fed condition.

Fasting isn't entirely the same as caloric restriction either because you can be consuming fewer calories but still not enter into a fasted state.

During World War II, they conducted a study called the Minnesota Starvation Experiment[128] on a group of lean men who reduced their calories by 45% for 6 months[129]. Their diet consisted of primarily carbohydrates which comprised 77% of total calories and had very little protein to mimic starvation conditions. They ate potatoes, cabbage, macaroni, whole wheat bread while still maintaining their active lifestyle. After the experiment, the men showed a 21% reduction in strength, decline in energy and vitality. One of them started having dreams about cannibalism.

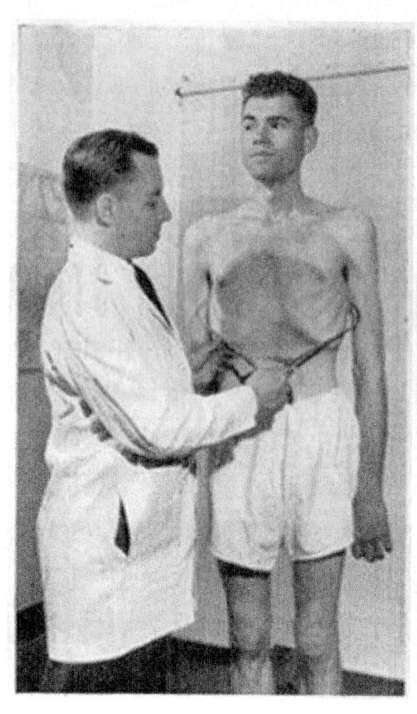

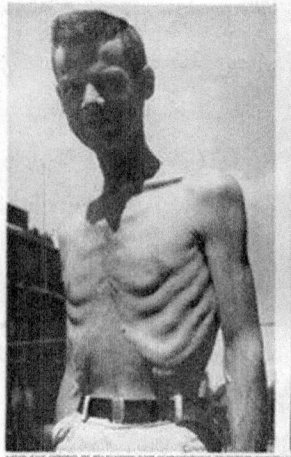

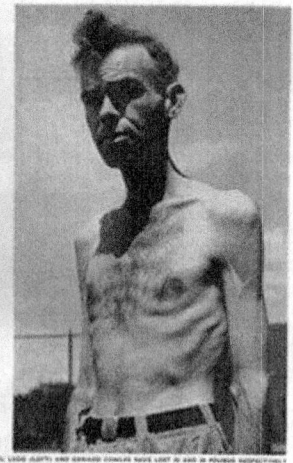

MEN STARVE IN MINNESOTA

CONSCIENTIOUS OBJECTORS VOLUNTEER FOR STRICT HUNGER TESTS TO STUDY EUROPE'S FOOD PROBLEM

Above:
Conscientious objectors during starvation experiment.
Life magazine - July 30, 1945. Volume 19, Number 5, p. 43.
Credit: Wallace Kirkland/Time Life Pictures/Getty Images.
Left:
Dr Ancel Keys measures the chest width of James Plaugher.

Figure 24 The Minnesota Starvation Experiment

The men in the Minnesota experiment were put under severe caloric restriction that resulted in starvation-like symptoms. Even though they were very malnourished, they weren't fasting because of still eating a significant amount of carbohydrates that kicked them out of ketosis.

Daily caloric restriction decreases metabolism, so it's easy to presume that this would be magnified as food intake drops to zero. However, this is wrong. Once your food intake stops completely (you start to fast), the body shifts into using stored fat for fuel (ketosis). The hormonal adaptations of fasting will not occur by only lowering your caloric intake. In the case of being fasted, your physiology is under completely different conditions, which is unachievable by regular eating.

Malnourishment happens when there is not enough nutrition to be found *i.e.* you go on a weight loss diet and restrict calories. While fasting, the organism is almost never fully deprived of essential nutrients, unless you lose all of your body fat. These fuel sources are mobilized from internal resources.

To prevent malnutrition and starvation, while restricting calories, you want to establish ketosis and autophagy as soon as possible. Even consuming small amounts of food will put you into a fed state. It doesn't matter whether you eat 200 calories or 1000, you'll still be shifted out of a fasted state.

That's why intermittent fasting is a lot better than caloric restriction. If you're feeding yourself, but in inadequate amounts, then your body will most definitely perceive it as scarcity. This will decrease your metabolic rate and creates a new set-point at which you can lose weight that's lower than previously. You'll be causing more damage than good. If you do it the wrong way, you'll end up like someone from the concentration camps.

Fasting isn't a mechanism of starvation because your metabolism will be altered. This shift won't occur entirely if you continue consuming food, even when you've reduced your calories to a bare minimum. It's actually a lot healthier way of losing weight, as you'll be burning only fat, not muscle. When on a restrictive diet you'll never make the leap and to keep your energy demands at a balance you begin to cannibalize your own tissue. When in a fasted state, this can be circumvented.

Instead, fasting is one of the best ways of escaping the hedonic treadmill and becoming immune to the ebbs and flows of nutrition. Not only will it improve your overall health but makes the relationship with food better. In the next chapter, we'll talk about what the benefits are.

Chapter III
Why Intermittent Fasting

"If a man has nothing to eat, fasting is the most intelligent thing he can do."

Hermann Hesse

Intermittent Fasting (IF) is quite well documented in science as well as history as an amazing dietary strategy with many health and anti-aging benefits.

Hippocrates, the father of modern medicine, said: *"To eat when you are sick, is to feed your illness"*. He recommended abstinence from food or drink for almost all of his patients.

Before Hippocrates, the infamous Greek god Hermes Trismegistus, or Thoth the Atlantean in Egypt, said to have invented the art of healing. He used fasting to unveil the secrets of the cosmos by cleaning his body and purifying his mind.

Plutarch said: *"Instead of using medicine, rather, fast a day."* In Ancient Greece, consuming food during illness was thought to be unnecessary and even detrimental, since it would stop the natural recovery process. There was also a widespread belief, that excessive food intake could increase the risk of demonic forces entering the body. Perhaps obesity and slothfulness so many are plagued by today?

The Greeks believed that abstention from food improved cognitive abilities as well. Pythagoras systematically fasted for 40 days, believing that it increases mental perception and creativity. He also wouldn't allow any of his students to enter his classes unless they had fasted before. So did Plato, Socrates, and Aristotle – all of the great philosophers. They frequently abstained from food intake for several days.

This would seem reasonably effective, as the only way for a hunter-gatherer to end their starvation was to get smarter and more efficient at chasing game. Toolkit complexity is linked with increased hunting and fishing practices[130], as traps and toggle-headed harpoons require more intelligence to make than simple digging sticks. Unlike plants, animals run away and if you want to eat them, you have to come up with better ways of catching them. This ever-imposed stress on hunting societies was probably one of the major driving forces behind the development of our species.

Fasting's been written about in many myths, legends, and religious texts as well. Christianity, Judaism, Buddhism, and Islam all instruct abstention from food in some shape or form. Similar teachings are preached in philosophical, moral and tribal codes. It's commonly thought of as a definite way of creating a communion between God or other divine deities. The only religion that prohibits fasting is Zoroastrianism, because of its belief that such asceticism will not aid in strengthening the faithful in their battle against the sources of evil. But intermittent fasting has extremely empowering effects, which actually make the practitioner stronger, healthier and sharper, especially mentally. Those traits are definitely useful for battling everything.

Siddhartha by Herman Hesse is a story of Gautama Buddha that talks about his life and path towards enlightenment. In the book, he is a monk and beggar who comes to a city and falls in love with a famous seductress Kamala. He makes his move on her but she asks: *"What do you have?"* One of the well-known merchants challenges him as well: *"What can you give that you have learned?"* Siddhartha answers the same way in both cases, which leads him to ultimately getting everything he wants. He said: *"I can think. I can wait. I can fast."* What does it mean?

- **I can think**: You possess good judgement and you're able to make good decisions.

- **I can wait**: You have the patience and perseverance to play for the long-game without expending unnecessary effort.

- **I can fast**: You are capable of withstanding difficulties and challenges. Fasting trains you to control your physiology and makes your mind more resilient.

These three traits are extremely useful for living life according to your own terms. This book will teach you how to cultivate all of them.

The Sumerian epic hero Gilgamesh sought to find immortality to revive his fallen companion Enkidu. He traveled the landscape for months and walked through the darkest of caves until he reached a sage. During his entire journey he hadn't eaten anything – day and night - and was thus considered enlightened and worthy of receiving a small potion from the Fountain of Eternal Youth. It didn't help him though, because it got stolen by a serpent while the man was sleeping. He reached immortality nonetheless, as his people kept telling his story and carried his achievement through centuries.

For health, fasting has been advocated by many, even since the Middle Ages. Luigi di Cornaro was a Venetian aristocrat living at the 15th and 16th century. By the age of 40, he had gathered severe illnesses thanks to excessive eating and drinking. No medication or physician was able to help him and his days seemed to have been numbered. However, there was one doctor who, contrary to popular practice, suggested periodic strict abstinence from food. Cornaro survived and rid himself from all his diseases. When he was 83 years old, he wrote his first treatise *Tratatto de la Vita Sobria (A Treatise of Temperate Living)* and died at the age of 102. Talk about a comeback!

The Swiss-German doctor Paracelsus was also in favor of this. He said: *"Fasting is the greatest remedy, the physician within."* In the medical community, Paracelsus is famous for revolutionizing the medical sciences, by utilizing empirical observations from nature, rather than referring to ancient texts. He also gave zinc its name, calling it *zincum* and noted that some diseases are rooted in psychological conditions. For all of his patients, he advised intermittent fasting. Not something you see in today's doctors who would much rather prescribe drugs.

In the mid-1800s, E.H. Dewey, MD published a book called *The True Science of Living*, in which he said: *"every disease that afflicts mankind develops from more or less habitual eating in excess of the supply of gastric juices."* Basically, consuming too much food too often, without letting the body repair itself. That's why proper autophagy is central to longevity.

Since the 20th century, as doctors became more knowledgeable about the human body, fasting became increasingly popular as disease treatment and prevention. Some methods lasted more than a month, allowing only the consumption of water, non-caloric beverages, exercise, and enemas. Other modified fasts allowed the intake of about 200 to 500 calories a day, which came from vegetable soup, broth, honey or milk.

There have been many non-obese persons who have recorded their prolonged bouts of abstention from food. Alexander Jacques fasted for 30 days in 1887 and for 40 the year after[131]. An Italian *professional faster*, Signor Succi, said that he had done at least 32 fasts of 20 days or more, with his longest ones lasting for 40-45 days[132].

For obesity, fasting has been used as an effective therapy for a long time. Overweight patients have been put on fasting regimens of up to 159, 200 and 249 days[133]. The longest recorded fast lasted for 382 days (1 year and almost 1 month). In 1965, a 27-year old obese man Angus Barbieri from Scotland asked doctors to help him to fast as long as possible. He lost 125 kg (275 lb) in the process by going from 375 lb to 100 lb[134].

Figure 25 Angus Barbieri fasted for 365 days and lost 275 pounds

However, the average person in the Western world hardly experiences the sensation of real physiological hunger nor do they enter a fasted state. This is a shame as it will not only prevent them from getting all of the benefits but also make them more fragile and dependent on their illusory food.

What I'm going to be doing now is go through all the physiological benefits of fasting and how it's going to affect your health. Starting with the most important component of the cell.

Fasting and the Mitochondria

The key to keeping mitochondria healthy is to maintain energy homeostasis and remove dysfunctional cellular components that are causing inflammation.

Time-controlled fasting prevents mitochondrial aging and deterioration. It can also promote the longevity of mitochondria by eliminating the production of reactive oxygen species and free radicals by dysfunctional organelles.

The study we mentioned earlier on nematodes in the journal Cell Metabolism specifically found that **fasting and caloric restriction substantially increased the lifespan of nematodes by promoting autophagic mitochondrial fission-fusion dynamics**[135]. One of the scientists William Mair said:

> *Our work shows how crucial the plasticity of mitochondria networks is for the benefits of fasting. If we lock mitochondria in one state, we completely block the effects of fasting or dietary restriction on longevity.*

Building new mitochondria is also critical for keeping yourself youthful and energized throughout your lifetime.

Mitochondrial biogenesis is the process of building new mitochondria through the activities of certain metabolic regulators such as PGC-1α[136] and AMPK[137]. AMPK produces new mitochondria and controls mitophagy as well.

The key to growing new mitochondria is to signal the body to produce energy under energy depletion and in stressful environments. This causes cellular crises that need to be compensated for by building new power plants.

- **Fasting increases AMPK which promotes fatty acid oxidation which produces ketone bodies**[138]. The mitochondria run a lot better on ketone bodies because they can get into the mitochondria faster via the electron transport chain and they'll yield more ATP than glucose.

- **Fasting increases FOXO proteins**, which regulate longevity through the insulin/IGF-1 pathway and mitohormesis. FOXO1 and FOXO3 promote mitophagy[139].

- **Fasting increases sirtuins**. Sirtuins regulate fat and glucose metabolism in response to physiological changes in energy levels, thus they're crucial determinants of energy homeostasis and healthspan of the cell[140]. SIRT1 regulates mitochondrial biogenesis and PGC-1α[141]. Suppressing SIRT2 restricts fatty acid metabolism, reduces mitochondrial activity, and promotes obesity[142]. SIRT3 is the major mitochondrial deacetylase and it protects against oxidative stress[143] through its anti-oxidant properties.

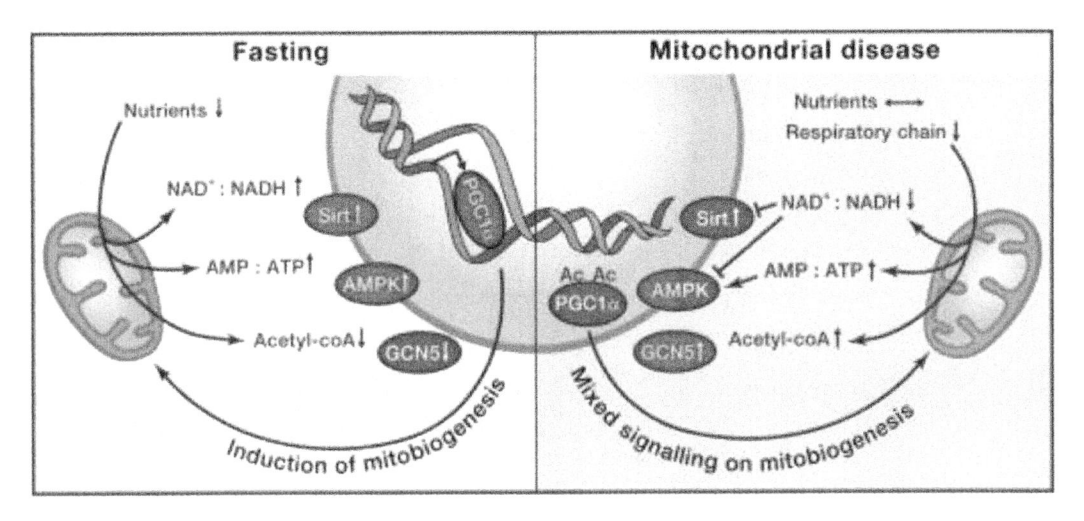

Figure 26 Fasting increases mitochondrial biogenesis through increased SIRT, AMPK, and PGC-1alpha activity

The key to increased mitochondrial biogenesis and longevity is to prime the body towards a more fat burning metabolism. This increases your cells' ability to produce energy from its own internal resources (autophagy) and lowers insulin levels (less oxidative stress).

Intermittent fasting benefits the mitochondria both ways – in protecting against free radical damage as well as enhancing energy production. But IF increases mitochondrial functioning in many other aspects as well.

Fasting and Mitochondrial Density

Mitochondrial density refers to the cells' ability to produce more energy from fewer resources and become more efficient at it.

- **Fasting increases NAD+ levels, which is an enzyme that helps with energy production** and promotes longevity.

    - NAD+ support mitochondrial functioning during youth and restore it in later life[144].

    - NAD+ protects the cells against oxidative stress with the help of sirtuins[145]. NAD+ activates sirtuins which then help to grow blood vessels and muscle[146].

    - NAD+ replenishment improves lifespan and healthspan through mitophagy and DNA repair[147]. NAD+ supplementation can promote DNA repair in mice[148].

- **Burning fatty acids and ketones cause less damage to the mitochondria as well.** Glycolysis, which is the process by which mitochondria burn glucose, causes more oxidative stress and the creation of free radicals, which in turn will speed up aging.

- **Reactive oxygen species and oxidative stress activate FOXO pathway to adapt to the stress.** Fasting causes mild stress that makes the body adapt to it through hormesis. Inactivity of FOXO factors accelerates atherosclerosis and compromises stem cell proliferation[149].

As you age or when you experience high levels of stress, you become more prone to mitochondrial dysfunction and accelerated aging. Mechanisms mediated by fasting such as increased NAD+, sirtuins, and FOXO proteins make your cells more resilient against environmental stressors and energy depletion.

In nature, animals would have to face high levels of stress only when going through some drastic seasonality or climatic change that would jeopardize their food resources. That inevitably imposes more bouts of fasting on them and thus make them react in a protective manner. If we were to replicate this response in the contemporary setting, then, as Herman Hesse pointed out: *„the most intelligent thing to do would be to not eat."* This can have a direct effect on one's brain functioning and cognition as well.

Fasting and the Brain

Comprising about less than 2% of your total body weight, the brain consumes roughly 20% of your daily calories[150]. It's one hungry beast and for a reason.

Keeping the brain energized and well is one of the main priorities of your body. Without enough higher executive functioning or learning, you would die in dangerous environments and you wouldn't maintain consciousness.

The growth of our brain has been one of the main drivers of our evolution as a species, skill acquisition, language, memory recall, social co-operation, tool crafting were all facilitated by improved cognition and getting smarter. That's why you want to provide the brain with the essential nutrients and other foods it needs for thriving. However, there are also many benefits to not eating and depriving yourself of those things with intermittent fasting.

Here are the benefits of fasting on the brain:

- **Does fasting grow new brain cells? You bet.** Fasting boosts brain power by increasing brain-derived neurotrophic factor (BDNF), which helps to grow new brain cells and synapses. It also promotes serotonin, which regulates synaptic plasticity with BDNF[151].

- **Fasting can boost BDNF by 50-400%![152]** Even 16:8 style intermittent fasting promotes neuroplasticity and stimulates the production of new brain cells[153]. BDNF also has anti-depressant benefits and it protects against stroke.

- **Fasting protects the brain against neurodegeneration.** During autophagy, fasting helps to clear out beta-amyloid plaques and lowers oxidative stress on neuronal tissues[154]. Fasting and the ketogenic diet are very commonly used to treat epilepsy.

- **Fasting boosts growth hormone that provides neuroprotection and regeneration[155].** Growth hormone not only protects against muscle catabolism but also prevents brain cells from dying.

- **Fasting gives the brain ketones which lower inflammation and maintain stable energy levels[156].** The ketogenic diet has BDNF increasing properties.

- **Fasting increases mitochondrial biogenesis,** which helps to produce more energy. There's a lot of mitochondria in the brain and other vital organs.

- **Fasting helps to lose weight**, which can improve brain function. Studies link a higher BMI with decreased blood flow to regions in the brain that are associated with attention span, reasoning, and higher executive functioning[157].

But there's a lot of people worrying about fasting causing starvation in the brain. Is it true?

The brain can use about 120 grams of glucose a day and if glucose levels fall below 40mg/dl its functioning begins to suffer[158]. However, during fasting, the brain can get more than enough energy from other sources:

- Ketone bodies are derived from fatty acid metabolism and after keto-adaptation, they can cover 50-75% of the brain's energy demands.

- Astrocytes in the brain and spinal cord can produce ketones that can be for neuronal metabolism and they have neuroprotective properties[159].

- Fatty acids from your body fat can also be converted into glucose through a process called gluconeogenesis which breaks fat molecules into 3 fatty acid chains and glycerol. Glycerol can contribute up to 21,6% of daily glucose production[160].

- Lactate can also give the brain energy during intense exercise. The brain prefers lactate over glucose when both are available[161]. Lactate gets produced during anaerobic metabolism.

- Recent research *in vitro* has shown that fuel alternatives to glucose improve neuronal efficiency and oxidative metabolism[162].

There's plenty of energy sources the brain can use while fasting and they all improve your cognitive functioning.

If you look at what excess glucose does to the brain, then it's obvious why there's a lot of neurodegenerative diseases on the rise. Alzheimer's disease is now being referred to as type-3 diabetes[163], as it's caused by an energy crisis in the brain. Insulin resistance in the brain contributes to the development of cognitive decline[164] and people with type-2 diabetes have an increased risk of Alzheimer's of 50-65% and higher.

Fortunately, fasting may help your brain with cognitive decline as well.

- Autophagy clears out the beta-amyloid plaques that begin to accumulate with cognitive decline and Alzheimer's progression[165]. Alzheimer's is also linked to obesity and insulin resistance.

- One of the main ketone bodies, β-hydroxybutyrate, actually blocks part of the immune system that regulates inflammatory diseases like arthritis and Alzheimer's[166].

- Ketone bodies also raise BDNF and lower oxidative stress. Ketones can also reduce too much excitement in the brain caused by excess glutamate and not enough GABA[167].

Being in a fasted state of higher ketone production sharpens your mind and prevents it from getting dull because of the mild stress response.

Fasting clears brain fog unlike anything else and it will boost your brain performance as well. A lot of people say they experience a heightened sense of awareness, mental clarity, focus, improved attention and motivation while fasting. This is caused by the rise in BDNF and other neurotrophic factors and hormones that increase your acuity, such as adrenaline, norepinephrine, cortisol, and other endorphins.

In my opinion, fasting ketosis is the most potent and easiest nootropics at your disposal. You can literally shift your mind into a higher gear of functioning by simply not eating and going into deeper ketosis. One study found that a short 24-hour fast had no impairing effects on cognition, attention span, reaction time, or memory recall[168].

I do all of my cognitively demanding tasks in a fasted state, including writing this book, and it helps me to stay productive throughout the day. Constantly eating may cause laziness, sleepiness, brain fog, and distractions, which is why I prefer to eat only once a day. This makes perfect sense from evolution again – to get your next meal, your brain had to become extra sharp and alert so that you'd increase your chances of eating. If you were to be dull and tired despite not having eaten for days, then our ancestors wouldn't have survived.

The biggest problem that may arise is hypoglycemia while fasting. If your blood sugar levels drop below a safe margin, you'll begin to experience shivers, fatigue, forgetfulness, and potentially pass out.

Hypoglycemia is a response to an energy crisis in the brain. Your entire body prioritizes fuel usage for the brain and whenever your blood glucose drops, you'll become hypoglycemic as to motivate you to find sugar ASAP. Usually, it happens when a person's blood sugar drops below 60 mg/dl or 3.5 mmol/L (See Figure 27).

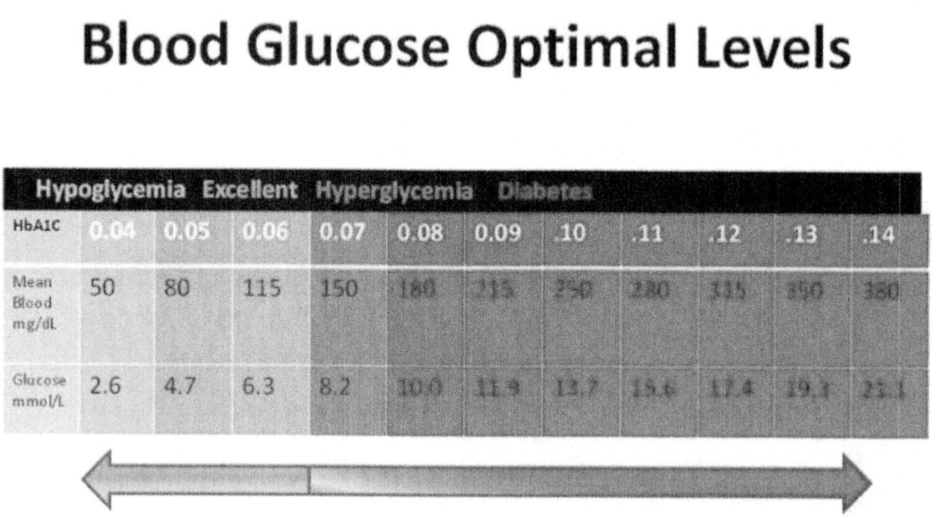

Blood Glucose Optimal Levels

	Hypoglycemia		Excellent		Hyperglycemia			Diabetes			
HbA1C	0.04	0.05	0.06	0.07	0.08	0.09	.10	.11	.12	.13	.14
Mean Blood mg/dL	50	80	115	150	180	215	250	280	315	350	380
Glucose mmol/L	2.6	4.7	6.3	8.2	10.0	11.9	13.7	15.6	17.4	19.1	21.1

Figure 27 Optimal Blood Glucose Levels

However, having low blood sugar while fasting shouldn't become an issue as long as there are ketones present. During my own 5-7 day fasts, I tend to have blood sugar around 50 mg/dl, which is hypoglycemic. Fortunately, my ketones soar up to 3-4 mmol-s, which prevents me from passing out and sustain a stable energy source for the cerebral tissue.

Therefore, if a person experiences hypoglycemia and gets the symptoms of such, then their brain is simply unable to use the other fuel alternatives. Ergo, when the body is keto-adapted enough, it's going to prevent any energy crises in the first place. The process of keto-adaptation will be covered in Chapter XIII where I'll walk you through how to start using ketones for fuel without necessarily sacrificing all glucose in the brain.

Let's now carry on with the first line of defense against the environment – your immune system.

Fasting and the Immune System

Your immune system is the most important line of defence against the outside world as it helps to deal with foreign intruders, infections, and other environmental stressors. If you hate being sick, want to have more energy, be healthier, and live longer, then it's vital to keep your immune system strong.

Fasts that last for 48-120 hours reduce pro-growth signaling and enhance cellular resistance to toxins[169]. They also trigger stem cells, which help to reinvigorate old cells and promote their youthfulness.

One study in particular done by one of the leading researchers of fasting Valter Longo *et al* showed that you can reset your immune system by fasting[170]. Mice and chemotherapy patients who didn't eat for several days saw a significant reduction in white blood cell count. This then turned on signaling pathways for hematopoietic stem cells (HSC), which are responsible for the generation of blood cells and the immune system. Longo said:

> *When you starve, the system tries to save energy, and one of the things it can do to save energy is to recycle a lot of the immune cells that are not needed, especially those that may be damaged.*

This study shows that extended fasting has a profound impact on the way your body can self-heal itself and strengthen its accord. There are also a lot of potential applications for this, starting with bolstering your immune system against the cold and ending with providing a healthier option for chemotherapy. But what causes these beneficial reactions? How does fasting reset the immune system? Valter Longo was thinking the same thing:

> *We noticed in both our human work and animal work that the white blood cell count goes down with prolonged fasting. Then when you re-feed, the blood cells come back.*

While fasting, your body starts mobilizing a lot of its internal fuel sources, such as body fat stores, stem cells, glycogen, and old cellular debris. Some white blood cells also get broken down as a means of throwing away unnecessary material.

Longo found that in order for the stem cells to be turned on, an enzyme called cAMP-dependent protein kinase A (PKA) needs to be shut down (See Figure 28). PKA inhibition signals the stem cells to become activated and start regenerating the immune system.

Prolonged fasting has also been shown to lower blood sugar, insulin levels, and other hormones such as mTOR and IGF-1, which are all growth factors that prevent the body from healing itself using its internal resources.

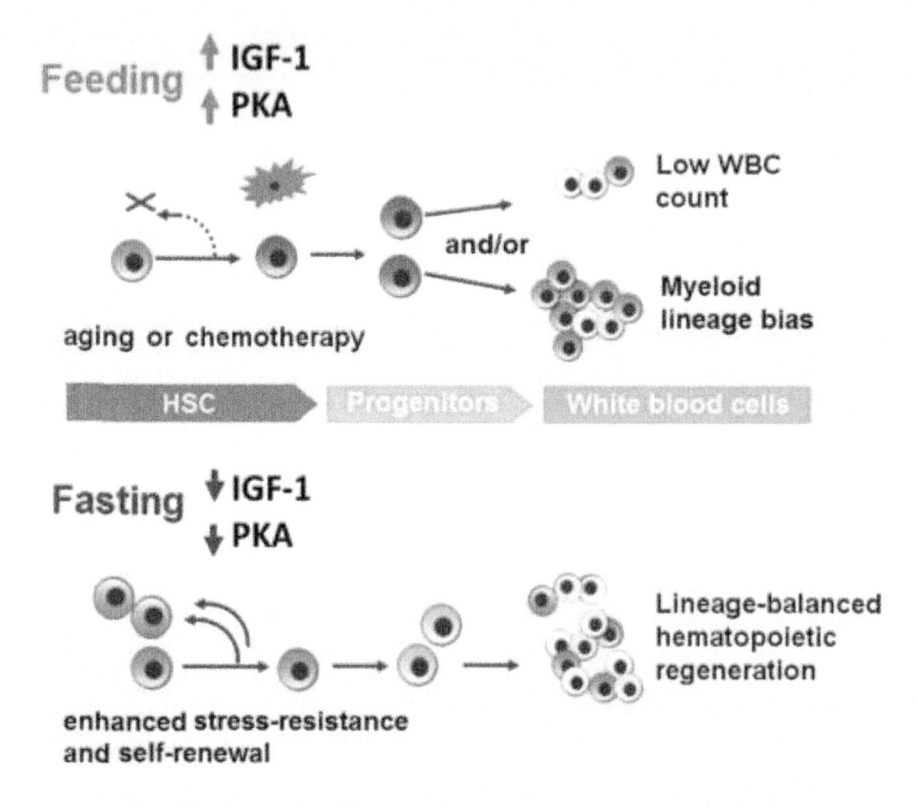

Figure 28 Fasting shuts down IGF-1 and PKA, which triggers stem cell regeneration

Fasting can weaken your immune system only if it becomes an overbearing stressor on your body. It's like any other physiological stressor your immune system has to deal with. If you're fasting for 5 days, having mad CrossFit workouts, not sleeping enough, controlling 3 screaming kids in the mini-van, and being stressed out, then, of course, you're more prone to getting sick. For the beneficial effects of hormesis to sink in, the stressors have to be taken at the right dose. Too much of the good stuff will still be bad and you can't expect to be a champ from day one.

It's important to expose yourself to different stressors to make yourself more resilient against them as well. However, you have to start from where your current ability is at. As you start practicing more intermittent fasting, your immune system gets stronger and you become less affected by the ebbs and flows of stress.

It's important to note that fasting boosting your immune system works both as disease treatment as well as prevention. If you want to keep your immune system strong, then you should practice some form of intermittent fasting continuously and adopt other hormetic lifestyle habits.

In Longo's study, it took about 3 days of fasting to reset the immune system. This was so because of shutting down PKA, lowering IGF-1, and elevating ketosis. For a complete reboot and strengthening of the immune system, you should have extended fasts of 3-5 days at least 2-3 times per year. I aim for about 4 and I also have 24-48 hour fasts sporadically every month as well. I'm also doing intermittent fasting daily and I usually fast about 20-22 hours every day so I don't have to have these very long 7-10 day fasts as my body isn't that damaged. Arguably, you can get a much stronger effect faster by fasting on the ketogenic diet that's already characterized by lower levels of liver glycogen and insulin.

PKA is a collection of enzymes who regulate glycogen, sugar, and fat metabolism through a molecular messenger called cAMP (cyclic adenosine monophosphate). PKA and cAMP get activated by several things but they're caused by the body mobilizing its energy stores. During the fast, once glycogen runs out, you'll start converting body fat into energy and ketones. That's why it takes such a long time to inhibit PKA and activate the stem cells – the body has to burn through the glycogen first before it goes for the fat. Having lower levels of liver glycogen and IGF-1 signals the body that it's okay to release stem cells and turn on cAMP.

You won't be able to have a complete reboot on your immune system or activate autophagy with eating the ketogenic diet throughout the day because high amounts of fat and calories still signal the presence of excess energy in the body. However, you can definitely get into the immune system reset zone much faster with an already depleted glycogen reserve when eating low carb. Combine that with eating one meal a day and you'll be in mild autophagy even on 1-day fasts.

Fasting Mimicking Diet and the Immune System

Valter Longo has also come up with his own way of replicating these benefits with what he calls the Fasting Mimicking Diet (FMD). Here's how it works:

- You eat about 500-1000 calories every day
- Your daily macros are low protein, moderate carb, moderate fat
- You eat things like a nutbar, a bowl of soup, and some crackers with a few olives or something
- Day One you eat about 1000 calories – 10% protein, 55% fat, and 35% carbs
- Day 2-5 you eat about 500-700 calories – 10% protein, 45% fat, 45% carbs
- Day 6 you transition back to a normal caloric intake with complex carbs, vegetables, and minimal meat, fish, and cheese

Studies on the fasting mimicking diet have shown that it lowers cholesterol, C-reactive protein, blood glucose, IGF-1, and blood pressure[171]. However, it's possible these effects simply came from the caloric restriction. FMD is often prescribed to elderly people or someone who can't handle fasting.

To get the full benefits of autophagy and stem cell growth, I'd say you still have to avoid all calories and fast for a longer time. The macronutrient ratios of the FMD aren't optimal either as you'd want to lower the carbs and total calories to inhibit PKA. You're much better off by having a longer fast.

Whatever the case may be, you should adopt a fasting focused lifestyle as to promote longevity, youthfulness, and create an antifragile immune system. The next most important thing for that is to take care of your gut because that's where most of the immune system is located in.

Fasting and Gut Health

Diet and lifestyle play a huge role in modulating the gut microbiota, which has an enormous impact on health and lifespan. Microbial dysbiosis or imbalances are associated with increased risk of cardiovascular disease, obesity, metabolic syndrome, cancer, and autoimmune disorders[172].

Different macronutrients and foods reshape the composition of the microbiome and both short term and long term dietary changes can influence microbial profiles[173]. However, not eating and fasting have a profound impact on the gut as well.

Gut homeostasis plays an important role in longevity. Dietary restriction has been shown to prevent gut pathologies and extend lifespan in fruit flies[174].

- **Short-term intermittent fasting improves gut health and extends lifespan in fruit flies** independent of the TOR pathway[175].

- **Intermittent fasting promotes clearance of pathogens and infectious bacteria** and helps to heal the gut in mice infected with salmonella[176].

- **Fasting protects the gut against the negative effects of stress**[177], such as inflammation. Fasting activates cAMP, which further activates genes that promote intestinal lining integrity and strength[178]. The cognitive benefits of fasting on the brain will also improve your mood and improve stress resiliency which protects against leaky gut again.

- **Intermittent fasting promotes white adipose tissue browning and reduces obesity by shaping the gut microbiome**[179]. The gut microbiota influences adipose tissue browning and insulin sensitivity[180] by signaling the browning of white fat into brown fat[181].

- **Caloric restriction and weight loss increase the number of beneficial bacteria in the gut** called Bacteroidetes[182]. Obese people have less of these bacteria than lean people. Caloric restriction enriches phylotypes in the gut that are positively correlated with increased lifespan[183]

Fasting and time restricted feeding heal the gut by giving your intestines rest from breaking down food. Digestion requires about 25% of the calories from each meal. Being in a fasted state promotes anti-inflammatory cytokines and cellular autophagy that instigate healing.

Fasting also increases the activity of *the migrating motor complex* (MMC), which is a mechanism that controls stomach contractions in a cyclical manner over 2 hour periods[184]. The MMC cleans out the GI tract and helps to eliminate undigested food particles. It's regulated by feeding/fasting hormones such as ghrelin, serotonin, cortisol, and somatostatin. Eating inhibits MMC and not eating increases it. If you have longer breaks in between meals then you'll digest and assimilate your food much better and prevent any kind of small intestine bacterial overgrowth or SIBO.

Another study found that time restricted feeding restored a variety of beneficial strains of bacteria in mice[185]. Fasting has been shown to improve symptoms of irritable bowel syndrome (IBS) as well[186].

A group of researchers from MIT found that fasting for just 24 hours boosted the regeneration of gut stem cells in young as well as old mice[187]. One of the scientists Omer Yilmaz said:

> *This study provided evidence that fasting induces a metabolic switch in the intestinal stem cells, from utilizing carbohydrates to burning fat.*

Indeed, fasting activates a group of transcription factors called PPARs, that turn on genes involved with fatty acid metabolism. Another aspect of it has to do with shifting into ketosis that reduces inflammation and supports healing of the gut. However, fasting may cause some gut issues if done wrong.

- **Fasting and eating at night may lower your sleep quality**. Some species of bacteria like *Enterobacter aerogenes* are sensitive to melatonin (the sleep hormone), which influence circadian rhythms[188]. A disruption in circadian rhythms can disrupt the microbiome and thus negatively affect metabolic health[189]. That's why it's important to not overeat at night. Chapter XXII talks about circadian rhythms and fasting.
- **Prolonged restrictive diets may cause a lack of microbial diversity**. The gut microbiome can respond to changes in diet very fast and thus restructure the

microbiome according to that. Short-term consumption of either fully animal or plant-based diets alter the gut microbial status, which can cause trade-offs in carbohydrate or protein metabolism[190].

- **Fasting decreases the size of some digestion organs like the small intestine and liver, which can lower the capacity to consume food[191].** That's why you may find it more difficult to eat as much as you did before when breaking your fast. However, these organs will regrow themselves quite rapidly and they'll become more functional afterward.

All of these issues are easily solvable and can be fixed with simply adjusting your fasting schedule.

It might seem that, because of not eating, fasting will destroy all the healthy bugs in your gut and lead to autoimmune disorders in the future. However, this is not the case.

In a study on mice, scientists found that intermittent fasting didn't wipe out bacteria in the gut or destroy their diversity. Instead, intermittent fasting actually improved bacterial diversity[192]. This probably has to do with the life extension benefits of dietary restriction found across all species. When the bacteria are faced with mild starvation or fasting they live longer and promote energy efficiency through diversity.

Fasting promotes the diversity and dynamics of the microbiome, which is determined by feeding and fasting cycles of the host[193]. At the same time, it will still starve off some of the pathogens, viruses, and bad bacteria.

Time-restricted feeding has been shown to prevent weight gain, improve gut diversity, and circadian cycling of gut microbiome in mice despite being fed a pro-obesity diet[194]. Mice who were allowed to eat the same diet *ad libitum* whenever they wanted got obese and sick.

All living organisms have built-in circadian clocks that connect them with the day and night cycles. The microbiome functions like a signaling hub in communication between you and your environment[195].

In the 2017 Annual Review of Nutrition, it was said that prolonged intermittent fasting has sustained improvements in health and they may be caused by benefits on circadian biology, the gut microbiome, and better sleeping patterns[196]. Intermittent fasting already has so many health benefits on your metabolism, body composition, cognitive functioning, and longevity but it's a win-win for your gut microbiome as well.

Part of the reason taking care of your gut is so important has to do with how the microbiome affects your cognitive functioning. Over 70% of your serotonin – the feel-good hormone – gets produced in the gut and all the other circadian rhythm-related processes are also linked with that.

Science has now discovered that microbial life inside living organisms has played a crucial role in shaping the evolution of said organism. In fact, the mitochondria are actually descendants of bacteria that millions of years ago developed a co-existence with our ancestral bodies. This has led to the suggestion of the concept of a Microbe-Gut-Brain (MGB) Axis[197].

The MGB Axis is this network of biochemical signaling between the gastrointestinal tract (GI) and the central nervous system (CNS). It includes the enteric nervous system (ENS), the endocrine system, the hypothalamic-pituitary axis (HPA), the autonomic nervous system, the vagus nerve, the endocrine system, and the gut microbiome[198].

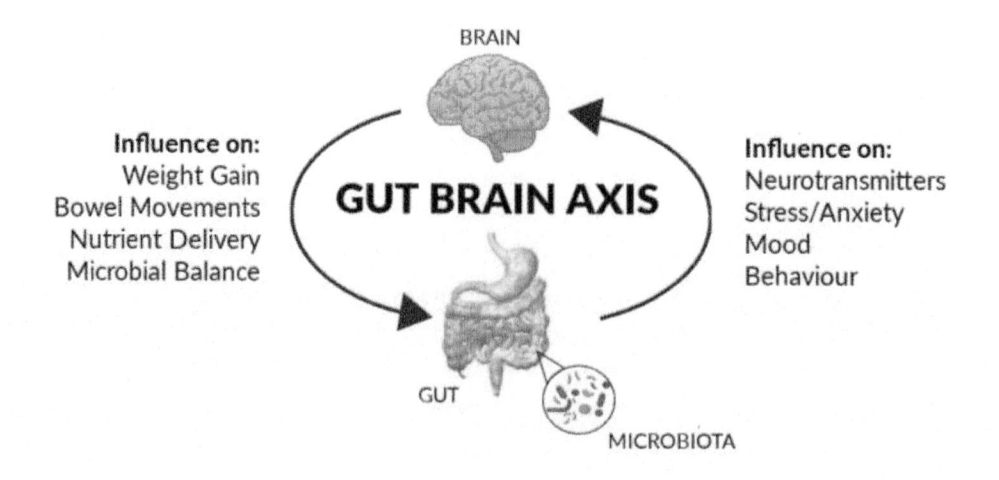

Figure 29 The Gut Brain Microbiome Axis

Your immune cells, muscle cells, cells of the gastrointestinal tract are all mediators of the neuro-immuno-endocrine system that are influenced by both the brain and the gut microbiome. In fact, it's been thought that the microbiome plays a much more influential role in the state of your being than the brain[199].

It's like a super-complex ecosystem that consists of trillions of microorganisms and bacteria. The 'thing' you call 'I' is literally comprised of this collective consciousness of many living organisms and cells inside your body. Hence the importance of taking care of your gut and mitochondria with fasting.

Fasting and Fat Loss

Fasting is also the healthiest and easiest way to reverse obesity. Before you can burn fat, you have to first "release" the fatty acids into your bloodstream through a process called lipolysis. Then they get transported to the mitochondria where they'll be oxidized into energy. See Figure 30.

During rest, our muscles start to use more fatty acids for fuel. When fat burning increases so does the amount of Uncoupling Protein-3 in our muscles. As little as 15-hours of fasting enhances the gene expression for UCP-3 by 5-fold[200]. We'll be using ketones to feed our lean tissue more effectively.

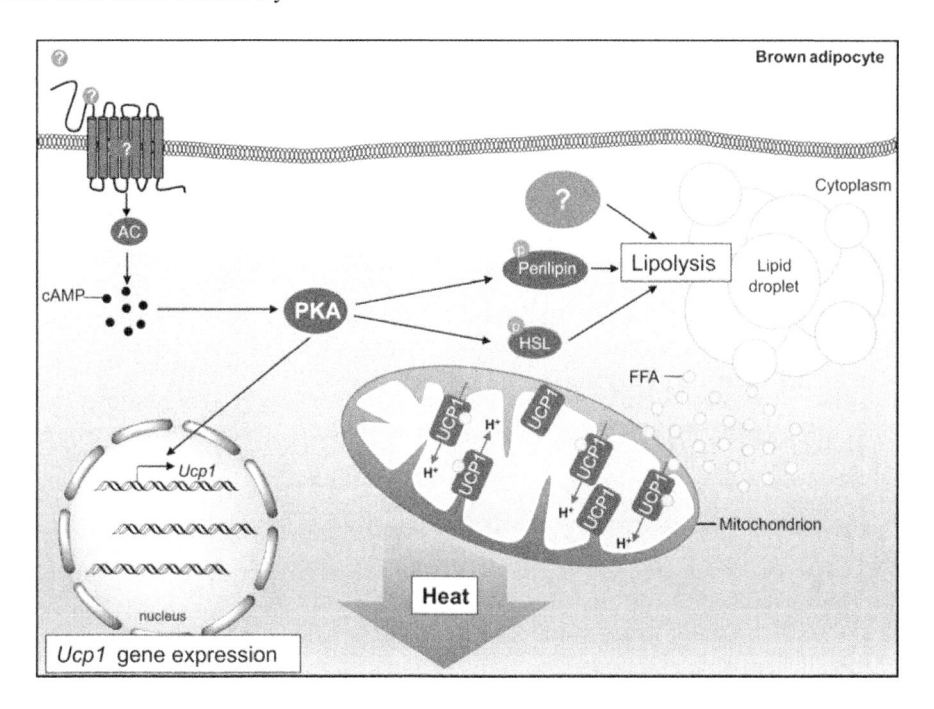

Figure 30 The process of burning stored body fat - low insulin raises cAMP, which leads to lipolysis that releases free fatty acids for the mitochondria to burn off as heat

Contrary to popular belief, intermittent fasting doesn't slow down the metabolism but actually increases it by 3.6% after the first 48 hours[201]. Even further, 4 days in, resting energy expenditure increases up to 14%. Instead of slowing down the metabolism, the body revs it up and puts it into higher gear. This is probably caused by increased adrenaline so that we would have more energy to go out into the savannah and find some food. The scarcer calories become the more detrimental it is to succeed in hunting.

People think that if they skip breakfast the body will hold onto its own body fat and store every calorie in the next meal. Think about it. Does your body really think it's starving after not eating for a day or is it simply your primal mind playing tricks on you?

In a fasted state, we actually become more efficient with the food we eat, instead of storing it all. With the lack of calories, especially carbohydrates, we become more insulin sensitivity[202], meaning that we need less of it to lower our blood sugar levels back to normal. Fasting can actually reverse insulin resistance and reduces overall blood sugar levels.

There's no reason to be concerned about malnutrition during fasting because our fat stores can deposit almost an infinite amount of calories. The main issue is rather micronutrient deficiencies. Potassium levels may drop slightly, but even 2 months of fasting don't decrease it below a safe margin. Magnesium, calcium, and phosphorus remain stable because 99% of them are stored in our bones. The man who fasted for 382 days maintained such a lengthy fast with no harmful effects on health thanks to taking a simple multivitamin. That's all you need to survive for that long.

Fasting and Growth Hormone

Another anabolic mechanism that gets increased is human growth hormone (HGH). After 20-24 hours of fasting, it does so by 1300-2000% (See Figure 31)[203]. It not only promotes tissue repair, body composition, and metabolism but also preserves youthfulness.

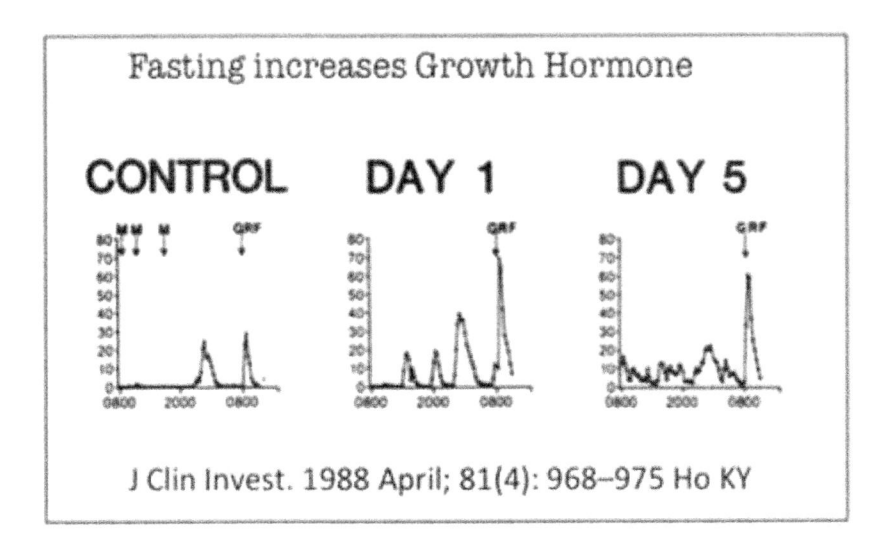

Figure 31 Fasting increases Growth Hormone Exponentially

Growth hormone plays a key role in the metabolism of all macronutrients. Its normal secretion fluctuates throughout the day and increases significantly during the first hours of sleep at about 11-12 PM.

After 3 days of fasting, HGH increases dramatically in non-obese individuals, but flats out after day 10[204,205]. In obese people, there is little to no reported rise after fasting from 14 to 38 days [206]. Hypothetically, this happens as a response to preserving lean tissue. I would suggest that beyond that point the body simply becomes extremely well keto-adapted and reduces both the overall energy demands as well as increases the efficient use of ketones as fuel.

What goes hand in hand with HGH is insulin-like growth factor (IGF-1). It's one of the major growth factors in mammals which together with insulin is associated with accelerated aging and cancer. Just 5 days of fasting can decrease IGF-1 by 60% and a 5-fold increase in one of its principal IGF-1-inhibiting proteins: IGFBP1[207].

Potential Side-Effects

There may also be some negative consequences to fasting. Headaches, dizziness, lightheadedness, fatigue, low blood pressure and abnormal heart rhythms are all short-term. Some people may experience impaired motor control or forgetfulness. But these are all symptoms of withdrawal, not fasting. Because most people rarely get to use their own body fat for fuel, they become too dependent on glucose. It's like an addiction that makes them crave more sugar.

When I first started practicing intermittent fasting I experienced some hypoglycemia but nothing serious. I simply got a bit lightheaded whenever I stood up too fast. After going on a ketogenic diet those signs have disappeared completely. Any mental hindrance is caused by an inner energy crisis. Once the body adapts to utilizing fat for fuel, the brain will accept ketones and will also reduce hunger.

Fasting may cause some flare-ups of certain medical conditions, such as gout, gallstones or other diseases. This is yet again not because of fasting directly but because of the overall high amounts of toxins in the body. The adipose tissue is more than a caloric pantry. It also stores poisons and infections that we digest. Once you start breaking down triglycerides, those same toxins will be released into your bloodstream again and need to get flushed out. There may also be some nervous stomach, irritable bowel of diarrhea. That's why fasting is an effective detox tool, as it cleanses the organism completely.

In comparison to all of the empowering health benefits of fasting, these few side-effects are minute and not guaranteed. They may or may not happen. What's certain is that they will be alleviated after time.

Why then have we been lead to believe that fasting is bad for us? Medical doctors and supplement companies all preach the consumption of 6 small meals a day. Why? I'm not going to be pointing any fingers or calling anyone out but simply put: there's no money to be made from healthy people who fast. How do you prescribe a pill of fasting that's completely free?

Intermittent fasting is truly the 80/20 dietary strategy in terms of improving your overall health and life expectancy. Unfortunately, it's never been implemented that thoroughly in the long-term. There aren't many people I can think of who've practiced time-restricted feeding over the course of decades, starting even from early adulthood.

When I first started I was 18 and I still have many years ahead of me. Although I'm very young and in my 20s at the moment, I don't feel like I'm getting any older. Of course, I stick to a consistent exercise routine, eat only healthy food, get enough sleep, and do other biohacks but I attribute most of my wellness to intermittent fasting.

By subtracting from your life and applying positive restriction with fasting you will actually gain exponentially more freedom and results. It would be unwise to not take advantage of such a simple yet powerful strategy as simply not eating. Instead of looking for another shortcut or trying to find the next magic supplement that will increase your lifespan, you can just stop eating for a while and see how your body reacts.

IF is one of the central cornerstones of Metabolic Autophagy because one of the few known ways of actually activating this procedure requires fasting. There are a few nuances to it but you'd get the biggest bang for your buck if you were to just fast. The next chapter will delve a bit deeper into the physiology and mechanisms of this process.

Chapter IV
What We Know About Autophagy So Far

„Life lives on life.
This is the sense of the symbol of the Ouroboros, the serpent biting its own tail.
Everything that lives lives on the death of something else.
Your own body will be food for something else.
Anyone who denies this, anyone who holds back, is out of order.
Death is an act of giving."

Joseph Campbell

In the early 1990s, a Japanese biologist Yoshinori Ohsumi was studying the transport of ions and small molecules in the cells' vacuole, which is a membrane-bound organelle in the cell. At that time, most of the other scientists were researching the plasma membrane and considered the vacuole just a garbage can with not much importance.

Cell differentiation requires a lot of protein degradation. Dr. Ohsumi decided to study the lytic function of the vacuole to observe any morphological changes in cells undergoing such degradation. He used a microscope to look at vacuolar proteinase-deficient mutant cells, which cannot sporulate as normal cells do when under conditions of nitrogen starvation. After 30 minutes of observation, he saw many vesicles appearing and accumulating in the vacuole. From the bunch, he identified formations of some autophagosomes, which sparked his interest in starting to research autophagy in yeast.

In 1991, one of Dr. Ohsumi's graduate students was observing some mutant yeast cells under a microscope and found the first autophagy-defective mutant named APG1-1, which is now called ATG1. ATG is an abbreviation of autophagy-related gene or autophagy-related protein[208]. In total, the researchers discovered 14 ATG mutants. Except for ATG1, all of the ATG genes turned out to be novel genes with their amino acid sequences not being very clear. Eventually, Dr Ohsumi's lab realized they had an entirely new pathway in their hands and published their results in the journal *Nature*[209].

On the 3rd of October 2016, Dr. Yoshinori Ohsumi was rewarded the Nobel Prize in Physiology or Medicine for *"discoveries of the mechanisms for autophagy."* Ohsumi's research is done in simple organisms like baker's yeast but the autophagy-related pathways have been found to be similar to autophagy in mammals as well.

Part of the reason why intermittent fasting is such a powerful thing for longevity has to do with how it primes the body to heal itself from within. This chapter will delve deeper into the mechanisms and benefits of autophagy.

What Is Autophagy

Autophagy or *autophagocytosis* in fancy science language translates from the Ancient Greek word *autóphagos* which means self-digestion or 'eating of self'.

Autophagy is a metabolic process during which cells disassemble and remove their dysfunctional components. You're basically recycling cellular debris and taking out the trash.

Initially, it was thought that autophagy was a hormonal response to starvation but recent research has shown autophagy to have several other roles in biology. There are many benefits to autophagy, such as reduced inflammation, improved immunity, prevention of genotoxic stress, anti-aging, suppression of cancerous tumor cells, and elimination of pathogens.

Compromised autophagy pathways will lower the body's ability to eliminate and heal the organism from inflammation, accumulation of toxins, and parasitic infections. Inability to cause autophagy makes rats fatter, less active, have higher cholesterol and impaired brain function[210].

The word 'autophagy' was coined at the Ciba Foundation Symposium on Lysosomes in London 1963. There, one of the scientists, Christian de Duve, who is a 1974 Nobel Prize Laureate and the discoverer of the lysosome, proposed the term autophagy to illustrate how lysosomes self-eat themselves[211]. That process is called macroautophagy but there are several types of autophagy that basically describe the same thing.

At the end of the 20th century, autophagy had been established as an important function in mammalian cells. However, there wasn't much proof to autophagy in morphological studies, which is why Dr. Ohsumi's work was so groundbreaking. His lab gave important insight into the mechanisms of this complex process. By now we don't know everything about autophagy but we do know something. So, how does it work?

When autophagy gets activated, the organelles of your healthy cells start to hunt out dead or diseased cells and then consume them.

Autophagy is mediated by an organelle called the *autophagosome*, which combines with other cellular components like the endosome and lysosome to form a double membrane around the cell that's going to be eaten. The autophagosome then dissolves the cell that are sentenced to death and converts it into energy[212] (See Figure 32).

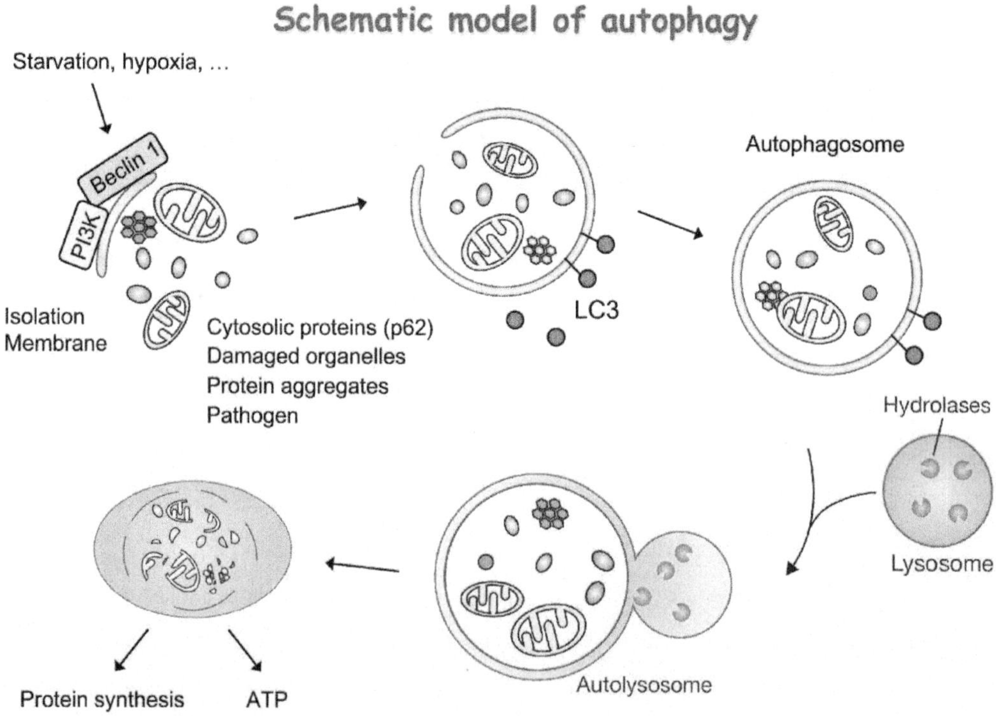

Figure 32 The Process of Autophagy

Where and how the autophagosome gets formed is currently a mystery for researchers. In yeast, it's been identified to occur when many ATG proteins converge at a site that's called *"pre-autophagosomal structure (PAS)"*[213]. Some equivalent structures have been located in mammalian cells as well but detailed information about the PAS is still unknown.

Autophagy gets triggered mostly by nutrient starvation:

- In yeast, starvation of nitrogen and other essential factors like carbon, nucleic acid, auxotrophic amino acids, and even sulfate can activate autophagy to some degree[214].

- In plant cells, nitrogen and carbon starvation can also trigger autophagy[215]. These were the points of interest for Dr. Ohsumi in yeast as well.

- In mammals, autophagy happens in various tissues in different degrees. There's macroautophagy in the liver, brain, muscle, mitophagy in the mitochondria and Chaperone-Mediated Autophagy (CMA) (See Figure 33). Depletion of amino acids is a strong signal for triggering autophagy but that depends on the type of cell and amino acids because amino acid metabolism differs among tissues.

- In vivo, it's thought that autophagy is regulated mostly by the endocrine system, particularly by insulin. Insulin suppresses liver autophagy by raising blood sugar and signaling the presence of nutrients. Glucagon, which is the counterpart of insulin, releases liver glycogen to be burnt for energy, and that increases autophagy.

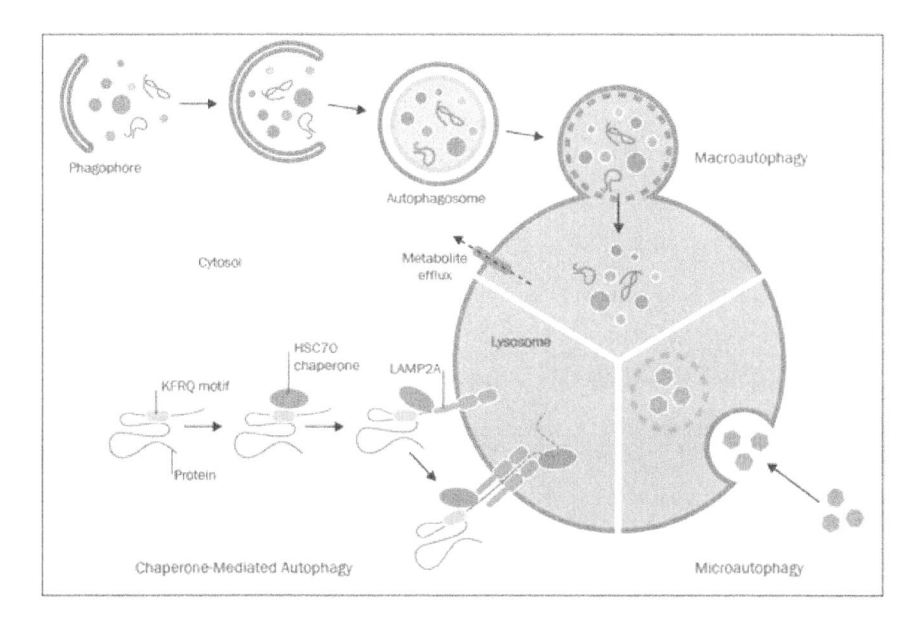

Figure 33 Different levels of autophagy

Both amino acids and insulin-like growth factors regulate the nutrient-sensing of mTOR. Suppressing TOR with things like rapamycin and CCI-779 has been shown to induce autophagy in yeast[216] and other animals[217]. However, not all of the autophagy signaling happens through mTOR as some amino acids can suppress autophagy independent of mTOR[218].

Recent reports have shown many other factors to be involved in autophagy regulation, such as Nf-kb[219], reactive oxygen species[220], calcium[221], AMPK[222], and many more. So, what I propose is to not look at autophagy as a binary on-and-off switch but more like a degree dependent state that's mediated through how depleted and deprived the organism is from certain nutrients.

The main inhibitor of autophagy in muscles is a kinase called Akt. It can regulate autophagy mainly in two ways: (1) a rapid regulation of mTOR activation, and (2) a slower response of gene transcription via FoxO3[223]. FoxO3 controls the transcription of autophagy-related genes, such as LC3 and Bnip3, which mediate the effect of FoxO3 on autophagy. Akt activation blocks FoxO3 and autophagy.

Although there's still much to learn about autophagy, current research is showing that most of the signaling happens through the pathways of mTOR and AMPK. When under nutrient deprivation, AMPK starts to inhibit cellular growth by suppressing the mTORC1 pathway, which in turn forces the body to catabolize its weakest components.

Autophagy is a catabolic pathway that makes you break down old cells. Although you're causing protein breakdown, autophagy is needed for muscle homeostasis. With poor autophagy functioning, your body wouldn't be able to maintain lean tissue. It improves your body's ability to deal with catabolism and atrophy by promoting protein sparingness. A weakened or inadequate state of autophagy may contribute to aging, and muscle wasting through sarcopenia[224]. That's why it's incredibly vital for not only living longer but also to age slower.

However, defective as well as excessive autophagy can lead to substantial muscle disorders and loss of lean tissue[225], which in turn can promote premature aging.

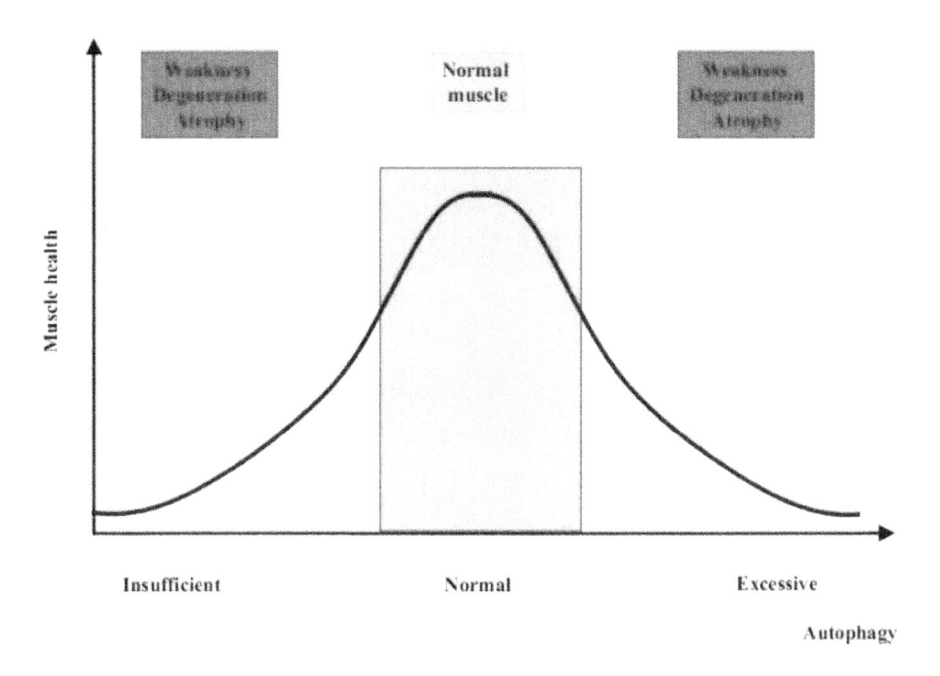

Figure 34 Too much, as well as too little autophagy, is bad for you

You need these anabolic pathways like mTOR and insulin to build new tissue and keep your cells alive. That's why strength athletes and bodybuilders are so focused on supporting anabolism and preventing muscle catabolism by taking different amino acid supplements and consuming foods that make them grow more. They're probably over-doing it but that's a whole nother story with potential consequences on their long-term health.

With that being said, **a constant supply of nutrients and access to energy inhibits the body's ability to induce autophagy and protect against catabolism.** In fact, a continuous circulation of both macro- and micronutrients all the time inhibits their usage and uptake by making the cells less responsive. It means that to actually absorb the nutrients you're feeding yourself, you need to go through periods of mild deprivation as you'll be more sensitive to those nutrients afterwards again.

Autophagy is essential to support skeletal muscle plasticity in response to endurance exercise[226]. To trigger the activation of autophagy during exercise, the activation of AMPK is also needed[227].

AMPK regulates both protein synthesis and breakdown pathways. AMPK has a vital role in skeletal muscle homeostasis. It's activated by conditions of energy stress, including nutrient deprivation and vigorous exercise.

Exercise performed in a fasted state shows a higher increase in LC3B-II level compared with a fed state, which suggests exercise done while fasting to have a better autophagic response[228]. This makes sense because you'll be tapping straight into your body fat for energy instead of burning through the food you ate. When activated, AMPK increases the flux of glycolysis and fatty acid oxidation while at the same time inhibiting gluconeogenesis and fatty acid and cholesterol synthesis.

In addition, AMPK has been recently shown to be a critical regulator of skeletal muscle protein turnover[229]. Protein turnover is the balance between protein build up and protein breakdown over the course of the day.

- If your protein synthesis exceeds the amount protein's being broken down, then you're in a more anabolic state.
- If you're breaking down more than synthesizing, then you'll be more catabolic. Or autophagic.

Your body is always in a flux between anabolism and catabolism – growing and degrading. Both of these ends of the spectrum are vital for a healthy life – you want to promote the growth and repair of your vital organs and muscles but you also want to eliminate and break down the old worn out cells and metabolic debris.

How Long Until Autophagy?

But how much time do you need to starve to activate autophagy? That's a difficult question because it happens in various degrees almost all the time. It's not as binary as you'd think but it's definitely not something you can hack easily either.

In general, suppressing mTOR and insulin will already begin to elevate autophagy in a dose-specific degree. Low blood glucose levels and depleted liver glycogen stores are indicators of energy shortage in the body. As energy depletion continues, the body upregulates its metabolic pathways that are associated with burning stored fat for fuel.

Eventually, this leads to a ketotic state with elevated levels of ketones in the blood. Chaperone-mediated autophagy can be maintained with the elevation of ketones and fatty acids thanks to suppressed insulin and low blood glucose.

Although autophagy is usually accompanied by ketosis, you can still be in ketosis without activating autophagy and you can activate autophagy without being in ketosis. The reason is that of nutrient signaling and mTOR. Fats and exogenous ketones have a negligible effect on insulin but they may still raise mTOR if you consume them in copious amounts and under the wrong circumstances. Likewise, you can see trace amounts of autophagy even while having fasted for 24 hours on a carbohydrate-rich diet if you do things right.

To really gain the benefits of autophagy, you'd have to be fasting for over 48 hours to allow the stem cells and immune system to do their work. That's why I recommend everyone to fast for at least 3-5 days 2-3 times per year. These extended fasts not only make you burn a lot of body fat very easily but they'll also recycle the weak cells that are simply dragging you down and giving you potential issues.

Contemporary eating habits like 3 square meals a day, snacking, high carb foods and processed ingredients are all anti-autophagy and they inhibit the recycling of old waste material.

Even people who eat „clean foods" but don't go through nutrient starvation may potentially be walking trash cans. There are many other sources of toxins and inflammation all of us get exposed to starting from air pollution, water, GMOs, plastics, heavy metals and who knows what else. What looks good on the outside doesn't mean that everything is okay on the inside. Having your autophagy pathways live and active is even more important for living in the modern world.

However, autophagy can do both good as well as harm to the organism. It has a dark side...

The Negative Side Effects of Autophagy

The entire process of autophagy and self-eating is called autophagic flux, which includes (1) the formation of an autophagosome, (2) fusion with lysosomes, and (3) the degradation of the autophagosome.

- **Autophagy controls inflammation and immunity by eliminating inflammasome activators**[230]. Removal of pathogens by autophagy is called xenophagy[231], which has many immune strengthening benefits. However, some bacteria like *Brucella* use autophagy to replicate themselves[232]. That may cause some bacterial overgrowth or at least prevent its death.

- **The essential autophagy gene ATG6/BECN1 encoding the Beclin1 protein has been found to suppress tumors in cancer**. However, it's not been found to be that big of a tumor-suppressor as previously thought and sometimes it can even promote cancer due to the self-replicative process[233]. Self-eating can enhance tumor cell fitness against environmental stressors[234], which makes them more resilient against starvation and chemotherapy. It may be that autophagy is better for cancer prevention rather than treatment.

- **It's not clear whether autophagy prevents or promotes apoptosis or programmed cell death**[235]. The outcome turns out to depend on the stimulus and cell type[236]. Blocking autophagy enhances the pro-apoptotic effect of bufalin on human gastric cancer cells, which is a Chinese medical toxin used for tumor suppression[237], through endoplasmic reticulum stress[238]. In this example, less autophagy led to more cancer cell death because the cancer cells were weaker whereas with autophagy they became stronger.

Essentially, when malignant tumor cells are put under nutritional stress through calorie restriction, autophagy may prevent them from dying by inhibiting apoptosis[239]. So, it's not all black and white with autophagy – some viruses and pathogens are eliminated by it whereas others hijack its mechanisms and replicate themselves. What's more, autophagy has various differences in different tissues[240], such as the brain, liver, muscle, and fat. Sometimes it's good, sometimes it's not. You want the beneficial autophagy in the liver and brain to clear out plaques but you don't want to self-eat your lean tissue and muscles.

The point of all this is that cells undergo constant deterioration and slow degradation because of entropy. You easily end up with suffering from the dysfunctional apoptotic or necrotic processes that may potentially leave the body in a lesser state than before. That's why it's important to not only get rid of the bad but to also keep the good, which is how well-functioning autophagy should work.

How to Measure Autophagy

Autophagy can be monitored by two different approaches: (1) direct observation of autophagy-related structures and their fate; and (2) quantification of autophagy-lysosome-dependent degradation of proteins and organelles.

To accurately estimate autophagic activity, it is essential to determine autophagic flux, which is defined as the amount of autophagic degradation. Monitoring autophagic flux is still complicated even in cultured cells and model organisms. and is currently impractical in humans.

At the moment, there is no established method to measure autophagic flux in humans; therefore, it's very difficult to know exactly how autophagy works in humans. However, studies done in other species have found some similarities and mechanistic principles that regulate autophagy.

As we know by now, autophagy is regulated most by the mTOR and AMPK pathways. To trigger autophagic cell death you need a catabolic catalyst that would increase AMPK and cause cellular stress. Being anabolic and growing will inhibit autophagy by raising mTOR through the insulin/IGF-1 signaling pathway.

To know whether or not you're more anabolic or catabolic or more mTOR or AMPK activated, you can measure your insulin to glucagon ratio (IGR). Both insulin and glucagon are important for your body's homeostasis and survival. They will either make you store energy and repair vital tissues (anabolism) or break down backup storage so that you'd survive (catabolism). In general, an increase in IGR is associated with more anabolism – weight gain, muscle growth, fat storage, hyperinsulinemia, and a higher risk of hypoglycemia. A reduction in IGR promotes catabolism, fat loss, and prevents hypoglycemia.

Everyone's average blood glucose levels vary but the consensus is that the normal range for non-diabetics while fasting should be 3.9-6.0 mmol/L (70-100 mg/dL). During the day it would fall around 5.5 mmol/L or 100 mg/dL. However, in my opinion, a healthier person would fall slightly lower than that between 3.0-5.0 mmol/L-s. Deviations of 5-20% will either release insulin to lower blood sugar or increase glucagon to raise blood sugar. It fluctuates throughout the day.

To know what's your insulin to glucagon ratio you can take blood tests for insulin as well as glucagon from your medical doctor. Here's what research has found to influence your IGR:

- A 1:1 insulin to glucagon ratio: 1.0

- While fasting you have lower insulin and more glucagon. Fasting + No Food: ~0.8

- While eating the Western Diet with higher carbs there's more insulin. Carbs + Eating: ~4.0

- On a Low Carb Diet, there's fewer carbs and less insulin. Low Carb + Eating: ~1.3

- Consuming protein while fasting causes a drop in insulin by raising glucagon-induced gluconeogenesis. Fasting + Protein: ~0.5

- Consuming protein on a low carb diet doesn't raise insulin and doesn't significantly affect glucagon. Low Carb + Protein: ~1.3

- Consuming protein with high amounts of carbohydrates spikes insulin 20x more than normally because of the anabolic effects. Carbs + Protein: ~70

Amino acids combined with carbohydrates produce a much larger anabolic effect and insulin response than just carbs or protein alone. That's a significant difference between macronutrients and their anabolic response. Calories in VS calories out just got much more complicated.

The Glucose Ketone Index

You can guestimate your general metabolic health and insulin-glucagon ratio at home by measuring your blood glucose and ketones with an ordinary ketone meter.

The Glucose Ketone Index (GKI) is a number between the relationship of your ketones and glucose levels. It can help to monitor your general health in relation to your blood glucose levels.

Here's the Glucose Ketone Index Formula: (Your Glucose Level / 18) / Your Ketones Level = Your Glucose Ketone Index

- Measure your blood glucose by pricking your finger and all that. Write down the number you got.

- Measure your blood ketones by pricking your finger again (*ouch*). Write down the number you got.

- Divide your blood glucose number by 18.
    - If your device is using mg/dl, then dividing that with 18 converts it over to mmol/l
    - If your device is already showing mmol/l, then you don't need to divide anything and can skip this step

- Divide your result from the previous step by your ketone numbers.
- The end result is your GKI.

GLUCOSE KETONE INDEX FORMULA

1. Measure Your Blood Glucose
2. Measure Your Blood Ketones
3. Divide Blood Glucose With 18
4. Divide That With Your Ketones

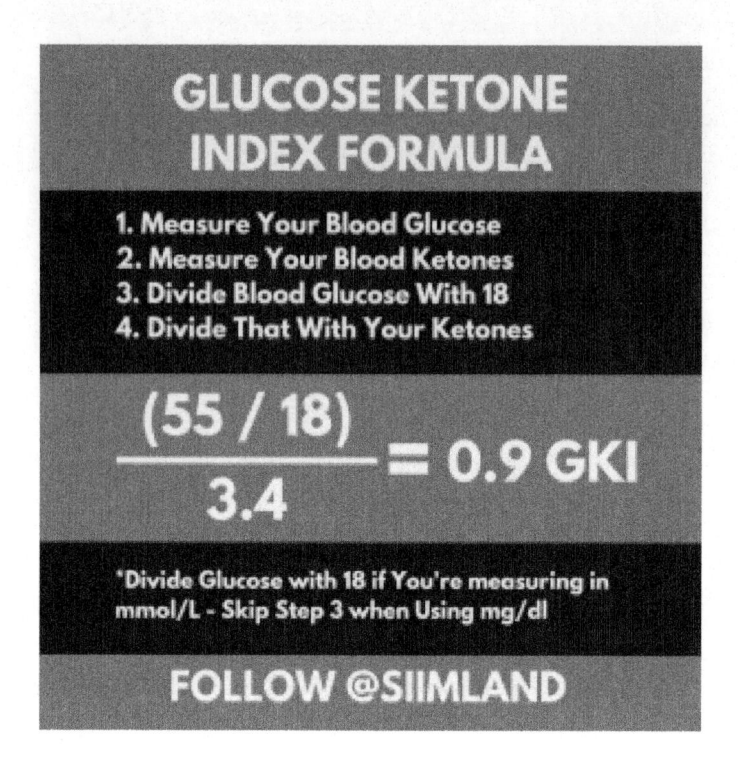

$$\frac{(55 / 18)}{3.4} = 0.9 \text{ GKI}$$

*Divide Glucose with 18 if You're measuring in mmol/L - Skip Step 3 when Using mg/dl

FOLLOW @SIIMLAND

Figure 35 Glucose Ketone Index Formula

Let's take an example from my readings during a 5-day water fast:

- My blood glucose was at 55 mg/dl which is 3.0 mmol/L (almost hypoglycemic!)
- My blood ketones were at 3.4 mmol/l (which is why I didn't have symptoms of hypoglycemia)
- Glucose at 55 mg/dl divided by 18 gives us 3.05
- Divide 3.05 by 3.4 and we get 0.9 rounded

In general, having a GKI below 3.0 indicates high levels of ketosis in relation to low levels of glucose; 3-6 shows moderate ketosis, and 6-9 is mild ketosis. Anything above 9 and 10 is no ketosis. Therefore, a lower GKI will reflect an estimated insulin-glucagon ratio by virtue of how glucose and ketones affect that relationship.

In his book, Cancer as a Metabolic Disease, Thomas Seyfried says that the optimal glucose ketone index range for cancer treatment and prevention is between 0.7-2.0, preferably around 1.0. As you saw with my 5-day fast glucose ketone index, it's quite hard to reach and keep your GKI that low. On a daily basis, my own GKI can fall somewhere between 5-13. If you're metabolically healthy and don't have any serious disease like cancer, then you don't have to obsess over your GKI score. It'd be a good thing to dip into the 1.0 range every once in a while during an extended fast but it's not a necessity.

These measurements aren't definitely indicative of autophagy or anabolism or something like that. However, they can give you a glimpse into your metabolic health and what state your body is in.

If you combine a lower insulin to glucagon ratio with a lower glucose ketone index while you're in a fasted state with depleted liver glycogen, then you can predict the degree of autophagy you're in. It wouldn't tell you about autophagy if you'd been eating because calories will suppress autophagy and you can be anabolic with higher mTOR while still maintaining a low insulin to glucagon ratio and vice versa. You can only predict it if you're in a fasted state because that's the surest way we know to increase autophagy.

You don't want to be in autophagy all the time either because it would prevent growth and repair of your body. Too much autophagy may lead to muscle wasting and dysfunctional cell death, which is why you want to balance catabolism with anabolism. Maintaining lean muscle is incredibly important for longevity and increased healthspan.

But is autophagy good or bad then?

Autophagy promotes cellular survival in response to stressful conditions such as hypoxia, starvation, mitochondrial damage, and infections[241]. That can be a double-edged sword as both the good and bad cells can get stronger from this.

Whether or not autophagy will be good or bad for you depends on the other lifestyle factors such as the circadian rhythms, overall biomarkers, the amount of body fat, muscle mass, methylation, microbiome and so much more. Fortunately, this book teaches you to optimize all of them. These components are dependent of autophagy to some degree which is why prevention is the best medicine.

Autophagy should be seen as a life-long process of maintenance and repair that needs to be conditioned and kept active so that you'd prevent it from becoming dysfunctional or inadequate.

Balancing Autophagy and mTOR

Both mTOR and Autophagy are amazing. It's fortunate that the human body has evolved with these metabolic pathways that promote survival and growth. However, they're not always optimal for our health and longevity in the context of a modern society.

The main idea of this entire book is that by regulating and controlling the expression of certain nutritional factors you can get drastic results in your body composition as well as expected lifespan. Let's take a look at the Protein Kinase Triad once again:

- mTORC1 (mammalian/mechanistic target of rapamycin complex 1) – it's the main regulator of cell growth and anabolism through upregulating protein synthesis
- mTORC2 (mammalian target of rapamycin complex 2) – it's a multiprotein complex of TOR that regulates the actin cytoskeleton, which is a network of filaments and fibers, extending throughout the cytoplasm of cells
- AMPK (AMP-activated protein kinase) – it's the main fuel sensor that helps to mobilize the body's internal energy stores in situations of energy deprivation
- ULK1 (Unc-51-like kinase 1) – it's a protein that's involved with activating autophagy and other catabolic reactions to amino acid deficiencies

These 3 pathways (mTOR/Autophagy/AMPK) sense the energy status of the body and determine whether your cells will be favouring anabolic processes of growth or catabolic processes of self-devouring and preservation.

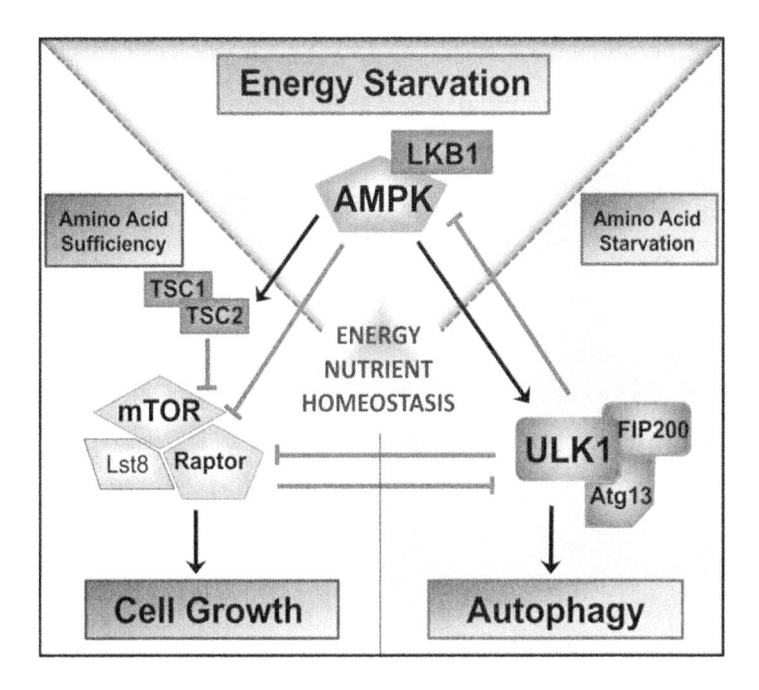

Figure 36 The Protein Kinase Triad that regulates nutrient signaling pathways

There's always an evolutionary trade-off between anabolism and catabolism as well as growth and repair.

- Being too anabolic and growing rapidly may speed up your biological clock so to say by causing oxidative stress to the mitochondria and simply making your other organs work harder. If you grow fast you'll inevitably age faster as well
- Being too catabolic and degrading at a quicker rate than you can repair will also damage vital cells and other processes in the body. In this case, you'll die sooner just by virtue of physical deterioration.

Both of them are necessary for health and longevity. You want to have enough lean muscle mass and bone structure to support vitality but you also want to be efficient with catabolizing your body's own internal fuel sources as to not waste unnecessary energy.

Being at a balance, on the other hand, is a myth because you'll never be completely stagnant and balanced. What you have to do is simply understand that certain things make you more anabolic and others catabolic so you can know how to structure your lifestyle based on what you want to accomplish.

Basically, **the expression of mTOR and its precursors is the most important signaling pathway in our body because mTOR will tell the body what's its energy status.** If your physiology is in a state of nutrient deprivation and starvation, your body recognizes that it's not the best time grow.

Whatever the research tells us, it's clear you don't want to be stimulating mTOR throughout the day because it's going to offset the balance towards too much anabolism. As I said, there is no such thing as being in balance as that would entail you're 50-50 on autophagy and mTOR. It'd be much better for longevity to be recycling your own cells most of the time and then when you stimulate mTOR you do it only briefly and with a larger effect. That's why I follow the ketogenic diet as well – because of keeping carbs low and protein moderate, I'm able to suppress mTOR the majority of time and trigger it only for building muscle.

Anti-Aging expert Ron Rosedale MD, who was one of the first people in the scientific community to point out the dangers of eating too much protein due to mTOR, has said that:

> *Health and lifespan are determined by the proportion of fat versus sugar people burn throughout their lifetime. The more fat that one burns as fuel, the healthier the person will be, and the more likely they will live a long time. The more sugar a person burns, the more disease ridden and the shorter a lifespan a person is likely to have.*[242]

But there's a caveat to the ketogenic diet as well because research has shown that some tumors and cancers can also feed off ketone bodies[243]. So, ketosis or the ketogenic diet themselves won't help you live longer. Neither will just autophagy. You have to know how to alchemically regulate your own metabolic pathways and be mindful of what direction you're heading with your dietary practices.

That's why intermittent fasting is still the most powerful component of this. Honestly, it's an irreplaceable part of life extension and balancing mTOR with autophagy. Any diet that doesn't include intermittent fasting is pro-mortality and it's very difficult to side-step this. You can gain some of the benefits of autophagy by eating ATG-boosting compounds and restricting protein but the effects of that aren't nearly as good you'd get from simply fasting. A therapeutic ketogenic diet may mimic a fasted physiology by keeping insulin and glycogen low but for the full benefits of autophagy, you need periods of energy deficit.

The process of autophagy resembles life itself – almost every organism on the planet is in a symbiotic relationship that more often than not entails eating each other. That is how predation occurs in the wild. That is how different ecosystems balance their species diversity. That's how cells within your body eat themselves and make them reborn.

In many mythologies of the world, there is a creature called the Ouroboros that represents the entire idea of autophagy perfectly. It's the snake biting its own tail – eating itself as to leave behind the old. Paradoxically, the serpent moves in a cyclical manner that entails constant death and rebirth.

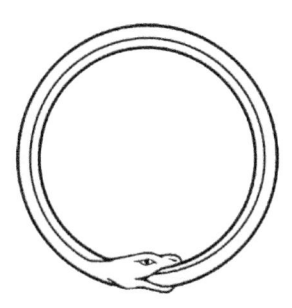

Metaphorically, the Ouroboros represents not only autophagy but any other form of personal transformation. In order to grow and become a new person, you have to leave behind the old. To promote the functioning of your healthy cells, you need to eliminate the dysfunctional ones. Inside your body, these processes happen inside a network of collective intelligence that includes your mitochondria, bacteria, cells, and physiological states. In your personal life, you have to be the one who'll bite their own tail and assimilate the weaker parts of yourself so you could become stronger and reborn into the greatest version of yourself.

Chapter V
Squaring the Curve

"I think you should always bear in mind that entropy is not on your side."

Elon Musk

When you look at the world around you, then you can see how everything functions in a cyclical manner with death and deterioration complementing life. The Universe itself seems to be moving towards greater complexity, chaos, and entropy, which may have potentially unwanted consequences on our subjective personhood.

However, entropy is never on your side, as the quote above pointed out. You're always on the verge of disappearing into nothingness. Whether or not that's a bad thing is a different question with many nuances. Instead of giving you immortality, this book seeks to help you live more fulfilled and have higher quality experiences.

In Chapter One, we talked about how age-related diseases have to do with a loss in mitochondrial functioning, decreased autophagy, cellular turnover, and pure physical damage.

Although the human body has its biological limitations that inevitably will reach its end, there are still many things you can do today to postpone that process. Aging doesn't have to be something that's guaranteed as most of it is relievable with lifestyle. It sure as hell doesn't have to cause you suffering but joy instead.

In the coming chapters, I'm going to be walking through how to promote longevity with resistance training, and how to maintain lean muscle tissue as you age. Take up your arms as we now begin to combat entropy.

Muscle and Longevity

"The best activities for your health are pumping and humping."

Arnold Schwarzenegger

Living organisms consist of many organic molecules and different tissues. Up to 60% of your body is made of water and it's found in muscle, organs, the brain, adipose tissue, and even bones.

A healthy body composition includes more muscle mass in relation to fat mass. About 30-40% of a fit person's total body weight is made of skeletal muscle tissue. This may increase up to 45% in advanced athletes and people who do strength training. Women tend to have less while men have more.

In 1972, Ancel Keys and others published a paper in the *Journal of Chronic Disease* where they coined the term Body Mass Index (BMI) to describe a person's ratio of bodyweight to squared height[244]. Keys said that the BMI is: *"...if not fully satisfactory, at least as good as any other relative weight index as an indicator of relative obesity"*.

BMI has been used for decades in attempts to measure an individual's weight status and categorize them as either underweight, normal, overweight, or obese. The formulas are as follows:

- Metric BMI Formula: BMI = weight (kg) ÷ height^2 (m^2)
- Imperial BMI Formula: BMI = weight (lb) ÷ height^2 (in^2) × 703

Let's take me as an example. I'm a 173 cm male who weighs 77 kilograms. This would lead us to calculate my BMI with these results:

- Convert my height into meters: 173 cm = 1,73 m
- Calculate my BMI: 77 ÷ (1,73)^2 = 25,6

That gives me the BMI of roughly 26. Common BMI ranges are: <18 underweight, 18-25 normal, 25-30 overweight, 30-40 obese, and <40 severely obese. Based on industry standards, I'm borderline overweight and in the yellow zone for increased health problems.

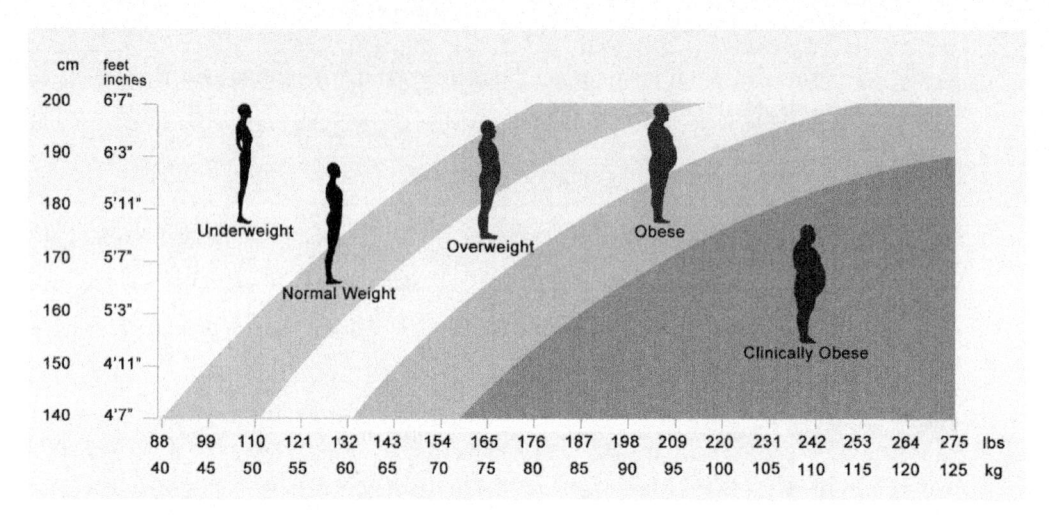

Figure 37 Congratulations! You may be borderline overweight or even obese if you happen to have some muscle

No doubt that overweight and obese people are at a greater risk of metabolic syndrome and dying prematurely. However, BMI isn't really a reliable indicator of lifespan. In a 2014 meta-analysis, people who were categorized as overweight and '*fat-but-fit*' had similar chances of dying as those who were in the normal weight-fit range[245]. Someone who's carrying a bit of extra fluff but exercises a lot may still be healthy in the long term.

However, **having higher body fat isn't definitely healthier nor optimal.** Excess adipose tissue, especially visceral fat, increases the risk of heart disease, diabetes, fatty liver, insulin resistance, and hypertension.

- **Visceral Fat** – The fat around your abdominals and between the organs. It's a bit harder to pinch and measure accurately. Often referred to as stubborn belly fat.
- **Subcutaneous Fat** – The fat under your skin. You can pinch it quite easily with your fingers and wobble around. *It's simple – if it jiggles, it's fat.* This one is less dangerous and used more for metabolic activities.

Visceral fat will continuously leach inflammatory cytokines and other metabolic substrates into your system that will promote inflammation and oxidative stress. Being overweight, experiencing too much stress, drinking alcohol, not sleeping enough, insulin resistance, and eating processed foods all contribute to visceral fat gain.

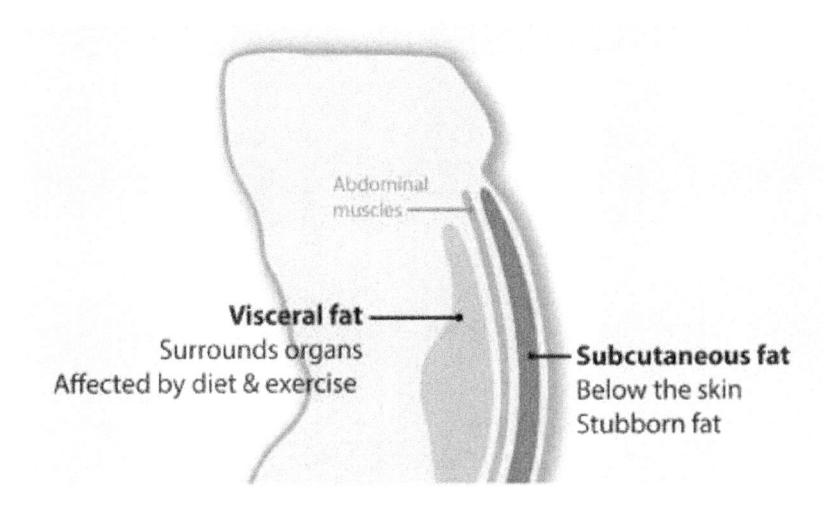

Figure 38 Difference between visceral and subcutaneous fat

Everyone has a combination of both types of fat in different proportions. To determine how much of each you have you can take DNA tests or full body composition scans at a medical facility. Naturally, if you're already carrying a bit of extra weight then you most likely already have a lot of both. However, even if you're skinny, you may still have some extra visceral fat, especially if you suffer from the *'skinny-fat syndrome'*. Having a wider more apple-like body shape is also indicative of visceral fat. To lose that you'd simply have to fix your diet, burn some fat, lower your stress, and allow your metabolism to heal.

Good body fat percentages range from 8-14% for men and 15-23% in women. Acceptable ranges fall between 15-20% in men and 24-30% in women. In men, you're considered overweight if your body fat % is above 21 and in women above 31%. Looking at your actual body fat % and relative fitness level is a much more accurate tool for predicting longevity than BMI.

Therefore, for optimal health and longevity, you would want to keep your body fat % always relatively low as to prevent obesity and other ailments. Being fat-but-fit shouldn't be an argument for staying fat because you'll always be healthier with slightly less body fat. There is no longevity benefit to having extra fat beyond a healthy range.

However, a lot more important factor for increased lifespan may be higher muscle mass. More muscle promotes chances of survival through many mechanisms but its main benefit comes from combating entropy.

The human body inevitably deteriorates with age. After the age of 30, aging is characterized by a progressive decrease in skeletal muscle. This process is called *sarcopenia* and it can happen at a rate of 3-8% reduction per decade. From the age of 40, lean tissue and strength get reduced by about 1% per year[246]. Keep in mind that even very athletic people's bodies consist of 30-40% muscle, which isn't that much. That makes maintaining your musculature an incredibly vital component of healthy aging and longevity.

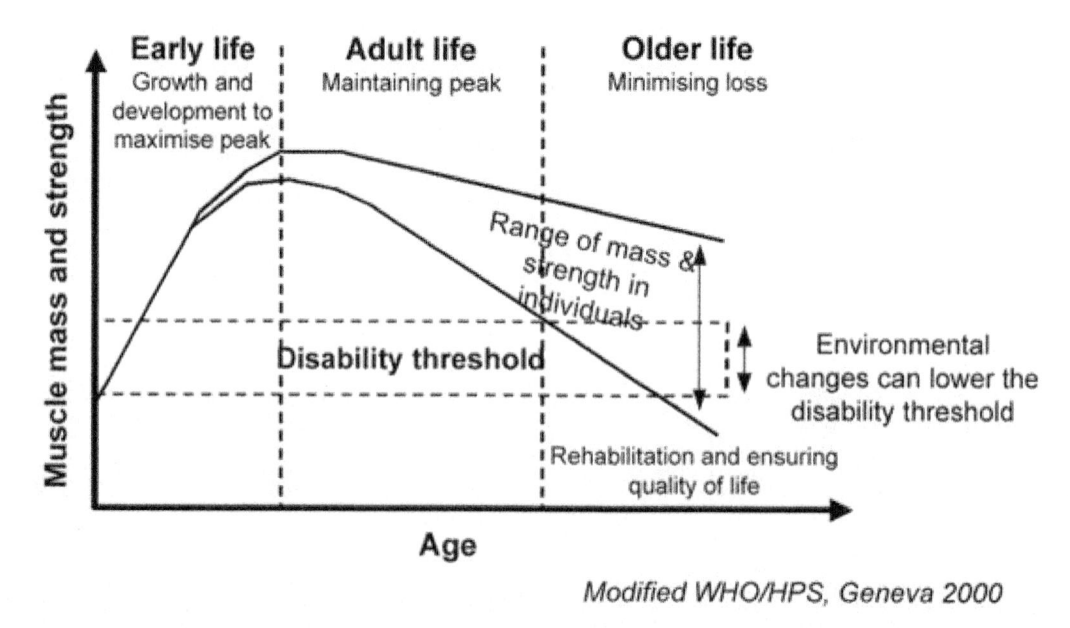

Figure 39 Muscle mass peaks during adulthood and drops in older life

Muscle wasting and a decrease in cardiovascular fitness are often attributed to disuse and sedentary living. Elderly people start to move less which makes their body lose its lean tissue. Low muscular fitness combined with inadequate nutrition is major risk factors for disease and mortality from all causes[247].

Loss of muscle is very well illustrated by hospital bed rest and weightlessness in space where astronauts or patients tend to lose a lot of weight. Most of it is valuable lean tissue.

This inactivity also decreases muscle protein synthesis (MPS), which makes the body more resistant to anabolic growth and hormones. Age also contributes towards resistance to protein intake.

On the flip side, it's well demonstrated that acute resistance exercise enhances myofibrillar muscle protein synthesis in both young and older individuals[248]. Even a single bout of resistance training can increase MPS by 2-3 times, which may be enhanced further with a protein-rich diet[249]. What you don't use you'll lose, especially muscle and strength.

For anti-aging and heart health, the American Heart Association (AHA) recommends aerobic activities, such as walking, jogging, swimming or cycling[250]. They also propose some strength training, but the vast majority of people aren't actively engaged in any form of resistance exercise. However, strength training may be much more important for longevity than just aerobics. There's a lot of research suggesting that muscular strength is inversely and independently associated with all-cause mortality[251]. Essentially, more muscle and strength reduces your chances of dying and can increase your lifespan, especially in older people.

A 2016 paper found that amongst a large cohort of 65 and older, mortality rates were significantly lower in individuals who did regular strength training[252]. This shouldn't be just encouraging for the elderly to continue exercising but it should also tell younger adults to keep working out as part of their everyday lifestyle.

The common idea of aging is that it's a gradual process of degradation that involves slowly becoming weaker, more forgetful, obsolete in society, a nuisance to the family, and miserable. How many people do you know who are already showing signs of deterioration in their 40s? *Oh...I'm not as supple as I used to be. I've got some back pain, I keep forgetting stuff, and I can't get out of bed easily. Must be old age catching up on me...*However, that's often the result of aging poorly and prematurely.

Aging shouldn't be thought of as something negative or seen with dismay. Of course, there is the inevitable physical entropy that accompanies the life cycle of your body but it doesn't have to be such a gruesome process. Like said, most people start aging in their 40s and they accept it as a natural thing. *Must be getting old again...*

Even if you did live over 100 years old, you wouldn't want to spend most of that time being disabled or enduring age-related diseases. I don't know about you but I'd prefer having a shorter life that's higher in quality VS living extremely long but suffering most of that time. Fortunately, you can get the best of both worlds.

'Squaring the Curve' describes the process by which you prolong your healthspan. Think of 2 lines plotted on a graph. Both start off at an equal level but in about 1/3 of the timeline, one of them starts to slowly decline. The other one maintains its heights and keeps going all the way until the last parts where it suddenly drops. That's how you'd want to age as well – to sustain a stable level of vitality and health throughout all of your years, not just the youth and then have a sharp and quick descent. It might seem horrible but in terms of total suffering and well-being, it sure does beat the weaning process of aging gradually.

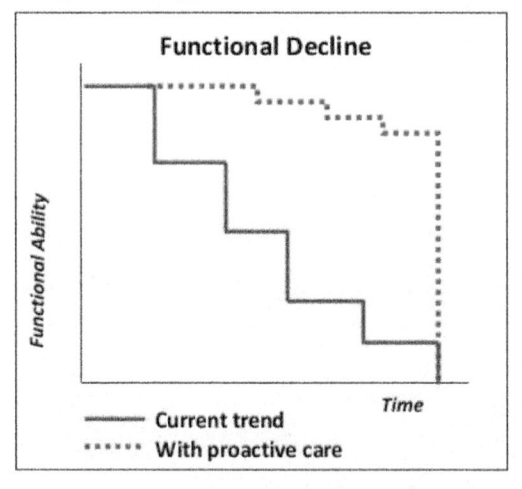

Figure 40 Squaring the Life Curve with Proactive Care

Most of the world is aging dysfunctionally – they're constantly slipping downward with their health and physical condition – like tectonic plates sinking into molten mantle. With a proactive approach to longevity and personal healthcare, you can avoid that deterioration and postpone it quite remarkably.

Funny enough, hunter-gatherers and some longevity hotspots tend to follow the same trend. Old people maintain high levels of physical activity and youthfulness all the way into late elderhood and then they die quite suddenly. There's nothing wrong with living like that as it's a sign of health and vitality. It's not even related to trying to live forever like a superhuman but simply improving the quality of life and extending healthspan. I'd prefer squaring the curve hard and quite rapidly rather than being subject to entropy throughout my entire life.

The Blue Zones

There are several of these regions across the globe called The Blue Zones where most of the population lives a very long life. The term was first coined by Dan Buettner in his 2005 National Geographic Magazine cover story *„The Secrets of a Long Life"*[253].

There are 5 Blue Zones: Okinawa in Japan, Sardinia in Italy, Nicoya in Costa Rica, Icaria in Greece, and the Seventh-Day Adventists in Loma Linda (California).

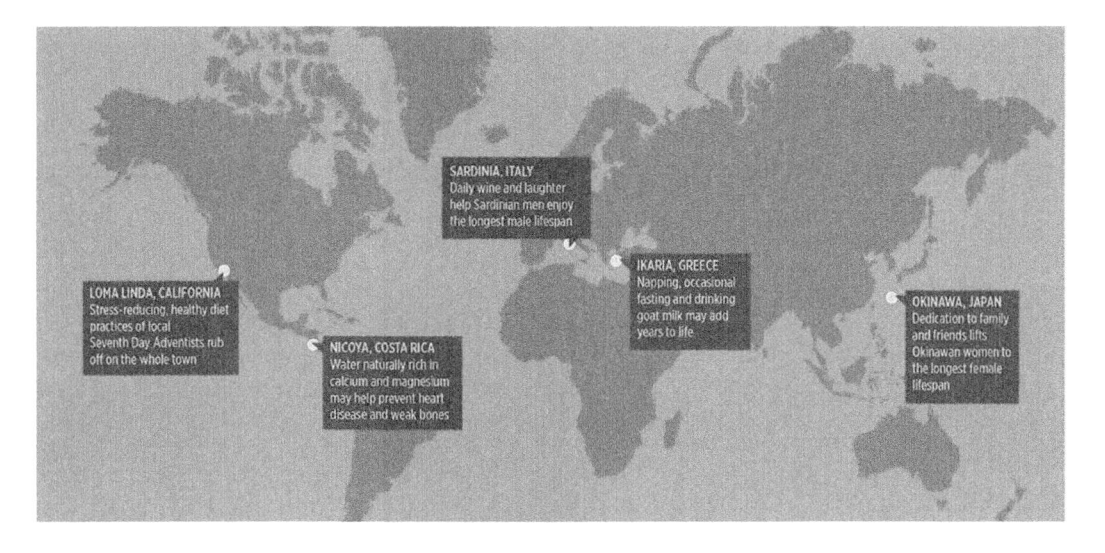

Figure 41 Longevity hotspots across the globe called The Blue Zones

These regions have the highest proportion of centenarians – people who live over 100 years old – they have less disease, increased longevity, and, in general, enjoy a good quality of life.

Given that these Blue Zones show up in a variety of locations across the entire globe, it's safe to say that the biggest contributing factor to their long lifespan is the environment.

It's epigenetics. *'EPI-'*, meaning *'on top of'* or *'outside of'*. Having certain genes plays only a minor role in determining your lifespan. A much more critical part is the environment in which those genes get expressed. Your environment is everything you get exposed to – the food you eat, the physical space, your natural surroundings, the thoughts and emotions you experience, the people you spend time with, and what overall interpretations are you making about all of it.

People in The Blue Zones are living in different geographical locations, they tend to eat different diets, they tend to follow different traditions but there are still a lot of things they have in common.

There are about 6 characteristics shared amongst these people that help them to live longer:

1. They eat a whole foods based diet with a variety of vegetables, fruit, fish, some meat, tubers, legumes, and whole grains.

2. They engage in moderate physical activity most of the day by working, gardening, spending time in nature, and walking.

3. They have lower levels of stress and less work which decreases their cortisol and slows down telomere shortening.

4. They stick to the circadian rhythms by following day and night cycles, getting a lot of exposure to natural sunlight, sleeping well at night, and having several naps throughout the day.

5. They have a strong sense of community as all of the people are actively engaged with their families and others around them.

6. They eat moderately and never too much because eating more would mean they'd have to work longer.

These are all the pillar stones to a healthy longevity-oriented lifestyle we can all follow.

The diets of Blue Zones aren't strict Paleo, vegan, or keto – they're actually very balanced and include a wide variety of food groups. These people eat meat but they also eat vegetables. They eat bread but they also eat fish. They eat dairy but they also eat fruit. This creates a more robust intestinal flora, which supports the immune system and prevents disease. Therefore, it's a good idea to mix up your diet every once in a while and rotate what you eat, especially if you're doing restrictive diets like keto or vegan.

However, a balanced diet is an oxymoron and it's not something you can quantify or prescribe to everyone – it's a matter of context and the person who's eating it. That's why the recommendations prescribed in this book will be more effective and quantifiable.

Blue Zones don't have a gym or treadmill parks – their exercise is enmeshed into their day to day activities. These people move around at a low to moderate pace most of the day by walking around the household, riding a bike to check up on the neighbors, doing gardening to grow their own food, hiking in nature, throwing hay or whatever the situation demands. They're not trying to burn their muscles or hit complete failure when lifting something – they're moving less intensely but they're doing it more frequently. It's important to stay mobile and active throughout the day so your mitochondria could keep producing energy.

In fact, having an intense 2-hour workout or a long bike run is still counterproductive if you sit in a sedentary position for the rest of the day. Frequency and total volume are much more important than maximum intensity when it comes to longevity.

The rationale in the Blue Zones is that eating more than you need requires more work and more stress. This problem is circumvented in the West because there's an overabundance of everything – calories are so cheap that people literally have to come up with more ways of burning them off. Combine that with not moving enough and you'll end up with the biggest obesity epidemic in history.

This same way of thinking in moderation applies to other areas of life as well – material possessions, working, sleeping, stress etc. The fact of the matter is that more is not necessarily better and you have to always keep in mind whether or not you need to have excess.

The most important thing for living over 100 is to optimize your environment. Naturally, we can't all go live on the sunny coast of Costa Rica, but what we can do is follow these best practices of longevity and improve our surroundings to the best of our ability. First, you have to start off with taking care of your own health and stress. Then that vitality will begin to ripple further beyond onto others around you as well and it becomes a feedback loop of happier people creating more happier people and eventually, you may start living in a Blue Zone yourself.

Why Does Muscle Promote Longevity

As mentioned already, more muscle and strength is a much greater predictor for longevity than just aerobic exercise. But why is that?

Resistance training can counter-act age-related sarcopenia. The worst thing about losing muscle with age is that you'll keep the fat. Sarcopenia decreases the amount of functional muscle fibers and proteins in your body while accumulating excessive intra- and extracellular lipids underneath the skin[254]. That's why people who do strength training and have more muscle tend to look healthier and are physically active even at an older age.

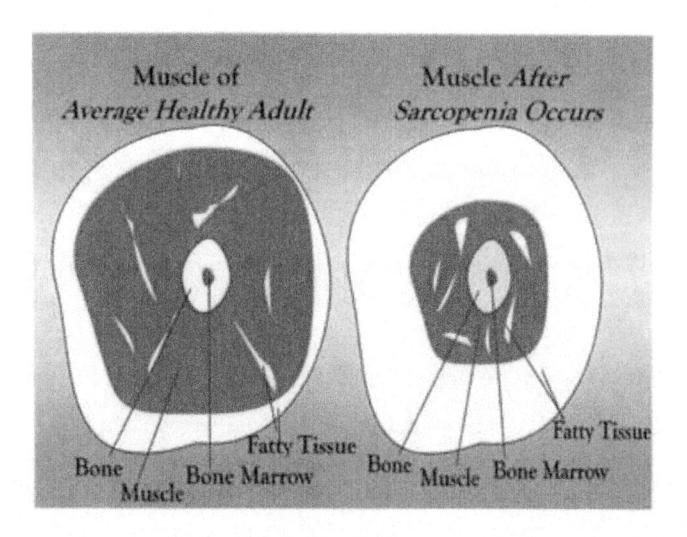

Figure 42 Healthy Muscle VS Sarcopenia

The best benefit of increased muscle mass is that it improves your insulin sensitivity[255], which will protect against diabetes, promotes glucose tolerance, and enables the person to eat more calories without getting fat. Skeletal muscle acts like a huge sponge for glucose and it comprises the majority of whole-body glucose uptake. That's why training and having more muscle can be preventative for insulin resistance.

To build muscle, the body has to have a reason for doing so. This entails applying mechanical stress that would stimulate muscle protein synthesis (MPS) and the mTOR pathway. Mild distress on the muscle preserves its mitochondrial function, which will prevent the age-related sarcopenia. It's another hormetic adaptation that prolongs lifespan. Muscle mitohormesis is thought to promote longevity because of repressing insulin signaling[256].

There is a hypothesis called *Uncoupling to Survive,* which suggests that increased mitochondrial uncoupling and thus increased energy expenditure might increase longevity by preventing the formation of reactive oxygen species (ROS)[257]. Expression of uncoupling protein 1 (UCP1) in skeletal muscle mitochondria increases lifespan considerably[258]. UCP1 or Thermogenin is an uncoupling protein found in the mitochondria of brown adipose tissue that's used to generate heat through non-shivering thermogenesis.

Fasting and resistance training promote UCP1, which in turn increases heat-shock proteins and lipid metabolism. Increased skeletal muscle uncoupling doesn't seem to reduce oxidative stress in muscle cells but actually increases endogenous antioxidant defence systems and redox signaling.

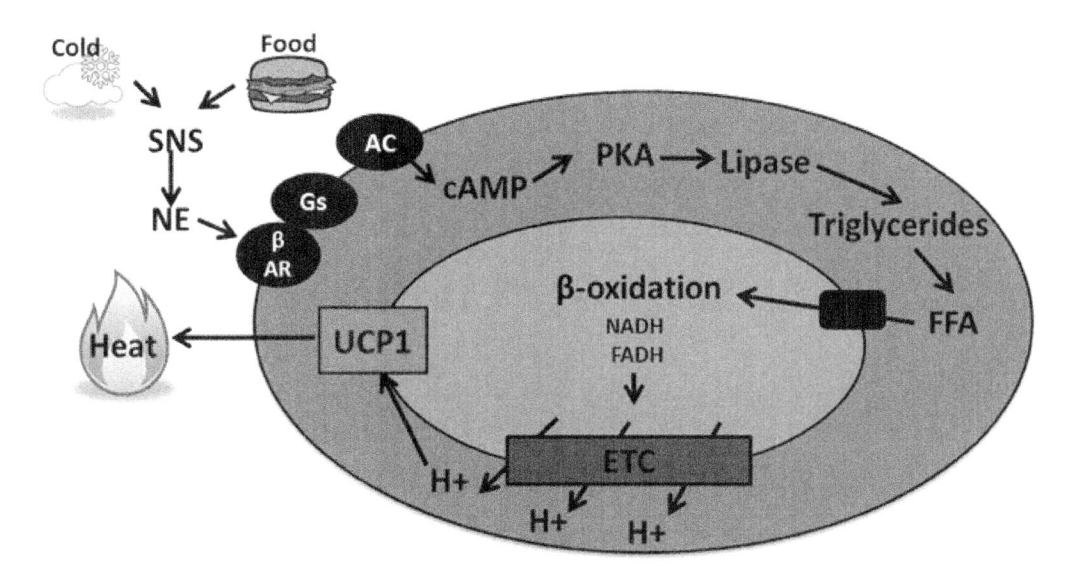

Figure 43 UCP1 results from beta-oxidation, which produces more heat and promotes longevity

There's evidence that skeletal muscle acts similarly to an endocrine organ, just like the adipose tissue does. Muscle cells (myocytes) produce certain cytokines called myokines that fight inflammation and maintain healthy physiological functioning[259]. That's another potential mechanism by which muscle uncoupling protects against chronic diseases. Being fit and muscular also enables the person to maintain a slightly lower resting heart rate and blood sugar, by reducing inflammation levels and burning off visceral fat.

Muscle mass, essentially, is like a solid pension fund that is associated with increased longevity, healthy aging, greater healthspan, and improved physical well-being throughout your entire life.

Unfortunately, not only are you more prone to losing lean muscle as you age but it also becomes increasingly more difficult to build it. As you get older, entropy begins to catch up and makes it harder for you to maintain fitness. For instance, rates of sarcopenia are often similar to the declines in growth hormone, testosterone, and other anabolic hormones (See Figure 44). All of those factors contribute to building and maintaining muscle, which is why you want to optimize your lifestyle.

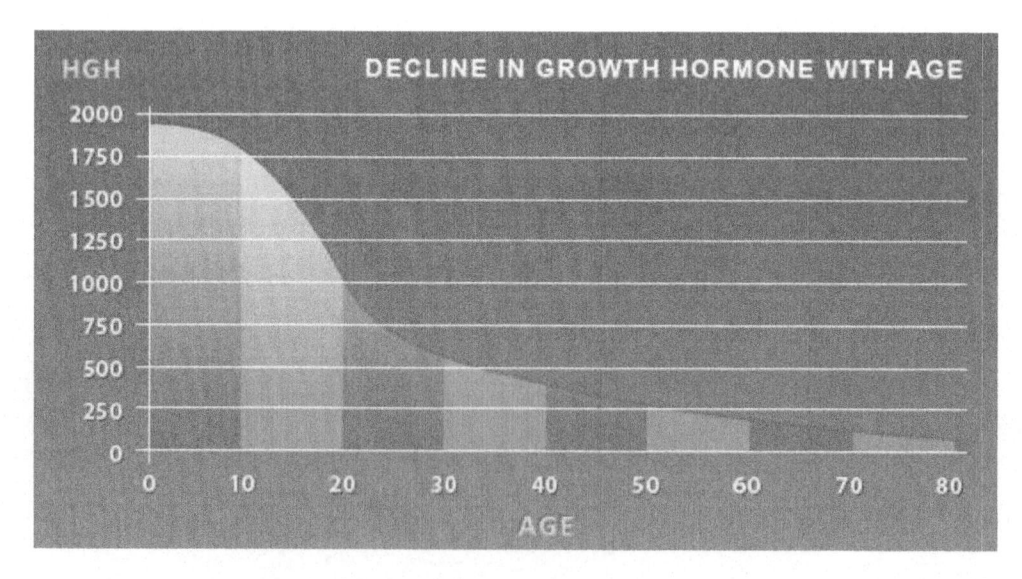

Figure 44 Decline in Growth Hormone with age

In my opinion, **part of the reason old people lose their muscle and anabolism is that they've become sedentary.** What you don't use you'll lose, especially strength and muscle. Those who are squaring the curve with a proactive approach to fitness and health will most likely be able to maintain more of their youthfulness whereas those who've simply given up will inevitably slip off into a gradual downward spiral of deterioration.

When looking at your family members who haven't made the healthiest choices in their diet and exercise you may indeed become depressed and want to postpone aging as much as possible. However, this should be encouraging and motivating because you don't necessarily have to suffer from the negative side-effects of aging with the right lifestyle habits.

What it should also tell you is that to increase your lifespan as well as your healthspan, you'd want to start working on it TODAY. It's a proactive process that requires an active approach.

What Kind of Muscle Benefits Longevity

As with everything related to the body, not all muscle tissue is created equally. There are several nuances to having more lean mass and not all of it is necessarily better.

In fact, it's thought that strength, not muscle mass *per se*, is associated with mortality[260]. Although losing muscle is a contributor to increased atrophy in old age, the elderly and obese people tend to have less strength and force production than healthier ones[261]. Maintaining muscle mass alone has not been shown to be enough to prevent loss of muscle strength, which suggests that muscle quality is equally as important for longevity.

Factors that underpin muscle quality include insulin sensitivity, motor unit control, body composition, metabolic status, aerobic capacity, fibrosis, and neural activation. All of them determine how well your muscle functions as well as the rate of decline with age.

Muscle quality is closely connected with muscle strength and power. With age, you tend to see a progressive decrease of type IIb fibers also known as fast twitch muscle fibers. Part of this may be because of disuse and sedentism but another reason is probably due to pure atrophy of lean tissue.

Therefore, having simply bulkier muscles like a bodybuilder may not necessarily correlate with increased longevity and lifespan. Even though you may exhibit higher insulin sensitivity and growth hormone, the body will still wean off because of the excess weight it has to carry. A good example is *cardiomegaly* (enlarged heart) where the heart can't pump out enough force and thus compensates for it by growing larger. That's why some athletes often die prematurely - their heart has to work extra hard to sustain its large size. Bodybuilders who take steroids to build muscle also tend to have larger hearts than regular people.

The key is to have quality muscle that's able to carry heavier loads and protect itself against muscle dystrophy. Being able to exert more force with less muscle size indicates higher muscle quality and mitochondrial density.

Grip strength is a simple and inexpensive measure of overall muscular strength. Low grip strength is associated with all-cause mortality, cardiovascular events, myocardial infarction, and stroke. It's thought to be a stronger predictor of dying than systolic blood pressure[262].

A recent UK study on over 400 000 participants tested the association between grip strength, obesity, and mortality[263]. It was found that a stronger grip was associated with an 8% decreased risk of mortality. Adiposity measurements were inconsistent with mortality but a BMI over 35 and abdominal obesity were strong predictors of mortality, independent of grip strength. Therefore, even if you are a strong-but-fat person with good grip strength, it's not justified nor optimal to be obese. *Not fat shaming, just doing what's best for you…*

It's suggested that overall grip strength may act as a biomarker of aging across the entire lifespan[264]. That's why some old men who have done physical labor all their lives tend to be quite healthy and robust even without going to the gym. Strong hands require you to be working and lifting things on a consistent basis. Old farmers in The Blue Zones also tend to have good grip strength despite their smaller physiques. Farm work requires much more forearms than biceps and triceps like you'd train at the gym.

Grip strength is the ability to use your hands to hold onto something and not give in under heavy load. It signifies enduring through and not letting go. Those people who have stronger grips may live longer just because of possessing this notion of grasping on and refusing to die.

Another potential indicator of muscle quality is leg strength. In 2011, researchers found that leg strength was one of the most important factors for determining physical function and mortality[265]. The link was purely based on strength and longevity regardless the amount of muscle mass.

What are the best exercises for building grip and leg strength? Compound lifts like squats, deadlifts, barbell rows, and bench press are the best ones for developing full-body strength and muscle. More on this in Chapter Seven.

Furthermore, resistance training increases bone mass and lowers the risk of osteoporosis, which a lot of people suffer from. Elderly folks who don't exercise and have lost their muscle are prone to hip fractures and joint pains just because of lacking bone density. In fact, one of the causes for dying is that they get an injury or they fall on the ground and they get hospitalized or die altogether.

Fortunately, it's never too late to start working out. It's been shown that elderly sedentary people can gain more than 50% strength after 6 weeks of resistance training. They even had to work out only 2-3 times per week with about 70-80% of maximal strength[266]. Age is not an excuse to not train or be un-fit.

Lifting weights and bodybuilding are actually incredible for injury prevention and rehabilitation. Strength training's been found to be more effective in treating painful muscles and mobility issues[267]. On the flip side, things like Pilates do not seem to improve symptoms of back pain or functionality[268].

Compared to aerobic exercise, strength-based resistance training protects against aging a lot more. It's been found that strength training can have a lot of the same health benefits as cardio, such as reduced chronic inflammation and improved cardiovascular health[269].

A recent meta-analysis found that resistance training is more effective than cardio for fat loss as well as overall health[270]. In addition to that, both high intensity and resistance exercise tend to suppress your appetite a lot more than low-intensity cardio. That's another hidden contributing factor for longevity – you'll subconsciously eat less because of working out smarter.

In conclusion, strength and resistance training are the best exercises for anti-aging. It not only maintains functionality of the body but also builds lean muscle tissue, which will postpone age-related diseases, protects against deterioration, prevents sarcopenia, and combats entropy.

A lot of research shows that muscle mass should be considered a vital sign of health and aging. Low muscle mass is associated with several negative health conditions. Just exercising or doing a lot of cardio isn't necessarily better because you may end up losing lean tissue. The NET benefit of excessive aerobic work isn't better than focusing primarily on resistance exercise.

Endurance and Cardio

However, there are still some benefits to doing endurance based exercise and cardio. To protect yourself against aging and disease, you'd want to maintain aerobic fitness regardless of your physique goals.

Aerobic capacity, which is your maximal ability to use oxygen during physical activity, tends to decline after the age of 50[271]. Having more aerobic capacity helps muscles to use oxygen more efficiently, which will improve physical performance.

Improved endurance exercise includes increased capillary supply, mitochondrial biogenesis, and improved transport of electrons in the mitochondrial transport chain. With higher aerobic fitness you'll also use less muscle glycogen and produce less lactate, which allows you to perform at greater intensities.

Both aerobic and anaerobic training contributes to the development of cardiovascular fitness. It means you can improve your endurance even without necessarily training it. There are much more effective and healthier ways of training your aerobic system than just cardio.

Long hours of cardio are not that beneficial because of how the repetitive motions begin to tear down your tendons and promote joint pain. Too much aerobics training also increases the risk of oxidative damage in the muscles which may speed up sarcopenia, especially when dieting. On top of that, the prolonged stress hormones and free radicals in the blood may damage the mitochondria.

The worst thing about too much cardio is that it's not even the best way to lose fat or become fit. How many people you know who've been exercising for a long time doing different runs, spin classes, and cardio sessions but they haven't gotten any fitter? They're exercising like maniacs but they're still fat or worse skinny fat. That's because the body receives an inefficient stimulus that would enforce proper adaptation.

Easy-going endurance releases the chemical BDNF in the brain, which makes us feel good and more cognitively sharp. That's why people experience this runner's high after hitting the zone. Their mind literally tells them that cardio is good whereas it might not be. The danger is that it may also become addictive because of the same very reason. This is called *Black Hole Training*.

Black Hole Training

It's a nightmare exercise zone somewhere between a piece of cake and a Navy SEAL workout. The pace is vigorous but not painful, which is enjoyable for your mind. You get an endorphin rush, which makes you think *I'm getting a good workout* but it's still stressful for the body.

Basically, the Black Hole is a heart rate zone that exceeds your aerobic capacity just a tiny bit. Once you can't hold a conversation anymore and have to breathe through your mouth, then you're using more glycogen and less fat for fuel. For a few minutes, that's fine, but most people never go running for 10. They hit runner's flow because of the adrenaline rush and can easily empty their glycogen tank. Once this happens, the body still needs glucose to perform at such an intensity. As a result, it begins to break down the protein.

Intensity Zone	VO$_2$ (%max)	Heart Rate (%max)	Lactate (mmol·L^{-1})	Duration
1	45-65	55-75	0.8-1.5	1-6 h
2	66-80	75-85	1.5-2.5	1-3 h
3	81-87	85-90	2.5-4	50-90 min
4	88-93	90-95	4-6	30-60 min
5	94-100	95-100	6-10	15-30 min

Zones 1 & 2: 'Zone 1'; Zone 3: 'Zone 2'; Zones 4 & 5: 'Zone 3'

Figure 45 Heart Rate Zones

This running craze became popular in the 70s and 80s. Although the joggers ran themselves into adrenal fatigue, they couldn't stop because they had become addicted to the endorphin high. Nowadays, running and jogging is considered the healthiest form of exercise and it's the average person's first choice for working out. Unfortunately, they are mistaken.

Cardio is beneficial and it has its place but you shouldn't make it the focus of your training. At least you'd want to avoid the Black Hole as much as possible.

If you're doing cardio for 30+ minutes then you should stay aerobic for the majority of the time. That's when your heart rate is below 60-70% of your VO2 max. At that intensity, you're using fat not glucose as fuel. Going higher than that will simply make you more glycolytic. When you burn through your glycogen stores and you keep exercising, then at that point you may end up sacrificing a bit of the lean tissue by converting some protein into glucose. With keto-adaptation, you can protect yourself against that to a certain extent but it's simply not worth it.

I'm also not trying to judge people who love to do long hours of cardio just for this zen-running kind of pursuits. Instead, I'm trying to show you a more effective way of exercising and relieve you from the perceived obligation of doing primarily aerobics, which you might have.

The 80/20 to exercise is to focus on resistance training and strength because it's what gives you the most results in terms of longevity. Endurance work can also be good but not if done excessively or for too long. Of course, you don't want to get winded because of climbing a flight of stairs. There are other more effective and time-efficient ways of doing aerobic conditioning than just cardio.

High-Intensity Interval Training

In 1996, a Japanese researcher Dr. Izumi Tabata conducted a study on two groups to assess the differences between training intensities on aerobic fitness[272]. All the subjects used cycling ergometers to train on for 6 weeks.

- The control group did 60 minutes of moderate-intensity exercise 5 times per week at 70% of their VO2 max. It's called low-intensity steady state cardio (LISS).
- The other group did high-intensity interval training (HIIT) with 20/10 sessions repeated 8 times at 170% VO2 max. They basically ran through a wall. 20 seconds beyond-maximum exertion followed by 10 seconds of rest. In total, the workout lasted for 4 minutes.

Results of the study were quite astonishing. Over the course of 6 weeks, the control group trained for 1800 minutes vs the 120 minutes of the HIIT group. The subjects who did LISS increased their VO2 max from 53 +/- 5 ml.kg-1 min-1 to 58 +/- 3 ml.kg-1.min-1 but their anaerobic capacity didn't improve significantly. The individuals who did HIIT increased their VO2 max by 7 ml.kg-1.min-1 while their anaerobic fitness also improved by 28%!

The Tabata study shows that high-intensity training is much more effective in inducing physiological adaptations and is a more time-efficient way of exercising. It also makes sense from the perspective of the body. You won't get stronger by lifting a pillow off the ground for millions of times. Lift a 400-pound barbell just twice and you'll get a much greater response.

Strength and fitness are the result of adapting to necessity. There needs to be a very good reason for your body to build muscle or get faster. Making a hole in the wall with a spoon will take an eternity. Hit it with a sledgehammer and you'll get out of jail in no time.

However, there are some implications about the Tabata study that need to be covered.

- **First of all, the aerobic group did improve their VO2 max quite significantly.** They did get fitter because the body became more efficient at running on lower intensities and thus they could ramp it up a notch.
- **Secondly, although the HIIT-ers increased their mean VO2 max slightly more, they started off with a lower value and may have had more room for "newbie-gains".** The LISS group was already fitter and thus it was more difficult for them to see improvements. See Figure 46 Below.
- **Thirdly, the HIIT group still ended up at a lower VO2 point than the LISS group.** They didn't become exponentially fitter although their relative fitness increased more.

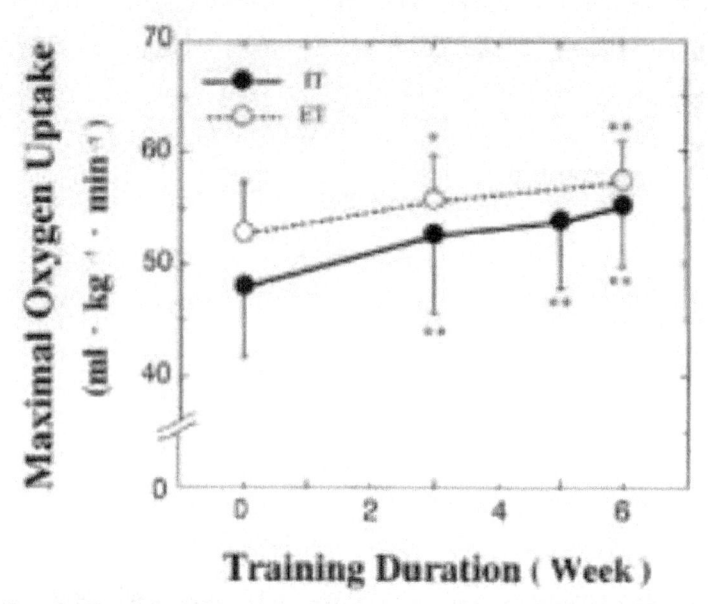

Figure 2-Effect of the endurance training (ET, experiment 1) and the intermittent training (IT, experiment 2) on the maximal oxygen uptake; significant increase from the pretraining value at *P < 0.05 and **P < 0.01, respectively.

Figure 46 Note that the endurance training (ET) group started off from a higher baseline level of fitness than the intensity group (IT)

Despite that, the HIIT group did improve their anaerobic fitness as well as their aerobic, whereas the LISS cyclers didn't. Unless you're training specifically for endurance, then you don't have to do hours of cardio. For an average person trying to be fit and healthy, doing high-intensity training is still a much more effective and time-efficient way of exercising. It saves time, preserves joints, and maintains a better metabolic rate.

High-intensity interval and Tabata training also coincide with resistance training. Both of them are more intense, they burn muscle glycogen, improve insulin sensitivity, and stimulate the sympathetic nervous system. In fact, HIIT may lead to similar physiological adaptations in terms of muscle growth than resistance training. Not entirely but it's not going to hinder the beneficial muscle building signals as much as cardio would.

Combining a lot of endurance with strength training may make you good at none of them because the body won't have enough time or resources to adapt properly. Confucius a Chinese philosopher said: *"The man who chases two rabbits, catches neither."*

That's why doing primarily resistance training and HIIT will give you more results and actually improve your overall fitness in the long term.

Moreover, it resembles the aboriginal hunter-gatherer way of living, where people would walk around at low intensities for the majority of the day and then for a very short amount of time face extreme exertion while running from a lion or chasing their own prey. This is what our primal physiology has become adapted to.

Next to fasting, exercise can also increase mitochondrial biogenesis and function through the same pathways of AMPK and FOXO proteins. High-intensity interval training increases the ability of mitochondria to produce energy by 69% in older people and 49% in younger[273]. Even just acute exercise increases FOXO1 phosphorylation, improves insulin sensitivity and promotes mitochondrial biogenesis[274]. Exercise also increases NAD+ and sirtuins[275,276].

There are many ways to improve your aerobic fitness but the main tenet is increased muscle quality. The best way to protect against sarcopenia and muscle-wasting is to combine both resistance training with HIIT cardio and then get a minimalistic dose on endurance. That's the main way to improve the quality of your life with exercise.

Chapter VI
HyperTORphyc Growth

*"You underestimate the Power of The Dark Side.
If you will not fight, then you will meet your destiny."*

Darth Vader, Star Wars Episode VI: Return of the Jedi

Cellular growth requires adequate amounts of energy and biosynthetic precursors such as essential amino acids and lipids or fatty acids. Because of that, your cells' energy and nutrient status are constantly monitored and regulated by the body wanting to maintain homeostasis.

One of the main mechanisms that makes the body grow and turn anabolic is the mammalian target of rapamycin (mTOR). It's mostly activated by nutrients and growth factors that activate muscle protein synthesis (MPS).

The process of growing new cells and tissue is called *hypertrophy*, which translates from Greek into 'excess-nourishment'. That's how muscles get bigger, that's how you grow as a child, and that's also how you conciliate entropy. Growth and expansion are just an extra step away from aging and death.

mTOR is connected to protein synthesis and cellular growth via many pathways and it affects hypertrophy at least indirectly. It's a piece of the puzzle and probably one of the most important ones but not the only one.

In this chapter, I'm going to give an overview of the main functions of the mTOR pathway and how it affects muscle growth and longevity. Let's begin with the basics!

Enter TOR

Mechanistic Target of Rapamycin or Mammalian Target of Rapamycin or mTOR is a protein kinase fuel sensor that monitors the energy status of your cells[277]. It's involved in every aspect of cellular life and existence.

There are 2 mTOR complexes – mTORC1 and mTORC2. They stimulate cell growth, proliferation, DNA repair, protein synthesis, new blood vessel formation (angiogenesis), muscle building, the immune system and everything related to anabolism[278].

- **mTORC1 functions as a nutrient sensor that controls protein synthesis[279].** mTORC1 is regulated by insulin, growth factors, amino acids, mechanical stimuli, oxidative stress, oxygen levels, the presence of energy molecules (ATP), phosphatidic acid, and glucose[280]. It's a key factor in skeletal muscle protein synthesis[281].
- **mTORC2 regulates the actin cytoskeleton, which is a network of long chains of proteins in the cytoplasm of eukaryotic cells[282].** It also phosphorylates IGF-1 receptor activity through the activity of the amino acid tyrosine protein kinase[283].

mTOR is involved with anabolic pathways such as insulin, insulin-like growth factors (IGF-1 and IGF-2) and amino acids[284][285].

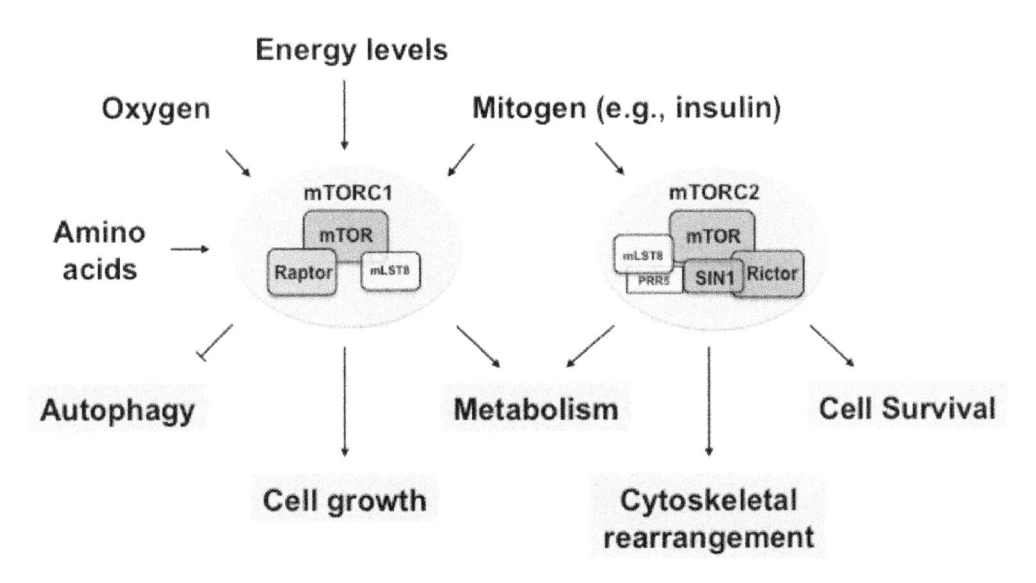

Figure 47 The mTOR complexes and their effects

The mTOR Pathway

The story of how mTOR was discovered is also quite an interesting one.

In 1975, scientists were collecting soil bacteria on the Pacific islands. On The Easter Island, they found that these bacteria called *Streptomyces hygroscopicus* secreted a particular compound with powerful immunosuppressing effects. They named the compound *Rapamycin* after the Polynesian name of the island Rapa Nui.

During the 1990s, Rapamycin was found to bind with a certain complex of proteins in the cells. That protein was named mTORC1. However, rapamycin doesn't bind to mTORC2[286].

mTORC1 detects many extra- and intracellular signals and growth factors[287], whereas mTORC2 is known to be activated only by growth factors[288]. Growth factors would be things like insulin, mechanical muscle stimulus, while nutrient factors would be things like amino acids and glucose that promote growth factors such as insulin and IGF-1.

After several studies, it was discovered that mTOR functions like an on and off switch for cellular growth. Simply put, if mTOR detects energy in the body, it's going to upregulate processes related to growth and ATP production. Elevated mTOR may not be the sole cause of hypertrophy but it's a catalyst nonetheless.

Here's a simple explanation of the mTOR pathway and how it works:

- **Whenever your body detects excess energy in the system, it'll try to direct it into the right places.**
- The mTOR complexes are activated by growth factors, primarily insulin and IGF-1, but also nutrient factors like amino acids and protein.
- Insulin Receptor (IR) and IGF-1 Receptor (IGF-1R) are in the class of tyrosine kinase receptors. Tyrosine is an amino acid. Activating these receptors leads to the phosphorylation of insulin substrate receptor proteins (IRS)[289].
- IRS activates a protein called phosphatidylinositol-3-kinase (PI3K) which further phosphorylates inositol phospholipids like PIP3. PIP3 interacts with proteins PDK1 and Akt.

- **Akt is thought to be one of the main upstream regulators of mTOR**[290]. Akt is a family of proteins that comprise of Akt1, Akt2, and Akt3. Akt1-2 are expressed in skeletal muscle while Akt3 is not[291].
- Akt inhibits protein breakdown by regulating FoxO proteins. FoxO proteins regulate protein breakdown and autophagy-related pathways[292]. They are inhibited by Akt, which prevents cell death[293].
- **When there are plenty of nutrients around, mTORC1 binds to ULK1, which is an autophagy activating kinase and inhibits the formation of autophagosomes which would initiate autophagy**[294]. When energy gets depleted, mTORC1 becomes inactive and releases itself from the ULK1 complex, thus freeing up the formation of autophagosomes.

Myostatin is a growth factor that inhibits muscle growth by suppressing hypertrophy[295]. It's basically your body's attempt to not build too much muscle as a way to preserve energy from an evolutionary perspective. Myostatin reduces Akt/mTORC1/p70S6K activity and protein synthesis, which blocks muscle cell copying[296]. Therefore, for muscle growth, you need the Akt/mTORC1/p70S6K pathway to be active.

AMPK, mTORC1, and ULK1 form *The Kinase Triad,* which maintains energy and nutrient homeostasis[297] (See Figure 48). These protein kinases and fuel sensors are in constant correlation and they're balancing each other out based on the energetic conditions of the body.

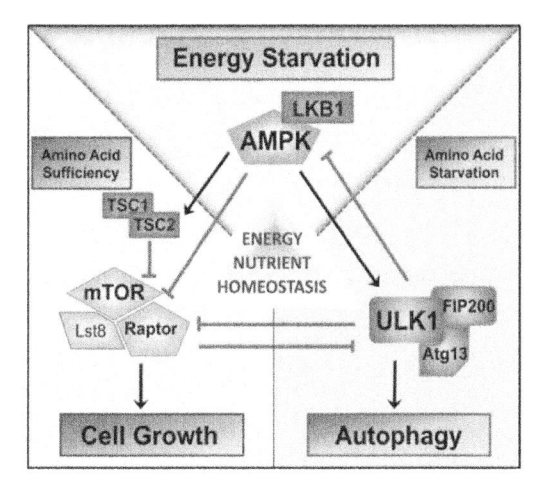

Figure 48 The Protein Kinase Triad of nutrient sensing

What Activates mTOR

mTOR regulation is mostly mediated through AMP-activated kinase (AMPK). AMPK monitors the energy status of the cells through their glycogen content and ATP to AMP to ADP ratios. It's called the adenylate energy charge that measures the energy status of cells.

Adenosine Triphosphate (ATP) is often referred to as the molecular currency of cells that's needed for energy production. Thus it's essential to life and anabolism. When ATP gets produced by the mitochondria, it converts into either adenosine diphosphate (ADP) or adenosine monophosphate (AMP). This process can be recycled several times, creating more ATP from its by-products.

Whether or not the body becomes anabolic or catabolic depends upon the energy charge. In an abundance of ATP, the body has more resources to conduct repair and growth. If ATP levels are low or depleted, ADP and AMP ratios increase and they'll get converted into ATP to maintain homeostasis. In that case, there isn't enough excess energy for additional growth thus the body remains either at a balance or, if energy depletion continues, becomes catabolic.

A reduction in energy activates AMPK which promotes catabolic pathways for maintaining energy homeostasis[298]. AMPK inhibits muscle growth by suppressing mTORC1.

Here are the other mTOR activating nutrients and factors:

- **Amino acids promote mTORC1 activity[299] without affecting mTORC2 activity[300].** Leucine specifically activates mTORC1 the most[301]. Some evidence also hints that leucine's by-product HMB may have a similar anabolic effect through the signaling pathway of mTORC1[302]. Although I'd say you'd get the same effect directly from leucine.
- **Mechanical stimuli from resistance exercise, especially eccentric contractions, increases the levels of mTORC1[303].** That's why cold exposure can sometimes also activate mTOR in muscles – you freeze up and contract the muscles. Phosphatidic acid gets regulated by exercise which activates mTORC1[304].
- **Phosphatidic acid enhances mTOR signaling** and resistance exercise-induced hypertrophy[305]. It's found the most in cabbage leaves, radish leaves, and herbs[306] or can be taken as a supplement.

- **Ursolic acid stimulates mTORC1 after resistance training in mice**[307]. It stimulates anabolism via PI3K/Akt pathways. Ursolic acid can be taken as a supplement or found in foods like apples, bilberries, rosemary, lavender, thyme, oregano and many more.
- **Creatine may potentially promote mTORC1 by increasing IGF-1 activity after exercise but doesn't further potentiate mTORC1 several hours after exercise**[308]. So, the best time to take it is with your post-workout meal. It's one of the most well-studied supplements that actually seems to work. Plus it's very cheap and effective.
- **Testosterone and androgens can also signal mTOR** and induce muscle hypertrophy[309,310]. Testosterone has many anti-catabolic as well as anabolic properties, which is why high cortisol tends to wreak havoc on this hormone.

These nutrient and growth signals are picked up by receptors on the surface of cells, such as insulin receptors, and by the energy availability throughout the cellular matrix via AMPK and its ATP:AMP:ADP ratios.

mTOR and Aging

Overexpression of mTOR or its dysfunction is often related to various cancers and genetic disorders[311]. Suppressing mTOR with diet or certain supplements like Metformin and Rapamycin are common ways of treating cancer and tumor growth.

- Rapamycin inhibits mTORC1, which is thought to increase life expectancy in animal studies.
- Disrupting mTORC2 with rapamycin may induce insulin resistance as well as symptoms of diabetes and glucose intolerance[312].

Increased glycolysis, which is the metabolism of glucose into lactate, is often found to be higher in cancer cells, also known as the Warburg Effect[313]. In 1924, Otto Warburg discovered that cancerous tumor cells primarily meet their energy demands from glycolysis. Akt regulates Hexokinase 2, which is thought to cause this enhanced glycolysis in cancer cells. mTOR promotes the activation of insulin receptors and IGF-1 receptors[314,315], which is in most cases accompanied by glucose and glycolysis. Hence the association of mTOR with cancer.

Inhibiting mTOR also promotes autophagy. High mTOR activity may promote tumor growth because of stopping autophagy from removing cancerous cells[316]. Patients with Alzheimer's disease also show dysregulated mTOR activity in the brain and connection with beta-amyloid proteins[317][318]. That's because of autophagy suppression.

Reduced mTOR expression has been found to increase lifespan in different yeast species, bacteria, and mice[319][320][321]. Whether or not it's related to insulin or mTOR specifically is something I'd pay more attention to.

Methionine restriction could be beneficial for longevity as well. SAM (S-Adenosyl-Methionine) is the 2nd most common cofactor in enzymes after ATP, which detects the presence of methionine-related nutrients in the body. One of the methionine sensors SAMTOR (S-adenosylmethionine upstream of mTORC1) inhibits mTORC1 signaling. Methionine restriction lowers SAM and increases SAMTOR, which improves glucose homeostasis and can promote longevity along the lines of caloric restriction. One of the reasons why it's thought that restricting calories and methionine extends life is because of decreased mTOR and insulin.

Due to its potent anabolic growth effects, mTOR signaling during early life is very beneficial and necessary for proper development. However, it may not be ideal as you become older. Likewise, people wanting to build muscle or improve physical performance would also want more anabolism. Unfortunately, most of the fitness and nutrition advice you hear on social media doesn't even consider the downstream effects of that nor what to do about it.

The free radical theory of aging states that reactive oxygen species (ROS) and oxidative stress damage the mitochondria, which decreases ATP production[322]. By lowering mTOR you preserve ATP that would otherwise be used for protein synthesis and thus have more resources for mitochondrial repair. Low TOR signaling also upregulates autophagy, which helps to remove ROS and recycle old cells back into energy thus slowing down aging.

mTOR may promote intestinal inflammation[323] as well as skin acne, which is more proof to how high mTOR all the time accelerates aging. However, this effect is probably due to a poor microbiome and other inflammatory lifestyle factors not necessarily mTOR itself. mTOR simply makes things worse in some cases because of its anabolic effects.

What Inhibits mTOR

Here are the mTOR inhibition mechanisms:

- **Dietary protein restriction lowers mTOR**[324]. Amino acid deficiency, in particular, regulates mTOR[325].
- **Calorie restriction lowers mTOR and promotes autophagy**[326]. Ghrelin the hunger hormone activates AMPK in the hypothalamus and inhibits mTOR[327].
- **Fasting lowers glucose, insulin, and suppresses mTOR while raising AMPK.** This is the most effective method of inhibiting mTOR. It also raises autophagy and promotes ketosis.
- **Ketogenic Diets are moderate in protein and low glucose, which lowers mTOR activity**[328]. Glucagon, which is a hormone that raises in the presence of low glucose and insulin, activates AMPK and represses mTOR[329].
- **Exercise inhibits mTORC1 in liver and fat cells**[330]. This is great because you'll be preventing fat gain while promoting longevity and muscle growth. The post-exercise time window, however, facilitates muscle protein synthesis in muscle cells because of the mechano-overload. That's another mechanism by which resistance training is great for increasing lifespan – more mTOR in muscles and less mTOR in fat cells.
- **Glucocorticoids and cortisol get elevated during physiological stress**[331]. Cortisol helps to mobilize glycogen and fatty acids. This shifts the body into a more catabolic state.
- **Metformin is a potent anti-diabetic drug that lowers blood sugar and insulin, thus lowering mTOR**[332]. Berberine is a medicinal compound that has similar effects.
- **Rapamycin is an immunosuppressing drug that lowers mTOR**[333]. It's been used to fight cancers and tumors in humans.
- **Resveratrol is a compound found in certain fruit and red wine that has a longevity-boosting effect**[334]. Part of it has to do with sirtuins and autophagy.
- **Curcumin inhibits mTOR signaling in cancer cells**[335]. Reishi fights tumors as well by blocking mTOR[336]. Rhodiola Rosea[337] and astragalus too[338].
- **Anthocyanins found in blueberries and grape seed extract promote AMPK and block mTOR**[339]. Pomegranate as well[340].

- **Alcohol activates AMPK and regulates the mTOR complex**[341]. It doesn't go to say that drinking is going to boost your longevity. Remember that mTOR inhibition is just a single piece of the puzzle.
- **Oleanolic acid contributes to anti-tumor activity**[342]. Main food sources of oleanolic acid are apples, pomegranates, bilberries, lemons, grapes, bilberries, and olives.
- **Carnosine inhibits the proliferation of human gastric carcinoma cancer cells by retarding mTOR signaling**[343]. Carnosine is an amino acid with anti-aging and antioxidant benefits that fights free radicals as well. Interestingly, it's found the most in animal foods and meat.

Although there's some evidence showing how mTOR can have negative side effects, it's still an essential pathway for cellular growth and maintenance. So, is mTOR bad for you?

Benefits of mTOR

Here's the light side of TOR:

- **mTOR is required for protein synthesis and skeletal muscle hypertrophy**[344]. Suppressing mTOR for too long or having too much autophagy leads to muscle atrophy and loss of lean tissue through sarcopenia, which can contribute to aging and metabolic disorders[345].
- **mTORC2 regulates the distribution of mitochondria and mTORC2-activated AKT is linked to mitochondrial proliferation**[346][347]. mTOR also promotes mitochondrial biogenesis by activating PGC1-alpha[348].
    - mTORC2 localizes mitochondria-associated endoplasmic reticulum (ER) and mitochondria-associated membrane (MAM)[349]. This localization is stimulated by PI3K growth factors.
    - mTORC2 deficiency creates a defect in MAM, which causes an uptake of calcium in the mitochondria. Probably not good for atherosclerosis and plaque formation.
- **mTOR can also help you lose weight and be healthier**. mTORC2 regulates glucose homeostasis via Akt. Akt promotes glucose uptake by increasing GLUT4 translocation to the membrane in adipocytes[350]. The same effect is also true because of increased muscle mass and insulin sensitivity that's accompanied by

muscle. If you have too low mTOR then you won't be able to build muscle thus having poorer metabolic flexibility and thus actually predisposing yourself to disease.

- **mTOR also contributes to neural plasticity and learning memory development**[351]. Neuroplasticity is a key factor in learning, skill acquisition, and memory retention. It seems that both too low levels of mTOR and overexpression of mTOR cause impaired learning and cognitive decline. Activating mTOR in prefrontal neurons by HMB inhibits age-related cognitive decline in animals[352]. mTOR also helps to grow synaptic connections.

All in all, mTOR signaling seems to be more problematic in people who already have a certain disease such as cancer, tumors, diabetes, or Alzheimer's. The reason is that mTOR inhibits autophagy which would help to fight the disease by clearing out the diseased cells. If you're sick and keep stimulating growth factors in the body, you'll keep feeding the disease while simultaneously stopping the healing processes. That's why strict fasting for an extended period of time seems to help treat a lot of diseases and cancers.

How Much mTOR is Too Much?

mTOR has its benefits for performance as well as some supportive aspects for longevity. However, it's not optimal to have it elevated all the time for obvious reasons. That's why knowing how to cycle mTOR is a vital thing for your long-term health.

You want to activate mTOR in muscle cells, brain cells, and mitochondria instead of fat cells and cancer cells.

- **Exercise activates mTOR in the brain and promotes skeletal muscle mTOR**[353]. Resistance training, in particular, will make you build muscle through the mechanistic stimuli of mTOR.
- **Time restricting your eating is probably the most effective and most critical thing for controlling mTOR.** Even though you may be eating a low mTOR diet you need autophagy as well if you want to promote life-span.
    - If you're fasting for longer periods of time and you're eating less often, then you need to make your meals more mTOR stimulating to counterbalance the catabolic effects of fasting and support muscle homeostasis.

- o   If you're fasting less and you're eating frequently, then it's indeed a better idea to keep your foods lower in mTOR as to avoid excess growth.
- **The mTOR pathway has many functions beyond just muscle growth and anabolism.** It regulates the immune system, fat storage, and with AMPK controls whole-body energy balance.

Timing is probably the most critical component to this. Activating mTOR doesn't guarantee muscle hypertrophy. It doesn't entail accelerated aging and carcinogenesis either. It's dependent on the body's entire nutrient status combined with the presence of growth factors or lack thereof as well as the overall energetic conditions of the individual. You can even add circadian rhythms and the status of the microbiome into the mix. What determines the effect of these pathways is vastly context dependent.

The same applies to AMPK. Elevated AMPK and increased autophagy won't always lead to lean tissue catabolism and atrophy. You can be effectively efficient at cellular turnover during some periods of the day without jeopardizing muscle or performance over the course of the 24-hour cycle. Likewise, you can have autophagy working in only some preferable regions, such as the liver and brain but not muscle thus preserving lean body mass while fighting disease.

That's why I'd say the best strategy is to maintain a state of low mTOR most of the day by fasting and then stimulating it in a post-workout setting. This ensures that you'll be turning on mTOR for its beneficial effects. If you're raising mTOR without needing to repair your body, then it's inevitably going to be worse for longevity than if you were to do it post-exercise.

Because of the same reasoning, activating mTOR after waking up isn't optimal as your body is in a state of deeper ketosis and mild autophagy after the overnight fast. mTORC1 does promote fatty acid storage by inhibiting lipolysis[354]. It also inhibits beta-oxidation and ketogenesis[355]. However, this may be because mTOR activity is often preceded by insulin and glucose. If you were to stimulate mTOR in a state of low insulin-glucagon ratios, then you'd probably avoid the inhibition of ketosis and fat burning. Nevertheless, turning on mTOR right in the morning may prevent you from going back into autophagy for the rest of the day.

Another benefit of restricting mTOR activation is that you're going to avoid the negative side effects of chronic mTOR elevation. In my opinion, all of the associations of mTOR with cancer and aging stem from having it elevated all the time *ala* Standard American diet 3 times a day with constant snacking or the high protein bodybuilder diet with 4-6 meals a day.

Intermittent Fasting and mTOR

There are virtually no studies done on people who follow a fasting focused lifestyle and actively time restrict their food intake. The effects of both mTOR and AMPK would be probably much different in individuals who fast and do resistance training simultaneously. That's why the arguments that mTOR will accelerate aging as well as that it's not that big of a deal both may hold water depending on the context. It's probably not a good idea to be anabolic all day and to optimize your longevity you'd want to express it in only certain situations.

mTOR isn't like a progressively increasing energy catalyzer, meaning that you're not going to keep growing more and more cells based upon how much protein or carbohydrates you ate in a meal. mTOR is more like a switch that opens up the valve for many other upstream anabolic hormones supportive of tissue growth and cell proliferation. If you have the valve open all the time then it's simply going to release more of these growth hormones into the body where they'll keep building everything they can, including the good and the bad.

It doesn't matter how much you restrict protein or carbohydrates because even small amounts of amino acids and energy will activate mTOR to a certain degree. That's going to have less of an effect on how much growth happens but it's still activated and thus inhibiting autophagy and catabolic processes. To effectively reap the benefits of low mTOR while eating you'd have to be eating a diet virtually near to starvation. However, that's just not optimal for muscle preservation nor performance. There are better and smarter ways of doing so.

If mTOR is suppressed most of the time like during fasting, then that's going to allow autophagy to actually kick in and give you the other lifespan-boosting benefits. It's much more effective and easier to fast and then stimulate mTOR only within a very small time frame.

I'd say that it's much more optimal for health and longevity to have a single large spike of mTOR and protein synthesis VS small and frequent blips of mTOR throughout the day because:

(1) Those small blips of mTOR won't make you build a lot of muscle.
(2) You'll still have to restrict your protein and calories to hopefully expect increased longevity if you eat very frequently.
(3) Restricting your calories and protein too much will lead to muscle loss and thus promoting aging and atrophy.
(4) A single high mTOR spike will make you build muscle enough and then helps you to go back into a state of autophagy faster for the next day.

I myself aim to fast about 20 hours every day and I've been able to build muscle with it. No skin issues, no health problems, no gut disorders, no signs of cognitive decline, supercharged mitochondrial function, energy all day, no hunger, and full vitality.

To be on the safe side, you may want to cycle being more mTOR dominant with periods of higher autophagy throughout the year. For instance, when training harder you naturally would want to be more anabolic. During periods of less activity, you'd want to take advantage of more autophagic conditioning.

Whenever you're trying to build some muscle you'd need higher mTOR and anabolism. Otherwise, you may find yourself swimming upstream and simply wasting your time. That's why sticking to a consistent intermittent fasting schedule is such a good idea. Such seasonality can be added to your yearly diet plans as well.

There are probably several ways of cycling mTOR and such but the main idea is that you have to know how to balance these anabolic and catabolic pathways. You can fast in different time frames or throughout different weeklong cycles but it's still best to focus harder on one end or the other without staying somewhere in the blimpy-zone of still having small amounts of mTOR but not being in autophagy.

Is IGF-1 Good or Bad

Next up, I want to look at another anabolic hormone that's considered to be bad for longevity and aging. Insulin-Like Growth Factor-1 (IGF-1) is connected to insulin signaling. So, is IGF-1 actually good or bad?

Most of IGF-1 gets mediated by growth hormone that's produced in the hypothalamus. When the anterior pituitary gland in the hypothalamus releases growth hormone into the blood, the liver responds by stimulating the production of IGF-1. IGF-1 activates the Akt pathway, which is a downstream activator of mTOR.

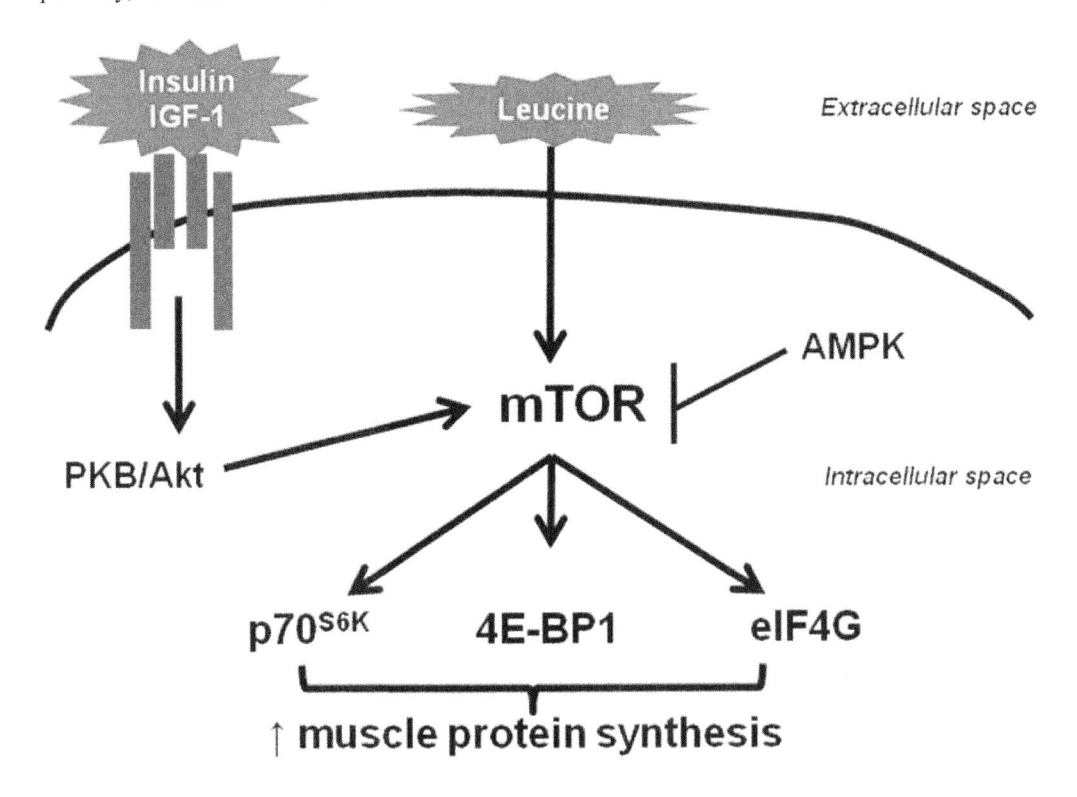

Figure 49 Insulin/IGF-1 signaling activates Akt, which turns on mTOR and growth

The role of IGF-1 is to promote the growth, survival, and proliferation of all cells, especially muscle, cartilage, bone, nerves, skin, and neurons, depending on what's needed. It's a critical component to childhood growth and it has many anabolic effects on the body.

There are many functional benefits to IGF-1:

- **IGF-1 supports muscle growth and protects against muscle wasting**. It also promotes bone growth and strength. Protein and amino acids increase IGF-1 levels as well independent of caloric intake. This supports anabolism and cellular growth.

- **IGF-1 regulates glutathione peroxidase, which is one of the most potent antioxidant pathways in the human body**[356]. These anti-oxidant benefits can protect against heart disease by clearing out plaques in the arteries[357].

- **IGF-1 fights autoimmune disorders** by increasing T-cells and boosts the immune system[358].

- **IGF-1 improves blood sugar regulation.** Lower IGF-1 is associated with metabolic syndrome and insulin resistance[359].

However, IGF-1 is also said to have its dark side...

The IGF-1 pathway has been found to contribute to some types of cancer[360]. People with Laron Syndrome, which lowers their IGF-1 expression, have a much lower risk of getting cancer. However, clinical drug trials have been unsuccessful and IGF-1s association with cancer is not completely understood

IGF-1 can protect cells against oxidative stress, which will prevent cell death and can protect against disease. However, if the person's already sick and has cancer, then IGF-1 will also prevent cancer cells from dying in chemotherapy.

IGF-1 and Aging

IGF-1 receptor signaling is thought to contribute a lot to biological aging and expected lifespan. Studies in roundworms show how genetically mutating the encoding of IGF-1R nearly doubles their lifespan[361]. IGF-1 inhibition also extends lifespan in mice. However, evidence in mammals is not clear and the results are inconsistent in humans.

Lower protein intake is associated with less IGF-1, a reduction in cancer and mortality rates in 65-year-olds and younger but not older people[362].

Both high, as well as low levels of IGF-1, are associated with cancer mortality in older men[363]. The same applies to the risk of dying from all causes. In fact, a meta-analysis of 12 studies with over 14 000 participants found that people with low IGF-1 were at a 1.27x risk of dying and those with higher levels were at a 1.18x risk. Lower levels of IGF-1 may actually be more detrimental as you age because you'll be more predisposed to muscle loss and bone fractures. Studies find an association with low IGF-1 and sarcopenia in older people[364].

The association of IGF-1 and mortality follows a U-shaped curve with both too high and too low levels of IGF-1 are linked to increased mortality risk[365] (See Figure 5050). IGF-1 improves the quality of life by strengthening your bones and promoting muscle growth.

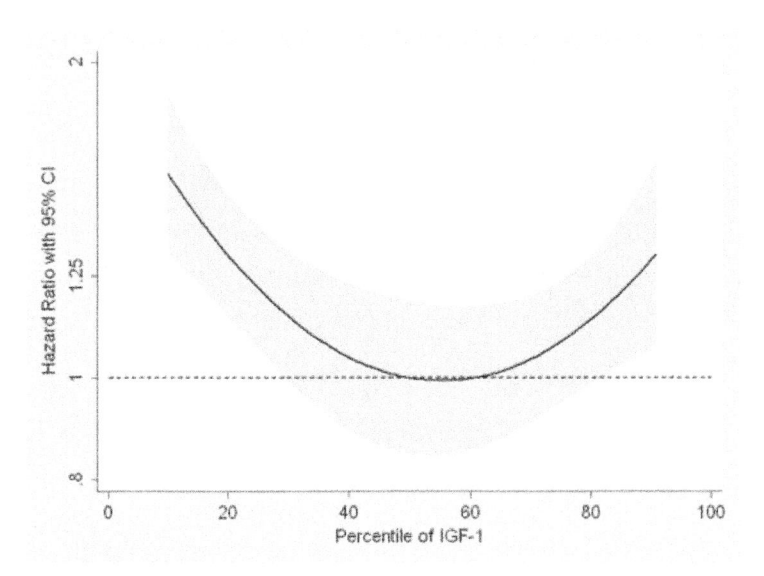

Figure 50 Both low as well as high levels of IGF-1 increase mortality rates

IGF-1 is correlated with longer telomere length which is an important predictor of longevity[366]. That's why there's this dichotomy between having enough IGF-1 for muscle and cellular maintenance VS not dying to the proliferation of cancerous tumors.

Furthermore, IGF-1 has many benefits on cognition and brain functioning. Administrating neurotrophic proteins of IGF-1 may potentially reverse the degeneration of spinal cord motor neuron axons[367]. IGF-1 helps to prevent cognitive decline by promoting new brain cell growth in rats[368]. It also improves learning and memory. In older men, it increases mental processing[369]. IGF-1 has anti-depressant and anti-anxiety effects.

IGF-1 prevents the accumulation of amyloid plaques associated with Alzheimer's disease in rats[370]. People with lower IGF-1 are more likely to have dementia and symptoms of cognitive decline.

IGF-1 deficiencies are linked to different types of growth failures such as dwarfism, and growth retardation. On the flip side, excess growth hormone production causes acromegaly or gigantism that leads to dysfunctional and excessive growth.

IGF-1 is also critical for healing and recovery. People with chronic inflammation tend to have lower IGF-1[371]. When IGF-1 is low inflammation tends to increase due to inadequate antioxidant and repair processes. IGF-1 stimulates collagen synthesis and prevents aging of the skin. However, too much IGF-1 and mTOR may cause acne and rashes.

In healthy individuals, IGF-1 expression would be balanced by the IGF-1 binding protein (IGFBP), which blocks IGF-1s effects. That's why IGF-1 is bad only if you have too much free serum IGF-1 in the blood. This may happen because the person is being more anabolic than their body needs to be whether due to sedentary living, eating too many calories, not doing proper resistance training, or some other metabolic mismatch.

That's why it's not black and white with IGF-1. You want to make sure you express IGF-1 at the right time in the right places, such as your brain cells, muscles, and bone matrix instead of fat cells, amyloid plaques, and malignant cells.

If you're not predisposed to some sort of a disease and you don't have cancer, then you don't have much to worry about dying from having too much IGF-1. Elevating IGF-1 above a safe limit is also very difficult unless you're taking growth hormone supplements, anabolic steroids, you have insulin resistance or you're eating copious amounts of excess protein.

IGF-1 is associated with greater risk of mortality and accelerated aging but at the same time IGF-1 helps to prevent those things and supports quality healthspan.

How to Inhibit IGF-1

You would still want to know how to inhibit IGF-1 just for the sake of being able to manipulate your body's anabolic and catabolic cycles.

- **Caloric restriction and intermittent fasting** can help to reduce cancer development, protect against cognitive decline, reverse diabetes and slow down aging. By now you shouldn't be surprised by the benefits of fasting.

- **Intense walking** affects serum IGF-1 and IGFBP3[372]. This is probably because physical movement helps to lower insulin and blood glucose, thus lowering IGF-1 as well.

- **Curcumin** lowers IGF-1[373] by activating AMPK and autophagy.

- **Luteolin inhibits proliferation of breast cancer cells** induced by IGF-1[374]. It's a polyphenol and flavonoid that stimulates AMPK.

- **EGCG from green tea** inhibits IGF-1 stimulated lung cancer[375]. Another polyphenol in action.

Limiting your protein and animal foods can also lower your IGF-1 levels but it may happen at the expense of other fat-soluble vitamins and minerals found from animal foods. Furthermore, chronic protein and mineral deficiencies will accelerate aging again by making you lose muscle and bone strength. That's why the most cost-effective way to lower IGF-1 again is to do intermittent fasting. You'll get a bigger effect with fewer long-term side-effects.

How to Increase IGF-1

In case you want to promote anabolism and cellular growth, here's how to increase IGF-1 levels. Let's start with some foods and supplements:

- **Deer antler spray extract** is said to contain IGF-1 and promote the production of testosterone.

- **Red meat, dairy products, and dietary calcium** are associated with higher IGF-1.

- **Dietary fat and carbohydrates** raise IGF-1 and lower IGF-1 binding proteins.

- **DHEA** is an endogenous steroid hormone that supports muscle strength and IGF-1.

- **Leucine and its by-product HMB supplementation** increase growth hormone, muscle growth, and IGF-1.

- **Low zinc** causes low IGF-1 as it's thought zinc potentiates IGF-1 actions.

- **Selenium and magnesium** are associated with total IGF-1.

However, IGF-1 is poorly absorbed by the intestines because of being broken down very rapidly in the gut. That's why you should focus on increasing IGF-1 through other means as well.

- **Resistance training increases the bioavailability of IGF-1** and supports bone density, especially in older people.

- **Sauna sessions can also boost growth hormone** and thus increase IGF-1. In fact, growth hormone may rise by 140% immediately after a sauna session[376].

When we return to the possibility of getting cancer from too much IGF-1, then it's safer to still primarily prime your body's growth hormone and strengthening pathways with exercise and fasting. These things are all linked to lower risk of mortality so it's your safest bet.

Whatever the case may be, for both optimal health, longevity, muscle growth, and performance, you want to balance IGF-1 and know how to cycle it.

Based on current research and understanding human physiology, you can say that the best range for IGF-1 is somewhere in the middle wherein you're not constantly under the effects of IGF-1 but you're not suffering from its absence either.

The best takeaway is that you want to keep IGF-1 relatively low and suppressed most of the time and activate it in only specific situations wherein the growth effects will contribute to your longevity by supporting muscle hypertrophy and cellular repair. That's why I'm practicing both intermittent fasting and resistance training. It's going to boost natural growth hormone production while giving me the other life extension benefits.

Chapter VII
Starting With Strength

"No citizen has a right to be an amateur in the matter of physical training. What a disgrace it is for a man to grow old without ever seeing the beauty and strength of which his body is capable."

Socrates

Although Metabolic Autophagy is mostly regulated by nutrition and diet, physical exercise and resistance training are equally as important. Building quality muscle and maintaining its function is essential for performance and longevity. In fact, you would probably live a very healthy and long life if you ignored all the nutritional advice in this book but kept lifting weights.

It's important to be actively engaged with some form of resistance training and other means of exercise because it's one of the most effective ways of squaring the curve. Furthermore, it's required to keep the signal of beneficial anabolism alive that would make the nutrients we eat partitioned more appropriately.

In this chapter, I'm going delve into the world of muscle building with strength training and resistance exercise.

Starting With Strength

Strength is the creation of muscular activation initiated by the nervous system. It's the ability of a given muscle to generate muscular force under specific conditions against a load or resistance.

The more muscle we have the higher our Total Daily Energy Expenditure (TDEE) will be because of the expensive cost of maintaining that tissue. It will also boost our metabolism and promote longevity through other means.

The body will by default always try to avoid putting on more muscle because of the high cost of maintenance. However, in our current contemporary environment, we're actually better off with more muscle than in the past when food was scarce. At the moment, calories are more than abundant and we have the possibility to become more muscular and healthier.

Building new tissue is an anabolic process that needs to be facilitated in some way. The first and most important thing is an adequate stimulus that creates the necessity for adaptation. Once we're facing the struggling situation of having to move a heavy object in any given direction, we're causing our muscle fibers to contract at an appropriate intensity according to the demand. The body then recognizes the need to build more muscle fibers so that we would be able to recruit more of them and do it more efficiently in the future.

Muscle hypertrophy is a phenomenon that increases the size of skeletal muscle by enhancing and growing its cells. The two contributing factors are sarcoplasmic hypertrophy and myofibrillar hypertrophy.

- **Sarcoplasmic hypertrophy** focuses on increasing muscle glycogen storage. It increases the volume of sarcoplasmic fluid in the muscle cells and isn't accompanied by significant strength gain. The amount of potential blood being stored in the muscles increases. This is the bodybuilder approach. As much size as possible. It's caused by several sets of 8-12 reps against a submaximal load.

- **Myofibrillar hypertrophy** increases the number of proteins necessary for adding muscular strength and will also cause small enhancements in size. This is the Olympic weightlifter and gymnast approach – as much strength with the least amount of weight. It's caused by muscle contractions against 80-90% of the one-repetition maximum for 2-6 reps.

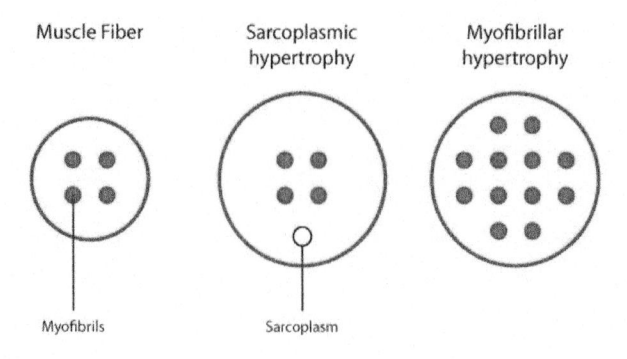

Figure 516 Different types of hypertrophy

Size doesn't necessarily mean more strength and *vice versa*. They can be built separately, as powerlifters are definitely a lot stronger than bodybuilders despite not being as bulky or with as big muscles.

Muscle growth gets facilitated by mTOR and it's down-stream pathways like Akt/mTORC1/p70S6K etc. When activated, mTOR signals the body to increase protein synthesis and gain size. When it's inhibited, muscle protein synthesis shuts down, eventually leading to muscle loss. The ratio between these two anabolic and catabolic states dictates whether we build or lose muscle.

The anabolic cycle has 2 stages. Firstly, we're stimulating growth that puts us in a catabolic state whether through training or intermittent fasting. Secondly, the effects get finalized, which leads to adaptation. Muscle and strength gain requires both of these steps and happens under the following conditions.

- **There is enough stimulus that forces the body to react**. This way the nervous system recognizes the necessity for proper adaptation to the stress encountered in the given environment. Too much, however, will lead to burnout and is difficult to recover from.

- **Hormones and pathways that stimulate growth are met**. Testosterone, HGH, IGF-1, mTOR are one of the more important ones for cellular repair and augmentation.

- **Protein synthesis is necessary for new tissue to be created**. With no building blocks to be found, we won't be able to construct anything.

- **Energy is the last condition, as all of this requires calories.** Under some rare conditions, the energy can be derived from the body's own fat store. But for optimal results, we want to help the process as much as possible with quality nutrition.

Even though sarcoplasmic hypertrophy can be created with no significant strength gain, we should still focus on building strength. Myofibrillar hypertrophy trains our nervous system as much as it does our muscles.

Strength determines the quality and quantity of our life. If you're stronger, you'll definitely be happier than if you were weaker. Having stronger muscles will make everything easier.

Constant growth and development resemble the essence of life itself. The Greek tale about Milo is the first story about the principles of strength training. Milo lifted a calf every day. As the calf grew larger, Milo also got stronger.

Figure 52 How Milo got strong

That's how we're supposed to train as well. We start from the level we're at, use proper form with our exercises and add a little bit of more resistance, as we get stronger. That's the secret to building natural muscle – strength.

Training Principles

The key to building muscle and burning fat isn't in doing hours of light or moderate aerobic exercise. Ever seen the difference between a sprinter and a marathon runner? The sprinter requires a lot more type IIb muscle fibers, which are primarily linked to muscle hypertrophy[377] whereas the endurance athlete trains for only type I fibers.

Long distance runners have a specific look to them because they're training certain muscle fibers and energy systems that elicit that kind of a hypertrophic response. Marathoners are generally much skinnier and less muscular than sprinters because (1) they don't train a lot of fast twitch muscles and (2) their sport prefers endurance based adaptations that include having less body weight.

Figure 53 Difference between sprinters and runners

Unfortunately, the man that chases two rabbits catches none...Doing cardio alongside resistance training decreases the positive effects of both. Concurrent strength and endurance training decrease gains in cardiorespiratory fitness, explosiveness, strength and muscle mass[378]. That's why it's not necessarily a good idea to combine a bunch of endurance exercises with hypertrophic ones. You can do your endurance work and aerobics as much as you like. However, whenever you're doing resistance training and trying to stimulate your muscles then you have to focus on primarily power and strength.

To create physical motion and contraction the body uses motor units, which are composed of a motor neuron and all of the muscle fibers that it supplies. One single motor unit can connect with many different fibers within a muscle, but only innervates one of the three types.

These motor units are on a similar spectrum as muscle fiber types. At one end we have low threshold motor units (LTMUs), which correspond with type I slow twitch fibers, and at the other high threshold motor units (HTMUs), that correspond with type IIb fast twitch fibers. Type IIa falls somewhere in the middle. Both of them get activated according to the force that's required to move an object. LTMUs for small power, such as lifting a cup, and HTMUs when the resistance is high, such as a near maximum deadlift.

Whether your activating LTMUs or HTMUs dictates the hypertrophy response and training adaptation you're creating. They can all be laid on a spectrum, which can be used as a guideline for the results you're after.

Variable	Strength	Power	Hypertrophy	Endurance
Load (% of 1RM)	80-100	70-100	60-80	40-60
Repetitions per set	1-5	1-5	8-15	25-60
Sets per exercise	4-7	3-5	4-8	2-4
Rest between sets (mins)	2-6	2-6	1-3	1-2
Duration (secs per set)	5-10	4-8	20-60	80-150
Speed per rep (% of max)	60-100	90-100	60-90	6-80

At near max effort, LTMUs are active along with HTMUs as well. This means, that to gain strength and cause hypertrophic adaptations, we want to be using heavy resistance and high intensity. You'd want to be recruiting as much muscle fibers and neurons as possible.

The 3 main variables of effective training you should know about are:

- **Volume** –Volume is the total quantity of movement performed during each exercise, training session and training cycle.

- **Intensity** – the amount of load, weight lifted, speed attained. Intensity is the qualitative component of training. It's about doing more work per unit of time.

- **Frequency** – the amount of density, how often you train, how many times per week you execute certain movements. This is another quantitative aspect but it is more concerned with recovery.

Together these 3 create a triad of training variables. The maximum capacity of each aspect is at the end-point. For instance, for volume, it would be 2 hours of training, for intensity 80-90% of near-maximum effort, for frequency it would be training every day 2 times.

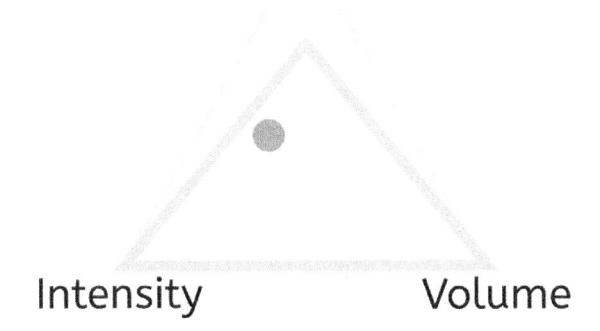

Figure 54 The Triad of Training Variables

You should strive for creating a balance between them. Otherwise, you would reach burnout. If you were to go for a high amount of volume every single day, then you can't be doing it intensely. Training only 2 times per week (low frequency) allows you to do a lot of work at greater loads as well

To reach your natural biological potential, you would want to focus on getting stronger and use sarcoplasmic hypertrophy as an addition. Some muscle fibers respond better to higher repetitions and we can adjust our training exactly according to our demands.

How Should You Train

Your main focus when doing any type of training is to promote health and longevity by building quality muscle because in the grand scheme of things they're much more important than just having shredded abs or running a mile under 4 minutes.

Basically, you want to target 3 main training regions: push-pull-legs. They're also divided into horizontal and vertical planes. To build muscle and strength, you need to increase the amount of muscle fibers in those muscle groups.

Whether or not you should follow an upper-lower body split/ do full body workouts or target a specific muscle group only once a week depends on your workout routine and how you prefer to train.

Current research shows that training a muscle twice a week leads to superior hypertrophy than once a week[379]. Therefore, you'd want to be targeting the main muscle groups like legs, chest, back, shoulders, and arms at least two times per week. However, there are reasons to believe that a higher training frequency may be more optimal in most people.

In 2015, Brad Schoenfeld *et al* published a study that showed how training a muscle group 3 times a week with full body workouts was better for muscle growth than training it once a week on a split routine[380].

More frequent muscle stimulation keeps protein synthesis more active and elevated. The window for growth lasts somewhere between 24-48 hours after training in advanced trainees.

- If you train your chest only on Monday, then the anabolic stimulus will be gone by the middle of the week and thus you lose out on a few days of potential growth.
- On the flip side, training a particular muscle 2-4 times a week will enable you to take advantage of this constantly elevated signal for building lean tissue. It's probably better for longevity as well because you'll be more sensitive to insulin and mTOR.

Higher training frequencies are also seen to be more effective in highly resistance trained men[381], which makes sense from a physiological perspective as well. The more you train and the stronger you get the more stimulation you need to facilitate further growth. Advanced trainees are more resistant to muscle damage and neuromuscular fatigue. They also show a blunted hormonal-anabolic response to training volume[382]. That's a good example of a hormetic adaptation outside of muscle hypertrophy.

Exercising frequently may actually inhibit mTOR a little bit by upregulating myostatin. That's why consistently working out frequently may be more beneficial for longevity whereas for pure hypertrophy purposes you'd benefit from taking breaks from training more often. Not exercising desensitizes the body to the anabolic stimulus again and potentiates further growth.

Increasing training frequency also lowers rates of perceived exertion (RPE) [383], reduces delayed onset muscle soreness (DOMS), and increases the testosterone to cortisol ratio[384]. Training more frequently requires you to scale down the volume or intensity of each workout, which enables you to train more often.

The main driver of muscle growth is total volume – how frequent intense growth signals are you able to send to a particular muscle throughout the week. Of course, there's a limit that the body can handle but generally working out more often facilitates increased hypertrophy because of increased volume.

It's not necessary to workout 4-6 times a week if it doesn't fit your schedule. As long as you put enough volume and intensity into your program, you can probably see as good results with a lower frequency and training only 2-3 times a week. In that case, you'd have to simply make your single workout sessions matter a lot more by increasing the amount of volume done with each exercise.

You would also want to focus on compound movements that move more than a single joint and tax the entire body, such as squats, deadlifts, bench press, pullups etc. Nevertheless, resistance bands are excellent in some situations as well. If you could also get some kettlebell conditioning then that would be great.

What routine you choose to follow depends on your preference. You can either do a full-body routine, upper and lower body splits or the push/pull/legs split. Train about 3-4 days a week and leave at least 1 day for complete recovery. On days when you're not doing resistance training, you can do some light aerobic cardio to keep yourself moving.

The most popular and simplest strength training program out there is *Stronglifts 5x5*. It consists of 2 full-body workouts. You train 3 times a week, alternating between A and B, with at least one rest day between workouts. See Figure .

- o   Workout A: Squat, Bench press, Barbell row

- o   Workout B: Squat, Overhead press, Deadlift

StrongLifts 5x5 Week 1

Monday	Wednesday	Friday
Squat 5x5	Squat 5x5	Squat 5x5
Bench Press 5x5	Overhead Press 5x5	Bench Press 5x5
Barbell Row 5x5	Deadlift 1x5	Barbell Row 5x5

StrongLifts 5x5 Week 2

Monday	Wednesday	Friday
Squat 5x5	Squat 5x5	Squat 5x5
Overhead Press 5x5	Bench Press 5x5	Overhead Press 5x5
Deadlift 1x5	Barbell Row 5x5	Deadlift 1x5

Figure 55 The StrongLifts 5x5 Program Overview

StrongLifts doesn't include any accessory work and it's not necessary for getting stronger or building muscle. However, I would add in some isolation exercises to promote sarcoplasmic hypertrophy in regions that require less intensity and will benefit from more volume.

- For your elbow tendons, wrist and forearms you can do extremely slow full range of motion pullups, pushups or biceps curls. Wrist curls and farmer's carry-s will also improve your grip strength and give you iron hands.

- For the legs and hips, you should do some walking lunges, Bulgarian split squats, hip extensions, and kettlebell swings. This will increase the range of your neuromuscular finesse and teaches you how to execute hip drive.

- Don't forget core work either, as it's important for structural integrity and a solid posture. Abdominal work will also help to define your six pack once you've reached a low body fat percentage. Do hanging leg raises, ab-wheel rollouts, dragonflags as your main exercises. Add hollow body holds, oblique twists and plank holds as accessories if you feel the need to.

This book isn't focused on maximizing muscle growth or becoming a professional powerlifter because it may have some downturn effects on longevity if taken too far. The general idea is that you should still incorporate resistance training into your exercise routine almost as a mandatory thing.

Next to diet and sleep, working out is the single most effective thing for improving your health. The workout section of this book may be somewhat less detailed as the nutrition part. Nevertheless, in the remainder of this chapter, I'm going to outline the most cost-effective and time-efficient resistance training exercises that will make you build muscle and strength.

Lower Body

The king of all lifts is the barbell back squat. It's our single most useful and powerful exercise in the gym and our most valuable tool for building strength and muscle.

The squat is literally the only exercise that directly trains *hip drive* – the active recruitment of the muscles in the posterior chain. This term refers to the muscles that produce hip extension – the straightening out of the hip joint from its flexed or bent position in the bottom of the squat. The best way to get strong *hip extensors* – hamstrings, glutes, and the adductors – is to squat heavy.

Here's how to execute the barbell back squat safely and correctly. Don't make the mistake of getting under the bar and just going for it. Merely winging it and doing something will lead to random results. If you do it wrong, you'll also hurt yourself in the process.

- Position the barbell on the lower part of your trapezius muscles. Keep the weight over your mid-foot, which is your center of gravity.
- Take a deep breath in and squeeze your glutes forward to unrack the weight. Take a few small steps backwards and let the bar settle in.
- Adjust your feet and position them about shoulder width apart, like you're about to jump. DON'T point your toes too far out. 45 degrees is too much. Keep them

at about 30 degrees, slightly apart.
- Before you squat, take a deep breath, brace your abdominal wall to activate your core muscles and maintain a neutral spine. Push out with your abdominal wall, as if you were to brace for a punch in the gut.
- Engage your glutes, start driving your hips back and sit between your legs. Drive your knees out over where your toes are pointed. DON'T allow them to collapse or shoot forward.
- Hit the right depth, just below the knees.
- As you come back up, think about squeezing your glutes and bringing your hips forward.

Good = Hips and shoulders rise at same pace

Bad = Hips rise faster than the shoulders

Squat depth is important, as it determines the safety of the movement. Full range of motion in the low bar squat involves the hips dropping slightly lower than the knees. Doing the weightlifting high-bar squat means that you'll be going almost all the way down. To maintain balance, you also have to set the bar higher on your upper traps and take a slightly narrower grip.

To squat properly, you have to use hip drive. Think of shoving-up the area above your butt and pushing it forward, as you come up. This way you're training the posterior chain.

Accessory work for the lower body include walking lunges, Bulgarian split squats and calf raises. They aren't necessary for muscle growth, as squats hit everything. You can still use them to bring some variation and really start carving your quads and hamstrings.

To develop speed and power in your legs, you should do low repetition plyometrics, such as broad jumps, vertical leaps and sprinting. Don't hit failure, as these adaptations are more taxing on the central nervous system. You have to be performing at your maximum to improve in these metrics.

The Shoulder Press

Pushing involves moving the body's center of mass or the weight away from the hands. In the upper body, it includes mainly the recruitment of your chest, shoulder and triceps muscles. At the same time, there's a significant amount of pushing involved in the squat and deadlift as well, as you're literally trying to screw your feet into the ground when executing these movements.

While doing any exercise, really, you have to use your entire body. Pushing involves some legwork as well, because, in order for you to maintain structural integrity, you have to tighten almost every muscle in the body at once. That's why compound movements are superior to isolation exercises – they stimulate mTOR and MPS in an all-encompassing manner.

The overhead press is the oldest barbell exercise. People have been picking things up and lifting them above their head ever since they developed the muscles to do so. Here's how to do it correctly and safely.

- Stand with the bar on your front shoulders. Keep your grip narrow and shoulder width apart, wrist straight, vertical forearms.
- Lock your knees and hips, keep your core and posterior chain tight.
- Raise your chest towards the ceiling by slightly arching your upper back. Think of touching your chin with your upper chest. DON'T bend your head downward

or upward but look straight ahead. DON'T arch your lower back either or fall too much backwards with your shoulders.

- Take a big breath, hold it tight. Core stability, push your abs out. Press the bar up in a vertical line. DON'T press it in front or behind your head. Press it *over your head.*
- While you're pressing the weight up, stay close to the bar. Shift your torso forward, once the bar has passed your forehead. Move your head slightly forward from your arms. Think of peeking out of a window with your hands above your head.
- Hold the bar over your shoulders and mid-foot to maintain balance. Lock out your elbows and shrug your shoulders up to the ceiling.
- Reverse the process, bring the weight back down to your front shoulders and repeat.

The dip works almost the same muscles as the overhead press and handstand pushup do, *sans* as much stress on the shoulders. Its main focus is put on the triceps, but if you adjust your body's position you can focus your chest and shoulders as well.

- Jump on parallettes or parallel bars and lock your elbows out.
- Keep your core tight and the legs straight, slightly bent forward. Maintain the hollow body position.
- Start lowering down vertically, while keeping your feet straight and flexed. DON'T flare your elbows out, but keep them in a vertical position, looking backward.

- Bend your shoulders slightly forward, until they get past your elbows. Look forward with your head.
- If you don't have enough strength to push yourself back up, then keep doing the negative as low as possible.
- If you're strong enough, push yourself back up again by slightly driving your chest forward. Use as little momentum as possible with your legs and back.
- Lock out your elbows and repeat the process.

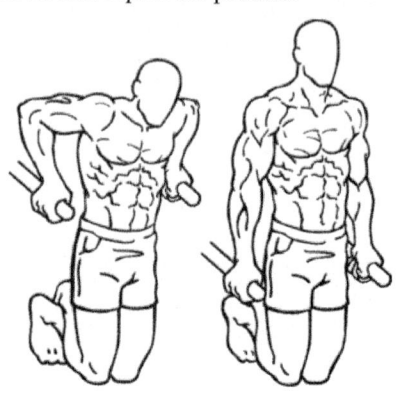

By bending more forward with your shoulders, you're putting more stress on the deltoids. If you keep it as vertical as possible, you're working the chest. Either way, the triceps will get a hell of a workout.

As you get stronger, you can go even further, by going lower and even doing the *"Russian dip."* You descent as low as possible and land your elbows onto the parallel bars. Then you drive forward with your shoulders again into the lowest dipping position and push yourself up again. Doing regular dips with weights between your legs or attached to a lifting belt increases the resistance.

The Push

The bench press is a full body compound exercise, that works your chest, shoulders, and triceps. It's also more effective for building upper body strength because you'll be lifting more weight than with the overhead press. Every gym that can call itself as such has a benching press. Here's how to do it safely.

- Lie on the flat bench with your eyes under the bar. Lift your chest up and squeeze your shoulder blades together. Keep your feet flat on the floor.
- Put your pinky on the ring marks of the bar. Hold the bar in the base of your palms and keep your wrist straight.
- Take a deep breath and make your core tight. You can arch your upper back by slightly lifting your lower back off the bench. Keep your feet planted to the ground.
- Unrack the bar by straightening your arms and move it over your shoulders. Keep your elbows locked out.
- Lower the bar down to your middle chest, while tucking your elbows about 75 degrees. DON'T flare your elbows out and keep them vertical. During that time hold your breath and keep the core tight!
- From the bottom, press the weight up again above your shoulders. DON'T lift your butt off the bench. Use your legs as driving force. At the top, lock your elbows out and breathe again.

The bodyweight equivalent of the bench press is the push-up. It's a great way to train your chest as a beginner, but can even be used to build strength and muscle as an advanced trainee.

- Get down on the floor into a plank position with your hands in front of you and your feet together.
- Lock out your elbows, keep your back straight and core tight! You know the drill.
- Slowly lower yourself down as low as possible. DON'T flare your elbows out and keep them vertical. DON'T arch your back either like a rubber band. Keep it straight and tight! Maintain proper form throughout the movement.
- Hit rock bottom with your chest and come back up again until you've completely locked out your elbows.

- Do them for as many reps as possible.
- As you get stronger, you can start leaning your shoulders more forward by standing on your toes. Eventually, you should be able to do pushups only on one of your big toes. This puts more stress on the wrists and forearms, as well as the delts.

hands slightly wider than shoulder-width apart

A.

flat back (in a straight line) no sagging, curving, or butt in the air

B.

C.

maintain a perfectly flat position when going down

Doing push-ups is a lot more difficult than it might seem if you do them right. You have to always have full range of motion and proper form. It's very easy to jerk around and start arching your back. In the military, we called it *screwing the ground.* Don't do that, nobody likes to see that.

The Deadlift

The deadlift is the best exercise for building back strength. If it were to include more legwork, then it would reign supreme over the squat and become the king of all lifts.

Deadlifting with proper form means you have to keep your lower back neutral. Rounding it during heavy lifting is very dangerous and will definitely lead to injuries. Actually, if you do it with proper form you'll increase your effectiveness. Moving the bar in a vertical line shortens the distance the weight has to travel, which increases the load you can pull.

- Walk to the bar and stand with your mid-foot under the bar. Take a hip-width stance, with your toes pointing out about 15 degrees.
- Put your hands straight in front of you, take a very deep breath. Put the air inside your abdominal wall and keep your core tight.
- Go down into the position with your arms straight, while still holding your core tight.
- Grab the bar, about shoulder-width apart. Your arms are vertical when viewed from the front and just hanging outside your legs.
- Bend your knees, until your shins touch the bar. DON'T move the bar closer, but keep it over your mid-foot.
- Lift your chest up, straighten the back. DON'T move the bar, DON'T drop your hips, DON'T squeeze your shoulder blades.
- Squeeze the bar as much as possible and start lifting it off the floor. Drive your feet into the ground, as if you're pushing the earth away from you. DON'T weaken your core. DON'T bend your elbows, keep them locked out and straight. Maintain a neutral spine, by keeping your chest up and looking up.
- As the bar passes the knees, engage hip drive and stand up straight. DON'T round your shoulders at the top. "Open" your upper body once you reach the top. DON'T lean back at the top, arch your lower back or shrug the bar.

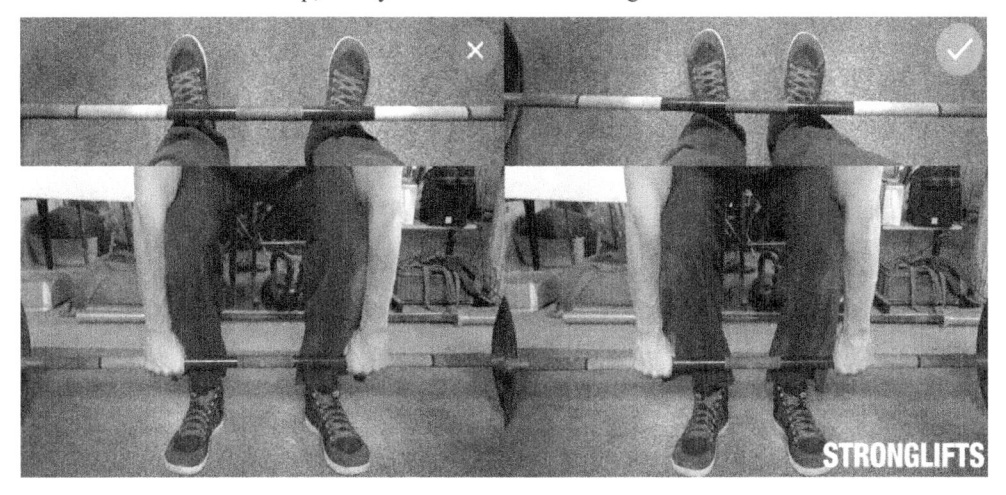

The deadlift is a full body exercise and it involves a lot more than pulling. If you do it correctly, then the actual pull part starts after the bar passes your knees. Your arms are there to simply hold the bar in place and don't get engaged in any other way. The initial part of the lift is all about pushing and generating torque with leg drive. It's as if you're screwing your feet into the ground.

Your form has to be impeccable for you to be able to lift heavy weights without damaging your spine and discs. False movement patterns will stick because of the neuromuscular aspect of training. If you deadlift with a rounded back, you'll do so with lighter weights as well. One day you'll reach down to pick up some books or a bag and you'll *snap your sh*t up!*

The Row

Rowing is the best exercise for developing a thick and wide back. It's an important skill to pull heavy objects towards you while maintaining proper form.

Barbell rows are also a full body compound exercise that works your entire back, hips, and arms. They're also great for building biceps, much better than curls. Like with the deadlift, it's important to keep a neutral spine throughout the motion. The bar should start from the floor and returns to the ground on every rep.

- Walk to the bar and stand with your mid-foot under it. Take a medium, shoulder-width stance with your toes pointing out.
- Grab the bar with a medium-width grip. It should be slightly narrower than on the

bench press but wider than on the deadlift. Squeeze the bar.
- Unlock your knees and keep them higher than on the deadlift. Bend the knees, but keep them back, so you won't hit them with the bar.
- Lift your chest up and straighten your back. DON'T move the bar towards you. DON'T drop your hips. DON'T squeeze your shoulder blades together.
- Take a big breath, hold it, keep your core tight and pull the bar against your lower chest. Lead with your elbows and pull them to the ceiling. DON'T raise your torso, or it will become a deadlift. DON'T use momentum to jerk the weight up and down.
- Drop the weight on the ground and repeat the process.

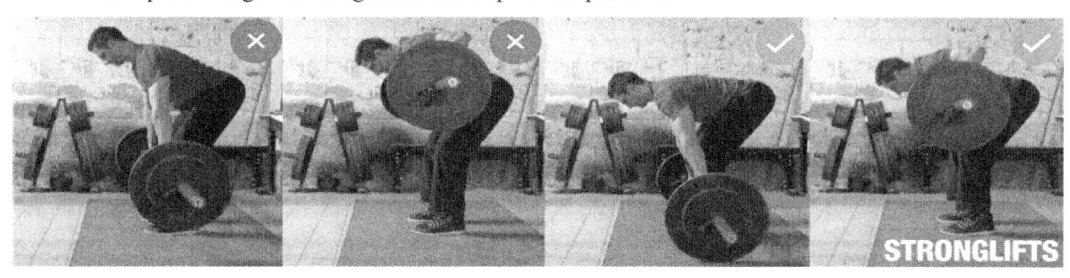

The king of all bodyweight exercises is the pull-up, because it works your entire upper body. It's also a great indicator of your relative level of fitness. There's nothing else but you and your own muscles. It's you versus gravity. Being big and muscular isn't noteworthy if you can't do at least a dozen dead-hang pull-ups.

- Jump up to a bar and grip it about shoulder-width apart. Leave yourself into a dead-hang with your hands completely extended.
- Take the hollow body hold and keep your core tight. Your elbows have to be locked out and your feet can be slightly in front of you.
- Pull yourself up by pulling your elbows down on the floor. Keep them close. DON'T swing yourself up or use legs as assistance. Maintain the hollow body position and a tight core.
- Pull yourself up until your chin passes the bar. DON'T do half reps. Lower yourself all the way down into a dead-hang again with your elbows locked out. Take a deep breath and pull-up again.

Once you get stronger, you can make them even more difficult, by doing L-sit pull-ups. Elevate your straight legs in front of you, parallel to the ground, by engaging your quads and hip extensors. You should look like a big "L". Do the same motion as you would with the regular pull up. Eventually, you can also start adding extra weights to a lifting belt to make it even more difficult.

Core and Abs

To get visible six pack abs you have to have a low body fat percentage, which is achieved by being in a caloric deficit. You have to burn more calories than you consume to lose fat.

The purpose of abdominal training isn't to get you a six-pack. Instead, it's about strengthening your core muscles that support your entire body. All compound movements engage the core and to maintain proper form you have to have integrity in your posture.

The Best Core Exercises:

- **Hollow body hold** – the most fundamental static hold in all bodyweight exercises we talked about earlier.
- **Ab wheel rollouts** – use a special ab wheel of a regular basketball.
- **Hanging leg or knee raises** – hang from a bad with your elbows locked out. Raise your legs or knees as high as possible.
- **Dragon flags** – lie down on a bench and grab hold of it with your hands. Raise your entire lower body up into the air and move it down as slow as possible. That's the exercise Silvester Stallone did in *The Rocky*.

Your core strength will determine how well you'll maintain good form during movements and will also give you a rock solid abdominal wall. It's literally the foundation to getting stronger in everything you do. DON'T neglect it.

Conditioning and Cardio

We already talked about the benefits of high-intensity training and cardio in Chapter Five. The main idea is to not spend hours and hours on low impact activities that yield insignificant responses. We want to focus on the 80/20 of most benefits, which is why HIIT is superior to LISS in most cases. Not only is HIIT better for anaerobic fitness but it's also more effective for fat loss. Dr. Doug McGuff writes in his book Body by Science:

> *There is an assumption that low-intensity exercise is necessary for fat burning and also that it burns more fat than high-intensity exercise. The reality is that no exercise, per se, burns a lot of body fat.*
>
> *The average person weighing 150 pounds burns roughly 100 calories per mile— whether the person walks or runs that mile. Since there are 3,500 calories in a pound of body fat, it would be necessary to run or jog for thirty-five miles to burn 1 pound of body fat.*
>
> *While both low- and high-intensity physical activity burn calories, high-intensity exercise does something that is highly important in the fat-burning process that its lower intensity counterpart does not: it activates hormone-sensitive lipase.*
>
> *When we're mobilizing glycogen out of a cell during high-intensity exercise, we're also able to activate hormone-sensitive lipase, which permits the mobilization of body fat. If insulin levels are high, even in the face of a calorie deficit, hormone-sensitive lipase will be inhibited, and mobilizing fat out of the adipocytes will become essentially impossible. This may explain why people who diet and take up either walking or jogging often find it difficult to lose much in the way of body fat.*

That's exactly my point – most people who are doing a bunch of cardio don't look very fit and they tend to have some health issues. It means that simply exercising won't work.

Doing 1 hour of cardio may indeed burn more calories than doing 10 minutes of HIIT. It can help you create a negative energy balance, but weight loss doesn't necessarily equal fat loss because you can also lose lean muscle mass. In this case, HIIT is more effective because you won't be tapping into gluconeogenesis during exercise and will torch fat burning afterwards. Plus there's the afterburn effect of HIIT that makes you consume more oxygen after exercise.

How to do HIIT:

- Pick an exercise that you can easily max out on, such as sprinting, burpees, kettlebell swings, push-ups, jumping squats etc.

- Warm up for 1-2 minutes with easy aerobic

- Then, go all out for 20 seconds

- Rest for 10 seconds

- Maximum effort for 20 seconds

- Recover for 10

- Repeat for 8-10 rounds

Because of the intense nature of this exercise protocol, you have to remain mindful of how fit you are. It's very effective but can be easily taken too far. Know your limits and medical condition before you try it out.

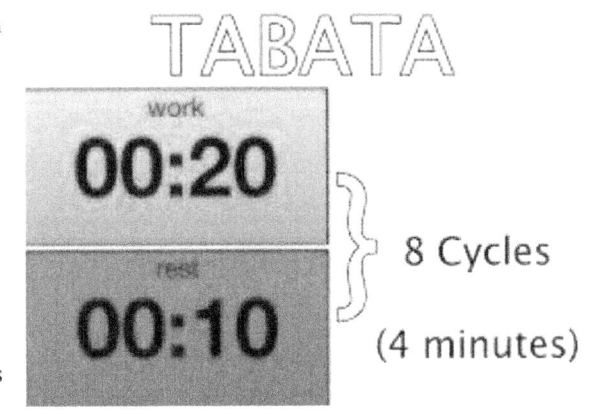

Doing 10 minutes of steady state cardio burns about 100 calories. Now, one single banana has about as many calories. Would it be easier to not eat it in the first place? Even worse, one slice of pizza and a can of soda has about 600 calories, which you can consume in less than 5 minutes.

Exercise is 1 step forward in the right direction. But a poor nutritional plan is 2 steps back. Simply put, you can't out-exercise a bad diet.

It's the treadmill effect – you have to keep trying harder and harder to keep yourself in one spot. You just have to keep on running and running. Otherwise, the ground beneath you will wipe you off. No matter how many hours you spend rolling inside the wheel like a hamster, you'll never reach your results if you make it all go down the drain.

Training Structure

When you're constructing your workouts, you should follow a few simple rules of hierarchy to maximize the amount of your training ability and efficiency. This should be a template upon which you structure your exercises.

- **Warmup.** The first thing you want to do is warm up. This will increase your core temperature, directs some blood to your muscles and puts you in the mood.
    - Do about 3-5 minutes of light cardio or something aerobic.
    - Spend at least 5 minutes doing mobility work. Do arm circles, deep lunges, squats, some pushups, hang from a bar, and get rid of all of the cracks you might have. Focus more on the body parts you're about to train.
    - DON'T do static stretching. This is the complete opposite to what we want to achieve with our training. Your muscles have to be tight and strong if you're lifting heavy weights. Stretching can lead to injuries because you'll soften your fascia. Do dynamic stretching with mobility instead.

- **Skill work.** This is the part in which you're practicing a technique of some sort. It's second because you'll still be fresh and ready to go. Do handstand holds, snatches, focus on perfect form and proper ranges of motion. Skill work is almost like an extended warm-up, as you'll be still priming your muscles for the actual work. Do this for about 5-10 minutes.

- **Strength work.** This is the core of your workout – the most difficult and taxing part. In here you'll be doing your key lifts, such as the squat, deadlift, pressing, rows or benching. All of your efforts should be directed into improving the weight you can move. Power and explosive work can also be included here, as you want to be as fresh as possible so that you could get stronger. Don't think about getting a cardio workout in this phase and focus solely on your lifts. This is the bulk of your training and should last for about 30-45 minutes, depending on how long your workout lasts in total.

- **Accessory/hypertrophy work.** After you've finished your compound lifts, you can also do some accessory work. Isolation exercises can add the extra benefit of sculpting your physique exactly the way you envision it to be. It's also a great way to build those smaller muscles, such as the forearms, calves and elbow tendons, that benefit more from higher reps. This is the hypertrophy part of your workout, that's actually necessary for increasing your key lifts as well. Do about 3 sets of 8-15 reps each and focus on the pump. Accessory exercises should complement the major lifts you did that workout. For instance, if you did squats, then you should do walking lunges, Bulgarian split squats or leg extensions, instead of biceps curls or dips. If you deadlifted, then do rows and pull-ups.

- **Metabolic conditioning.** To improve your cardiovascular fitness and burn more fat, you should also include some metcons. You can either do aerobics or HIIT, we know which one yields better results. Do 5 minutes of Tabata or about 10-20 minutes of LISS cardio. These exercises are done to take advantage of the state in which your muscles are at after resistance training. They're not as taxing on the nervous system as the main lifts. You can still have a good conditioning session after strength training, whereas it wouldn't work the other way around.

- **Prehabilitation/mobility and cool down.** Lastly, flexibility and mobility work are done at the end. These help your body to relax and help to prevent injury. Try to increase your mobility by doing deep squats, back bridges, splits and foam rolling. Work on your rotator cuffs, hips and elbows so that they would get stronger.

We can manipulate intensity by (1) increasing the difficulty of our exercise, by adding more weight to the bar or progressing in bodyweight movements. Volume can be modulated by (2) the amount of reps done per set, (3) the amount of sets per exercise, and (4) the total amount of exercises we do. To continually get stronger, we can't be repeating the same workout over and over again, because our body has already adapted to the stress. Therefore, improving at least in some of those 4 metrics indicates progress.

Immediately after a workout we actually get weaker. However, as our body heals itself and the nervous system recovers, we'll go through supercompensation and come back stronger. The training stimulus must exceed a certain threshold that causes good adaptations (to not undertrain), and it mustn't be too much that causes excessive damage from which we can't recover from. More is not always better and we should always be mindful of under which conditions our body is at.

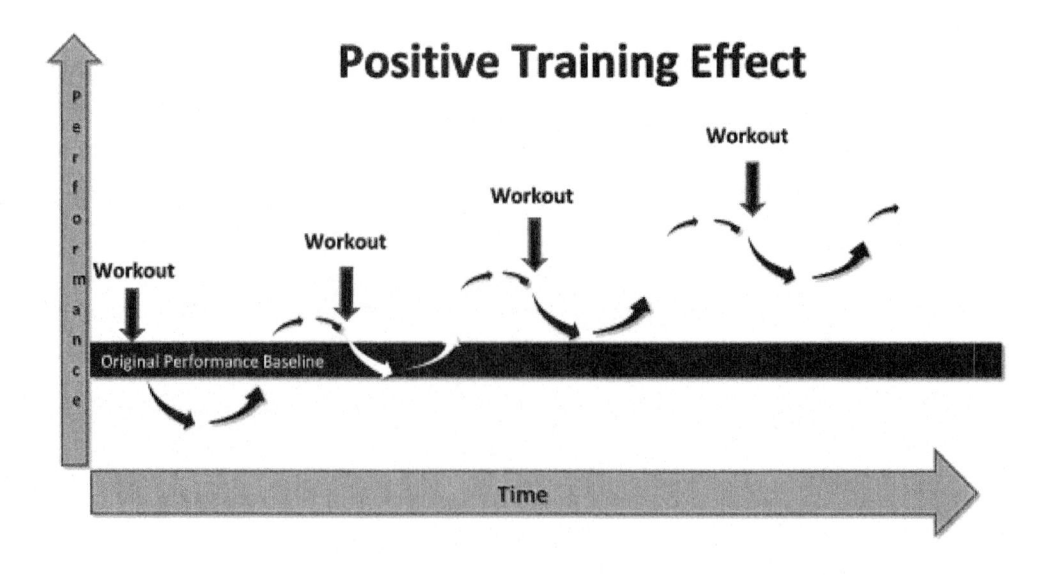

Figure 56 How you should progressively get stronger

Generally speaking, optimal recovery from workouts takes about 48-72 hours. If you didn't push yourself through the dirt like a maniac, then your muscles should be repaired by that time. However, the central nervous system may need up to 6-7 days of rest. That's why it's important to not overdo the intensity and volume. You'll feel it when you've fried your CNS. Your motor control and balance will decrease and you'll be more tired.

Periodization is about strategically designing our workouts by systematically manipulating variations in training specificity, intensity and volume.

The goal is to maximize your gains while reducing the risk of injury, staleness and overtraining. It will also address peak performance for competitions or events. An intelligently structured design will include several different chunks or periods of time across the entire year that each has its own priorities.

There are a lot of ways you can structure this, too many if you ask me. Anyone can benefit from clever periodization, but only the serious competitive athlete would need to dial down very deeply into the subject. If you simply want to get stronger then you don't need to go crazy with this because your career isn't dependent on this. If you're getting weaker then you won't miss out on anything if you take a few extra days for recovery. However, there should still be a few guidelines we should follow. I'm going to give you an example of very simple and basic periodization.

There are 4 weeks in a month. Each week represents certain aspects of your training you're trying to improve. Every workout ought to progress you towards what you're trying to accomplish. To let your body rest while still maintaining high output in your performance, you can follow a cycle of overreaching and recovery. The same cycle applies to weeks as well as workouts. You have a hard workout, followed by an easier one. After your hard week, you'll have an easier one. The first training session focuses on strength, whereas the next one on hypertrophy etc. This way you'll be able to hit all of your lifts hard and allow the supercompensation to kick in.

	Monday	**Tuesday**	**Wednesday**	**Thursday**	**Friday**	**Saturday**	**Sunday**
Week 1 Hard Week	Hard Lower Body	Easy Upper Body	Light Cardio	Hard Upper Body	Easy Lower Body	HIIT	Rest
Week 2 Easy Week	Hard Lower Body	Light Cardio	Easy Upper Body	Light Cardio	Hard Upper Body	Easy Lower Body	Rest
Week 3 Hard Week	Hard Upper Body	Easy Lower Body	Light Cardio	Hard Lower Body	Easy Upper Body	HIIT	Rest
Week 4 De-Load	Hard Lower Body	Rest	Light Cardio	Hard Upper Body	Rest	Light Cardio	Easy Full Body Workout

This is a very sustainable way to progress and can be adjusted according to your own preference. You can also use the push/pull/legs split with this. Just make sure you cycle between harder and easier days.

What I mean by hard is 80-95% of your 1RM maximum. In strength training, it follows the rep range of 2-6 reps with 4-8 sets. On easier days, the load should be slightly smaller and should fall somewhere between 60-80% with 8-12 reps and 3-4 sets each exercise.

You should also take into account the principle of *auto-regulative training*. Basically, it's about structuring your workouts based on how you're feeling on that day.

It's much wiser to back off when you feel like you're still too exhausted from your previous session. Adding more stress on top won't give you the desired results. Our pursuit towards excellence can be overshadowed by our type-A personalities that think that we're simply being lazy and need to grind through the pain. The ego is the enemy here and we should learn to listen to the signals our body is sending us.

Based on a scale of 1 to 10, start measuring how you feeling each morning and then act according to that.

- 10 would mean that you can literally run through a wall. In that case, go for a heavy workout with no regrets.

- Number 1 would mean that you can't even make it out of the bed and need to be hospitalized, which is a sign of serious overtraining.

- Number 5 and anything below that feels like you have some joint pain, too much muscle soreness, troubles finding balance, forgetting things, mental fatigue and shivering limbs. Back off and have a rest day.

- If you're between 6 or 7, have an easier day – you'll be feeling quite fine but don't have that explosive spring in your step if you know what I mean.

- 8 and 9 means that you'll feel great and are motivated to train. You're eager to push yourself hard and aren't afraid of squats, deadlifts or even HIIT. Have a heavy workout.

Don't let yourself go below 6 or 7. This compounding effect will make weaker and destroys your nervous system. Most importantly, don't neglect recovery. For the beneficial adaptations to actually sink in we need to give our body some time to repair itself. If we were to bust in another sledgehammer without having healed the previous impact, then the supercompensation and growth will never take effect. Frequency, volume, and intensity go hand in hand and we should always choose 2 of them.

Principles to Remember

I have to say that I'm not trying to force you into any given way of exercising because it all still comes down to what you're trying to accomplish. But there are still some key points you should keep in mind.

- **When training, have a clear idea of what you're trying to achieve.** Exercise, go for a walk, play sports and do yoga just for fun, but train with a specific goal in mind.

- **Structure your workout routines around only 2 of the 3 variables of the triad.** They are frequency, intensity, and volume. Don't try to scale them up all at once.

- **When you're aerobic and can breathe easily through your nose, then you're burning primarily fat.** If you start breathing heavily through the mouth, then you've reached the anaerobic zone and will be utilizing glycogen for fuel.

- **If you do cardio, then avoid the Black Zone.** Running and cycling for more than 30 minutes are supposed to be aerobic. Start slow and you'll be able to go faster as your heart rate improves.

- **Doing Tabata and HIIT is a much more time-efficient way of improving cardiovascular health for the majority of people.**

- **Resistance training is a lot more important for health, longevity, muscle growth and bone density.** The least you should be doing is 12 minutes per week but for more optimal growth slightly more is needed.

- **Nutrition is more important than exercise.** Whoever you may be, lowering your carbohydrate intake will be beneficial for your health and body composition. But low carb won't work if you don't have an idea of what you're doing. It would only keep you in the Black Hole of eating.

Instead of trying to avoid the momentary discomfort of working out, you'd have to muster enough guts and courage to occasionally push a heavy boulder up the hill. Once you do that, you don't have to compensate that much for your lack of insulin sensitivity with other miracle drugs or interventions – you're just fit to handle it. That's the idea of cycling anabolism and catabolism as well – sometimes you need both. The next chapter delves deeper into this topic.

Chapter VIII
Anabolic Autophagy

"The struggle itself towards the heights is enough to fill a man's heart. One must imagine Sisyphus happy."

Albert Camus

One of the ultimate pursuits of physique building is to build more muscle and lose fat. Preferably simultaneously and as fast as possible.

When it comes to burning fat and losing weight then intermittent fasting has been shown to be very effective. Not only due to the physiological effects of fasted ketosis but also because of the increased adherence and convenience.

Fasting is also one of the critical components to overcoming this dichotomy between mTOR and autophagy. It's the quintessential component that will flip the context of how these pathways manifest completely upside down.

What about muscle growth and bodybuilding? Can you build lean mass with intermittent fasting and time-restricted feeding? Or are you destined to be at the mercy of meal-prep, nutrient timing, and the anabolic window?

This chapter will delve into the realm of anabolism through resistance training, protein intake, nutrient partitioning, and intermittent fasting.

Can You Build Muscle and Lose Fat at the Same Time

Like mentioned in earlier chapters, being in a fasted state is a catabolic stressor to the body, which is the opposite to anabolism. If you're depleted of your endogenous resources, it's very difficult to build something out of nothing.

This coincides with physics and Newton's First Law of Thermodynamics, which goes like this as viewn through the anabolic window:

- To lose fat you have to be at an energy deficit i.e. burn more energy than you consume.
- To build muscle you have to be at an energy surplus i.e. consume more energy than you burn.

It might seem obvious and self-explanatory because energy can't just appear and disappear. However, muscle and fat tissue have distinctive roles and they're used differently by the body as well.

A surplus of calories is context dependent because it can cause a completely different NET gain effect on body composition, depending on the proportions of macronutrients and quality of nutrients. Furthermore, overall conditions on the body are equally as determining in what the person's going to look like. The same applies to a deficit of calories combined with training.

Calories aren't just calories because they can be partitioned differently according to the macronutrient ratios, quality of nutrients, hormone levels, training status, and overall energetic demands on the body.

Menno Henselmans, the founder of Bayesian Bodybuilding, is a scientist and physique coach who's written an amazing article about this on his website (https://bayesianbodybuilding.com/gain-muscle-and-lose-fat-at-the-same-time/). I'm going to reiterate a lot of what he's said there to convey the idea of how it's possible to build muscle and lose fat at the same time.

Calorie Partitioning and Muscle Growth

In 1982 Heymsfeld *et al* assessed the biochemical composition of muscle tissue in normal and semi-starved individuals[385]. This is what human muscle tissue is composed of (See Figure 57).

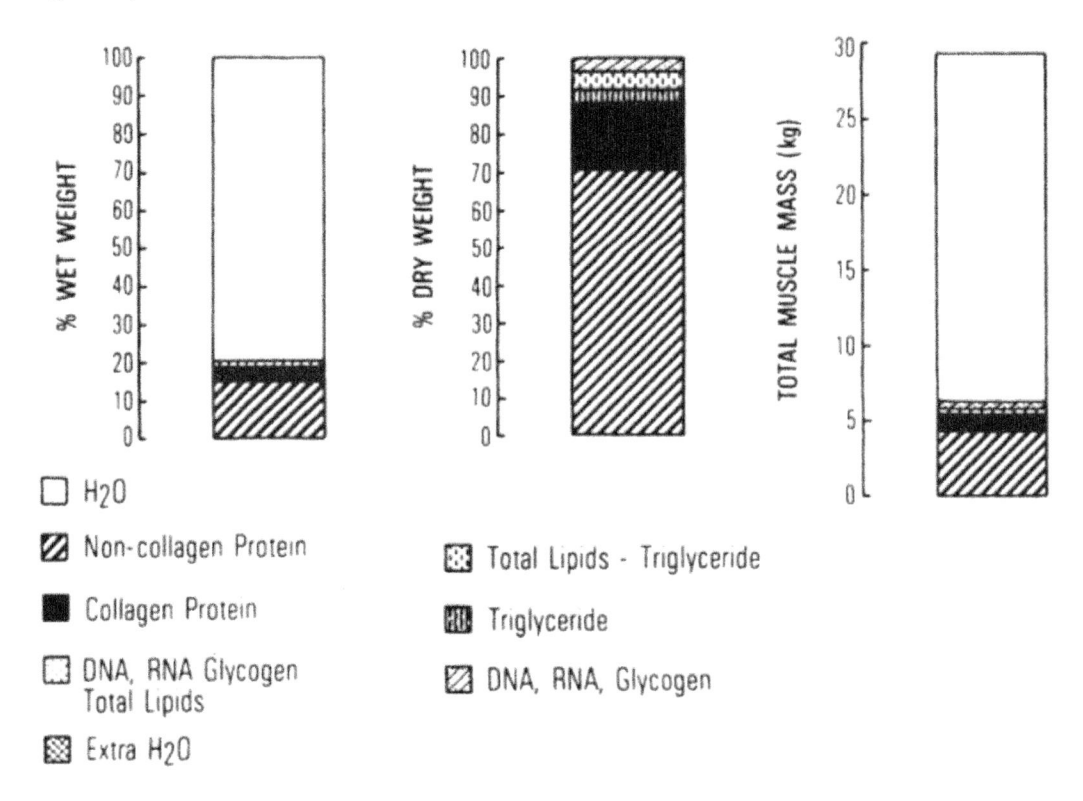

Figure 57 What muscle tissue is composed of

According to this, muscle is composed of a lot of water (H2O), different proteins, stored glycogen and triglycerides, which are energy substrates. This shows that you don't need a whole lot of extra energy to build muscle as long as there's an adequate stimulus for growth. Most of these „ingredients" can already be found inside the body.

If your workouts are stimulating the muscles enough and if you follow it up with adequate muscle protein synthesis by consuming enough protein, then the rest of what you need can be derived from stored body fat. Likewise, you can gain fat and lose muscle at the same time by doing a lot of catabolic exercise like chronic cardio for hours and over-consuming daily calories with very little protein.

It also shows that it's still possible to gain muscle and lose fat at the same time if you're clever with training load, nutrient partitioning, and meal timing. Therefore, energy conservation is irrelevant to caloric partitioning and how your body composition changes.

There have been several studies showing how it's possible to build muscle and lose fat at the same time:

- Overweight police officers with 26% body fat started weight training and lost 9.3 pounds of fat while gaining 8.8 pounds of lean mass in 12 weeks[386].
- Women who start resistance training lose fat and gain muscle[387]. Sometimes even when they're on sub-optimal diets.
- Elite level gymnasts on low carb ketogenic diets dropped their body fat percentage from 7.6% to 5% and gained 0.9 pounds of muscle[388]. They also increased relative strength and the amount of chest to bar pull-ups they could do while training up to 4 hours a day and eating less than 22 grams of carbs a day.
- Many powerlifters and strength athletes recomp their bodies consistently by losing fat mass and increasing their lifts[389].
- An extremely overweight person who hasn't trained before may start resistance training at a huge caloric deficit and they're more than likely to end up with losing body fat and building muscle

These are not anecdotal results or broscience as they've been replicated by both elite athletes as well as overweight people. It is possible to build muscle and lose body fat at the same time, resulting in improved body composition. You don't have to even eat anything to refill your muscle glycogen after working out. *Whaaat?*

It's true that carbohydrates immediately after resistance training exercise will replenish glycogen stores faster just by virtue of them being consumed. However, compared to just drinking regular water, there isn't much difference in the coming few hours.

Pascoe *et al* (1993) did an experiment on 2 groups of men who trained leg extensions in a fasted state[390]. Some of the subjects were given 1.5 g/kg of a carbohydrate solution and the others an equal amount of water. Total force production, pre-exercise muscle glycogen content, and degree of depletion weren't significantly different between the two.

During the initial 2-hour recovery, the CHO group had a significantly greater muscle glycogen re-synthesis compared to the H2O group. However, after 6 hours, muscle glycogen was restored to 91% of pre-exercise levels in the CHO group and 75% in the H2O group. Keep in mind that this was fasted – no food was consumed beforehand nor afterwards. The carbs did promote muscle glycogen resynthesis slightly more but the subjects who didn't consume any calories at all still resynthesized 75% of the glycogen they had lost during the workout.

That's just mind-boggling – **you don't have to eat anything at all to restore the glycogen you've lost.** Of course, having something to eat will promote recovery and muscle growth but the self-resourcefulness of your body is just phenomenal.

The 1.5 g/kg of carbs simply drank in a solution isn't something I'd volunteer for. For an average person weighing 70-90 kilos, that would entail over 100 grams of carbs just from the drink. That's not a really good trade-off in terms of the insulin spike and carb load. The effects weren't even that significantly better…

The carbs in Pascoe's study spiked glycogen resynthesis during the first 2 hours post-workout but by the 6-hour mark it had kind of flattened out a little bit. The higher glycogen rates in the CHO group were then primarily due to the ingested carbohydrates not because of carbs flipping some sort of a magic switch. Most of the work was done by the body itself. But how does this work?

The adipose tissue consists of stored triglycerides, which is an ester comprising of 3 fatty acid molecule chains and a glycerol backbone that holds them together. This single fat particle can cover most of the body's metabolic needs in at least the short term. Fat is fuel that most tissues and muscle can use.

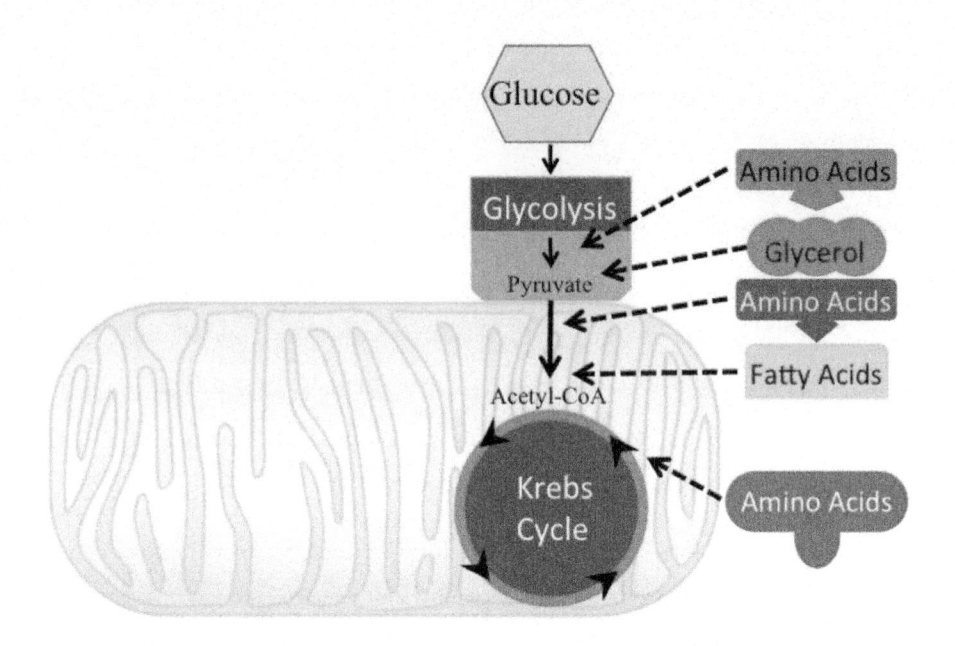

Figure 58 Different fuel alternatives the body can use to produce energy

However, the brain and some other organs still need a small amount of glucose because fatty acids themselves can't cross the blood-brain-barrier. During glucose deprivation, the liver will convert that glycerol backbone into glucose through the process of gluconeogenesis. The other 3 fatty acid chains will be metabolized into ketones and all of these substrates will be used to cover the brain's energy demands.

Even if you're not eating anything you can resynthesize the glycogen you've lost during exercise. Part of the restored glycogen will come from glycerol and fatty acid gluconeogenesis but a significant proportion will also come from lactate[391].

Lactate is the byproduct of glucose metabolism and it's been shown to contribute up to 18% of skeletal muscle glycogen synthesis after high-intensity exercise[392]. Basically, during high-intensity workouts, you're producing a lot of lactic acid by burning off your muscle glycogen. To eliminate the burn effect and restore the glycogen you lost, the body uses some amounts of that lactate for muscle glycogen resynthesis. *Whaatt...that's amazing*! Your body literally recycles the energy you burned off and then restores it.

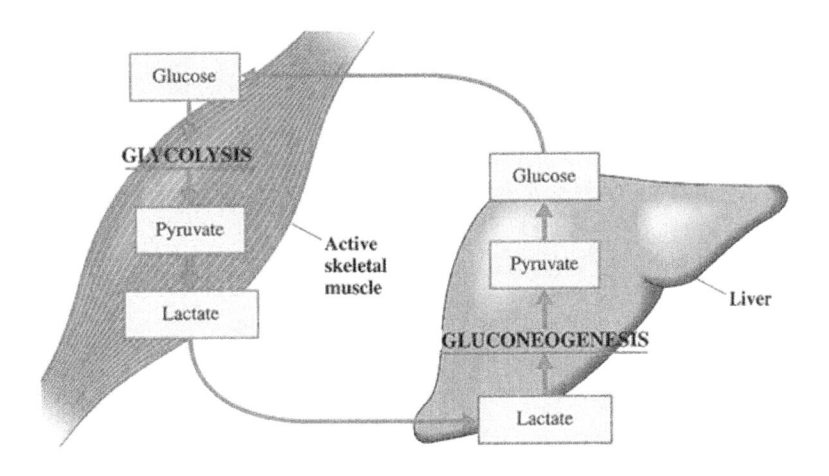

Figure 59 The conversion of lactate into glucose through gluconeogenesis and back again

That's just phenomenal – your body literally is a survival machine that can adapt to almost everything. It means that even if you don't consume any food at all, you could fuel a few good workouts with just the resynthesized glycogen from your own endogenous energy stores. While still not having eaten anything.

It means that even if you're eating a high carb diet you need far fewer carbs to replenish muscle glycogen than you think and you're wasting a lot of readily available fuel that's already produced by your body. With keto-adaptation, you're more likely to increase the rate at which lactate and glycerol contribute to muscle glycogen resynthesis without even eating any food at all because of limited glucose in the diet.

This is just an example of how proper fuel partitioning can make you gain muscle and lose fat at the same time. You can be in a caloric deficit and still build lean tissue as long as you train enough and provide the other components of muscle protein synthesis *i.e.* protein, amino acids, and leucine. The body has the energy – fatty acids, lactate, ketones, glycerol, and glucose – it just needs the building blocks from food. A clever and carefully orchestrated nutrition plan would be structured around optimizing lean muscle growth with virtually zero fat gain.

Furthermore, meal timing and protein intake will be even more important for making this kind of a recomposition possible. There are certain times the body needs more fuel and amino acids than at others. For instance:

- After working out, the muscles are more prone to shuttle the nutrients you've consumed into glycogen stores and to stimulate muscle protein synthesis. In that scenario, all the calories you eat would be primarily directed towards positive muscle growth rather than fat accumulation because the body prioritizes recovery rather than storage.
- On the flip side, eating a bunch of excess calories without having moved a flower will inevitably be more pro-fat gain just because the body isn't under such energetic conditions that would favor high energy intake. What the body doesn't need right away will be used for storage.

Of course, at the end of the day, what you do throughout the entire 24-hours is going to dictate the end result. However, it's safe to say that some moments are more important than others and they'll yield a more favorable outcome in terms of body composition.

In the case of intermittent fasting, you would inevitably see a much bigger lean muscle gain if you were to consume most of your calories after a resistance training workout. The dominos will all be set in line – the mechano overload from exercise, depleted glycogen stores, activated mTOR, and nervous system fatigue – everything is much favorable for building muscle as long as you stimulate MPS and bring in the building blocks. Eating that same food without having stimulated these anabolic mechanisms won't be nearly as effective. You may still gain lean muscle if you workout afterwards but it wouldn't be that well partitioned.

If you haven't worked out and the muscles aren't in need of recovery, it's better to not eat anything and continue to fast. At least that's what I would do. I mean, if the anabolic signal hasn't been set, then I'd prefer to continue fasting as to reap the benefits of more autophagy, get into deeper ketosis, and burn extra fat. The only time I want to eat is after having trained and stimulated the muscles as to make the calories more directed towards lean muscle growth.

Most people would see a more optimal body recomposition with more muscle and less fat if they were to backload the majority of their calories into the post-workout scenario. The thing is that there's only a certain amount of calories you need to build muscle and in order to lose fat you can't be eating an unlimited amount of food. Therefore, with limited calories, it's simply smarter to consume them only when there's an actual need for it.

Intermittent fasting and time-restricted feeding are such powerful tools for building muscle and burning fat at the same time.

- Whenever you're fasting, you'll be burning more fat, suppressing hunger, and promoting growth hormone that helps to maintain muscle.
- You don't need a bunch of calories or energy to do strength based resistance training. That's the main catalyst for muscle growth and it can be easily done with limited supplies.
- Eating food after working out will promote more muscle growth rather than fat gain. It will facilitate a more anabolic response while still eating fewer calories.
- Eating more food before working out may make you lose a bit more muscle if you're eating at a caloric deficit. The reason being nutrient partitioning and meal timing.

That's why even if you're not fasting that aggressively you'd still want to fast as long as you can every day and eat most of your protein and calories after working out. If you want to build lean muscle with virtually zero fat gain that is.

As long as there's an adequate stimulus there will be a sufficient response. The degree of how much or effective it is depends on many variables we've been talking about indefinitely by now. You just have to know it's possible, understand the principles, and use various tools to control the direction you're heading towards.

Losing Muscle While Fasting

Now that I've shown how it's possible to build muscle and lose fat at the same time, I want to turn to refuting some of the myths about intermittent fasting and muscle growth.

You might have heard from bodybuilding experts and fitness gurus that you need to eat every few hours or else... or else you're going to lose all your muscle mass. I'm going to tell you right away that yes that is possible if you don't know what you're doing. If you do things right you'll be actually able to build lean muscle while still losing fat.

One of the biggest reasons why fasting doesn't equal immediate muscle loss is because of growth hormone. Under normal conditions, your body has only one spike of growth hormone in the morning and another one at night.

Studies have found that when you're fasting, your body goes through these spikes of growth hormone several times during the day[393]. So, you experience surges of growth hormone more frequently when in a fasted state. Part of the reason has to do with the body trying to preserve more muscle despite being deprived of calories.

In fact, growth hormone increases exponentially by up to 2000-3000% at the 24-hour mark. #ARMEXPLOSION (See Figure 60)

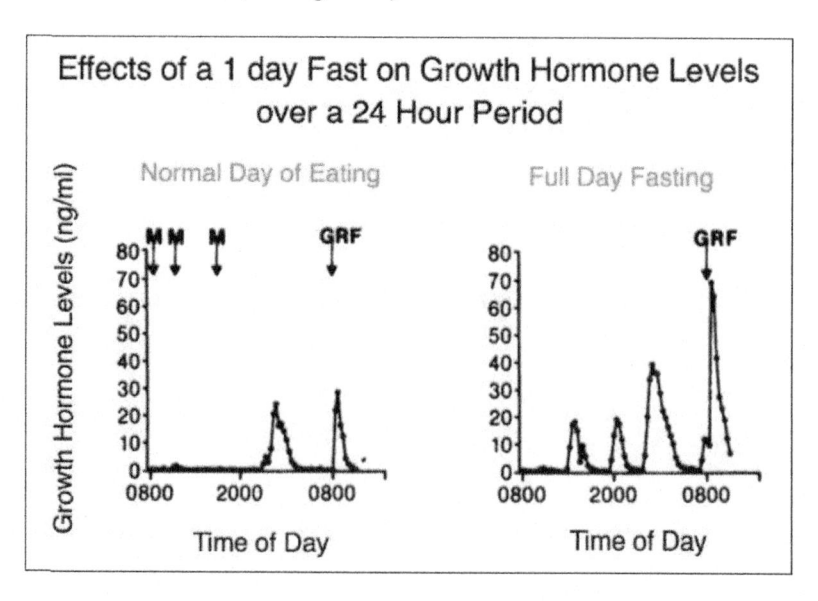

Figure 60 Fasting increases the amount of growth hormone surges during the day

When you look at the amount of GH increase then the micrograms aren't that large, but you wouldn't experience anything like that in a regular metabolic state. Not eating gives you something you wouldn't gain from regular eating. This boost is actually going to last for the upcoming several hours.

Another hormone that's going to help you build muscle is testosterone.

- Short-term fasting has been shown to increase Luteinizing Hormone (LH), which is a precursor to testosterone. In a study done on obese men, LH increased by 67% after 56 hours[394].

- Another study found that obese men saw a 26% increase in GNHR (Gonadotropin-releasing hormone), which is another testosterone stimulant[395]. The same study found that men who were working out saw a 67% increase in GNHR, which led to a 180% boost in testosterone.

So, there's plenty of hormonal benefits to fasting and I'd suggest that if you're suffering from low testosterone or you can't seem to build fat, then it's just because your endocrine system is unable to produce its own testosterone. Some careful and strategic intermittent fasting can help your body to not be so lazy so to say and jack up your muscle building pathways.

Nevertheless, there are many people saying that they're losing muscle while fasting. Why is that?

It's important to realize that not all weight loss or decrease in the equals muscle loss. Losing weight doesn't mean you're burning fat either. The goal is to lose weight at the expense of fat tissue and fill it up with muscle thus ending up with a positive NET gain.

With that being said, it doesn't mean you can't lose muscle while doing intermittent fasting. Of course, you can and you will if your body feels the need to do so. There are 2 reasons why you may begin to break down your lean tissue.

(1) Gluconeogenesis While Fasting

First is gluconeogenesis – the conversion of protein and fat into glucose.

Gluconeogenesis is driven by demand not by supply, which means that it happens when your body needs glucose for survival and the only source of glucose it can find is its own organs and muscles.

The reason you may trigger gluconeogenesis is that you don't have access to other fuel sources, like fat. Your body isn't keto-adapted to burning ketones yet and the next best thing it can think of is protein.

That's why you'd want to get into ketosis as soon as possible when you're doing fasting or even when maintaining a caloric deficit. The ketogenic diet will make you burn exclusively your own body fat while preserving muscle because ketones give more energy to the brain, which spares protein.

After you flip the metabolic switch of starting to use more ketones for fuel, you'll increase fat burning and protect your muscles because the body has access to an abundant fuel source which is your own body fat.

(2) Autophagy and Muscle Loss

The second reason why you may lose muscle while fasting is the inhibition of autophagy.

Studies have found that autophagy is needed for maintaining muscle mass[396].

- If you're doing a caloric restriction diet but blocking the effects of autophagy, then you're going to keep yourself in a semi-starvation state because your body will never switch into ketosis. This leads to gluconeogenesis and so on – the bad stuff.

- If, however, you allow autophagy to kick in, whether that be through strict water fasting or a fasting mimicking ketogenic diet, then your also stimulating the other growth hormones we've been talking about so far and it's going to preserve your muscle.

What it means is that if you want to burn fat, or if you want to prolong your lifespan, then you're actually better off by avoiding all calories whatsoever.

Even as little as 50 calories of 2-3 grams of leucine will stop autophagy and shift you into a fed state. It's going to be better for fat loss, for muscle sparing and for longevity, to avoid all calories during your fasting window. That goes back to the idea of timing your calorie intake more in the post-workout scenario wherein you've reaped the benefits of fasting ketosis and are now prone to recover.

That's why when dieting or staying at a mild caloric deficit, it's better to stay in ketosis for at least most of the time because it'll maintain more lean tissue while still enabling the burning of body fat.

One study in the 70s compared 3 diets with the same amounts of 1800 calories and 115g protein but with carbohydrate ratios of 30-g, 60-g, and 104-g[397]. After 9 weeks of dieting, the 30-g group lost 16.2 kg, the 60-g group lost 12.8 kg, and the 104-g group lost 11.9 kg. What's more, the 30-g group's weight loss came from 95% fat loss, for 60-g it was 84%, and the 104-g 75%. These findings show that lower carb ketogenic states can promote more fat loss at the preservation of muscle mass.

There are several possible mechanisms by which a ketogenic diet preserves muscle mass and prevents protein catabolism.

- **Low blood glucose stimulates the secretion of adrenaline, which regulates skeletal muscle protein mass.** Adrenaline's been shown to directly inhibit protein breakdown[398].

- **Ketone bodies provide a plentiful source of fuel to brain and muscle tissue, which suppresses protein oxidation and gluconeogenesis of muscle**. In fact, BHB has been shown to decrease leucine oxidation and actually promote protein synthesis in humans[399].

- **Dietary protein consumption also has a much greater muscle sparing effect than carbohydrates**. Eating protein increases protein synthesis by increasing the availability of amino acids in the blood. When eating at a caloric deficit, higher protein intake has also been shown to reduce muscle loss and promote fat loss[400].

One of the branched-chain amino acids leucine has also been shown to stimulate protein synthesis via the insulin signaling pathway without actually needing carbohydrates[401]. Leucine's metabolite HMB has also been found to have anti-catabolic effects[402]. High protein diets have been shown to increase muscle protein synthesis despite the reduced levels of insulin[403]. A ketogenic diet isn't necessarily high in protein but it's still can provide the individual with all the essential amino acids and complete protein sources.

Can Fasting Make You Build Muscle

But what about the opposite? Can fasting actually make you build muscle?

It depends on how hard you train and how many calories you eat during your feeding window when you're not fasting. I'm going to leave the nutrition aspect for the coming chapters because I want to cover the hormonal side of how it's possible.

Studies have found that fasting lowers the expression of mTOR and IGF-1, which are both needed for cellular growth by increasing one of their inhibiting proteins called IGFBP1[404]. Within 12-14 hours of fasting, SIRT1 gene regulation starts rising which will begin to suppress mTOR and AKT[405,406], thus down-regulating mTOR mediated protein synthesis.

TOR has quite a detrimental role in anabolism. Inhibiting mTOR blocks the anabolic effects of resistance training and prevents muscle growth[407]. mTOR is clearly anabolic but also anti-catabolic. It's going to protect the body against the harmful effects of cortisol and glucocorticoids on muscle tissue[408]. Fasting is one of the most anti-TOR things there is because of upregulating AMPK and depleted amino acid availability.

Although fasting increases growth hormone exponentially, it also decreases serum IGF-1 levels, which again decrease the body's anabolic state[409]. This may seem paradoxical at first because we know that the production of IGF-1 in the liver and muscles gets instigated by the release of growth hormone from the pituitary gland in the hypothalamus. What gives?

Truth be told, growth hormone is not as much of an anabolic hormone as it is an anti-catabolic one. GH has many fat burning properties that promote catabolism but its main role is to still preserve lean tissue not build it. The muscle building properties are triggered by IGF-1 and mTOR both of which get decreased while fasting. However, it doesn't mean you can't build muscle.

Muscle growth results from the positive balance between muscle protein breakdown (catabolism) and muscle protein synthesis (anabolism). See Figure 61. You can stay in a highly catabolic state most of the day (fasting) if you compensate for it with enough anabolic stimuli (eating). It's possible to build muscle with a time-restricted feeding schedule, even with eating just one meal a day. Whether or not it's the most optimal thing for muscle growth remains to up for debate.

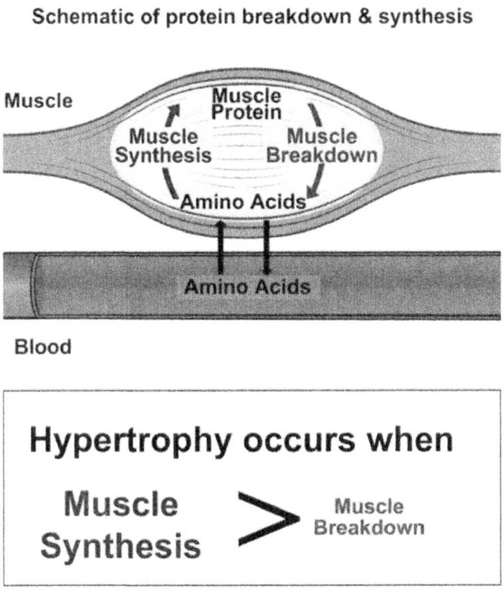

Figure 61 Muscle Hypertrophy Results from a Positive Gain of Muscle Protein Synthesis over Muscle Protein Breakdown

If you're trying to maximize your genetic potential for lean muscle mass, then it's naturally going to be easier to do so with a more frequent eating schedule. Likewise, it would be possible to go through periods of more anabolism followed by brief periods of higher catabolism that can lead to a positive muscle building condition because of practicing intermittent fasting.

The rationale of trying to build muscle and strength with intermittent fasting isn't oriented towards maximizing muscle growth or performance. It's about prioritizing longevity and not over-stimulating the anabolic effects of mTOR all the time.

That's why you can't build muscle in a fasted state directly – it's not even supposed to happen as the goal is catabolism. However, the anabolic state comes from the feeding stage wherein you nourish the body and drive quality lean muscle gain.

It may seem like pushing a boulder up a hill as Albert Camus described in his novel *The Myth of Sisyphus*. In Greek Mythology, Sisyphus was the king of Corinth who was sentenced to push a boulder up a hill because of his lavish lifestyle and deceit. Every time he started to reach the end at the top, the boulder would roll down the hill again. Sisyphus was made to repeat this process for an eternity, never getting anywhere. Camus used this story to depict the notion of the absurd where men seek futile meaning and clarity in an indifferent world devoid of an eternal truth.

Doing intermittent fasting for muscle growth may indeed seem like pushing a boulder up a hill because it doesn't send an anabolic signal for building lean tissue. However, it can still facilitate a positive increase in lean mass if done right. In fact, it's much more favorable in terms of longevity-oriented muscle growth and other health benefits. It's not as anabolic as taking steroids or eating high carb high protein 6 times a day by any means but the gains will be quality and leaner. The process may be slower but like Camus said: *"One must imagine Sisyphus happy"*. Which means you should still enjoy the journey despite the struggle and adopt the right long-term mindset.

I myself have been doing intermittent fasting since high school and have gained lean muscle mass and virtually zero fat. Throughout this period, I've never eaten breakfast, mostly skip lunch, eat the majority of my food in a single meal, have built an impressive physique, gotten stronger, and have never lost my six-pack abs. With deep keto-adaptation and proper nutrient timing, I'm able to build muscle at linear progression. The Metabolic Autophagy Protocol will teach you the same principles.

Chapter IX
Protein Absorption and Anabolism

"Part of my daily regime is my glucosamine and, of course, a multitude of multivitamins. Branched-chain amino acids, glutamine, of course, protein. I have one protein shake a day, and that is immediately after my training."

Dwayne 'The Rock' Johnson

We've been talking about protein for quite a while now and it's time to close the circle on this essential macronutrient. Although it might seem that fitness people and nutritionists over-emphasize the importance of protein, it really is that important. In fact, for body composition as well as longevity it's probably quintessential. That's why we're getting into such detail.

To maintain a certain amount of body weight and lean tissue, you have to support that with adequate protein intake. If you were to eat less protein than your body currently needs or if you were to fast over the course of a long period of time, you'll gradually lose some bodyweight because of the lack of amino acid building blocks and lower levels of protein synthesis.

We mentioned earlier that the process of gluconeogenesis is driven by demand not supply. In the case of not getting enough protein or glucose without being in ketosis, the demand is there and it's quite huge.

If the brain isn't using ketones or the liver hasn't even produced them, then the only option for the body to maintain its energy homeostasis is to sacrifice its muscle tissue and cellular debris by converting them into glucose. The reason is that muscle is very expensive to keep around during energy crises because it requires a lot more calories to sustain. When protein requirements are met and ketones are accepted, this process is circumvented and the body will use body fat stores instead.

Protein is the only macronutrient that can't be stored inside the body long-term.

- **Carbohydrates can be stored as liver and muscle glycogen (100-500 grams)**. Extra carbs will be burnt off as energy or converted into triglycerides and get stored as body fat.

- **Fat and extra carb can be stored in an infinite amount as body fat in the adipose tissue.** You can gain as much adipose tissue as you can possibly consume from too many calories.

- **Protein intake will be used for elevating muscle protein synthesis and activating mTOR which will help to maintain your current lean muscle mass.** To activate these pathways, you need only a certain amount of protein and more won't have a dose-increasing effect.

You can't really store protein inside the body beyond a certain necessary limit. To store protein it has to be converted into glucose through gluconeogenesis first. That glucose will then either be burnt off as energy or if you've already met your daily caloric needs, stored as fat.

The notion that protein can be converted into glucose is a double-edged sword. At times of over-consuming protein, it'll be simply turned into another fuel substrate that the body will then use based on its requirements. However, when you're not getting enough protein on a consistent basis and you're also at a caloric deficit, that same process of gluconeogenesis may turn to your lean muscle tissue and convert that into glucose.

Like I just said, gluconeogenesis is driven by demand, not supply, which is why you want to minimize the demand for glucose by becoming more keto-adapted and staying in ketosis while fasting.

In the short term, an influx of increased protein supply won't trigger gluconeogenesis of your own muscle tissue because there is no demand there. The body will have met its need for amino acids and thus doesn't require additional glucose. Temporary protein stores fluctuate throughout the day and they're connected to the feeding and fasting cycles[410].

How Much Protein Does Your Body Need

The amount of protein you need per day depends on many things and people need different amounts.

- **The more lean muscle mass you have, the more protein you need to sustain that amount of muscle.** Higher bodyweight requires more building blocks just because of having more mass. However, for optimal health and body composition, you'd want to focus on your lean muscle mass. The idea is to lose the fat and maintain the muscle not feed the fat with extra calories from unnecessary protein.

- **Being more active in general increases your protein demands because physical activity damages the muscle cells to a certain extent**[411].

    - If you do resistance training, you need more protein to support that training with enough protein synthesis and mTOR activation.

    - If you do primarily endurance training, you need slightly less protein because endurance training doesn't break down that much muscle tissue as resistance training does. Even if it does, the purpose of endurance training isn't to build muscles so the desired intake of protein wouldn't be higher either.

- As you age your ability to maintain skeletal muscle decrease and thus you need more protein as well.

The recommended dietary allowance (RDA) for protein is 0.36g/lbs of bodyweight which for an average individual weighing between 150-180 pounds would be 55-70 grams of protein per day[412]. However, this is not ideal for the majority of the population and most people actually need more, especially if you're exercising.

In general, the optimal amount of protein tends to be somewhere between 0.7-1.0 g/lbs or 1.5-2.0 g/kgs of lean body mass (LBM), which for the same average individual weighing between 150-180 pounds would be 110-160 grams of protein at a minimum. There are no seeming benefits to eating more than 0.8 g/lbs of LBM, even when trying to build muscle. You definitely don't need to be eating above 1.0g/lbs of LBM as you'll simply waste that protein.

How Much Protein Can Your Body Absorb in One Sitting

One of the additional reasons why it's thought you can't build muscle with intermittent fasting is that you won't absorb enough protein during the 24-hour period. To keep MPS active you have to spike it up with frequent protein ingestions because there's only a limited amount of protein your body can absorb in one meal. Is that true?

When you digest protein, it gets broken down into amino acids that will be transported into the bloodstream to be used as building blocks. There are a limited amount of transporter cells and receptors in the small intestine which restricts how many amino acids can be moved into the blood. Hence the theory that your body can only absorb a certain amount of protein in one meal.

Certain proteins are absorbed faster than others which allows the amino acids to be used more quickly as well. However, there are many other factors that determine protein absorption such as the pH levels of the gut, the permeability of the intestinal lining, protein sensitivity, and the presence of hormones related to gastric emptying[413].

The general consensus is that you can only absorb 30 grams of protein per meal and you need to spread your protein intake across 4-6 meals to maximize protein synthesis over the 24-hour period. However, it doesn't mean that eating fewer meals with higher amounts of protein would make you waste away that protein.

Amino acids and some peptides are able to self-regulate their time in the intestines. For example, the digestive hormone cholecystokinin (CCK) can slow down the contraction speed of intestines in response to protein intake[414]. CCK gets released when you eat dietary protein and it slows down your digestion as to absorb it better[415]. See Figure 62.

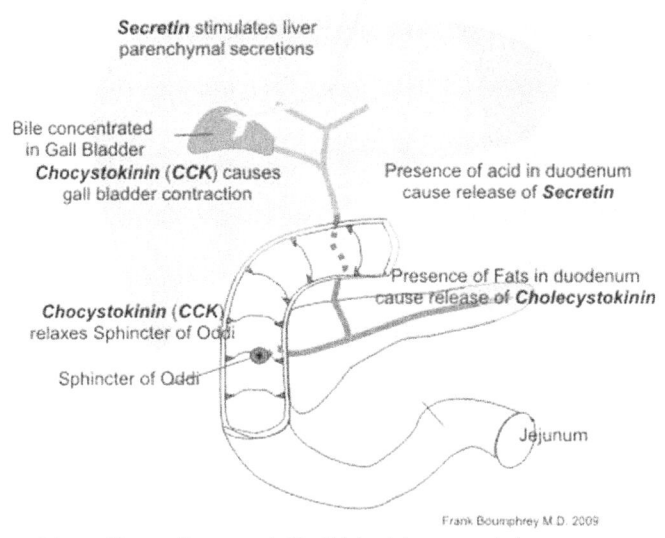

Figure 62 CCK controls gallbladder contractions

If you were to absorb protein too quickly, your liver wouldn't be able to maintain a steady stream of amino acids into the blood over the 24-hour period because you'll burn them all for energy.

Even if you've eaten a large piece of steak with over 60 grams of protein, you wouldn't be converting those amino acids into energy immediately anyway. Because of CCK and the generally slower speed of digesting steak, the protein from that steak will be digested over the course of many hours and your body will slowly assimilate those nutrients without wasting them away.

A Mayo Clinic study found that, on average, it takes about 24-35 hours for food to fully travel through the digestive tract and be completely absorbed[416].

- As soon as you consume something with calories, you're entering into a fed state and your body's going to be breaking down that food.
- After a few hours of digestion, your body goes into the post-absorptive stage, wherein the nutrients of the last meal are still circulating the bloodstream. This can last up to 8-12 hours and that's when you'll truly enter a fasted state.

Our intestines will contract according to the speed at which it can digest food. If they can't handle any more protein, then they won't waste this precious resource away but will simply slow down gastric emptying. After a few moments, when you've digested the protein you already consumed, the intestines will then move the remaining protein down the line so to say and continue absorption.

If you eat more protein than your body needs right now to trigger protein synthesis, it slows down the digestion of the excess and then gradually releases the amino acids into the blood over the course of the coming hours when your protein synthesis gets lower. Some amino acids can even be temporarily stored inside muscle cells for future use whether for maintaining amino acid homeostasis or for energy production.

The reason it's thought that you can only absorb 30 grams of protein in one sitting is that you only need about 20-30 grams of protein to trigger muscle protein synthesis and actually build muscle[417]. Any more than 40 grams doesn't stimulate MPS further. See Figure 63.

Triggering muscle protein synthesis is mostly regulated through leucine, which is the main anabolic amino acid. It requires about 2-3 grams of leucine to activate MPS and generally, you can get that amount of leucine from 20-30 grams of a complete protein. That's where this rationale originated from.

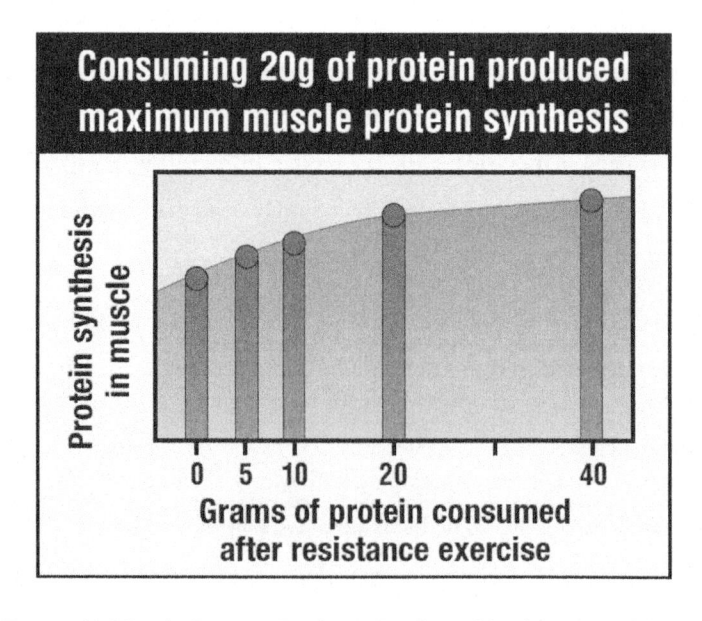

Figure 63 Muscle Protein Synthesis Peaks at 20-40 grams of protein

However, this doesn't really tell you much about how much protein you can end up absorbing in one meal. It just tells you that if you want to keep the muscle building signal activated more frequently then you'd have to spike muscle protein synthesis more frequently as well. There's no indication of how it affects muscle protein synthesis over the 24-hour period.

However, the stimulation for muscle growth after resistance training will remain elevated for a long period of time. Studies have found that the potentiation of exercise-induced increases in myofibrillar protein synthesis and Akt/mTOR signaling by protein consumption is sustained for at least 24 hours post-workout[418]. Even if you stimulate MPS twice within those 24 hours compared to 6 times you can still build muscle if you eat enough protein within that time frame.

Additionally, more frequent spikes of protein synthesis won't necessarily mean more muscle growth either because if you just eat 30 grams of protein then there aren't many amino acids in that small meal to build new tissue either. If you were to eat that large steak again with 60 grams of protein, you'll activate protein synthesis and you'll have more than enough building blocks as well. So, in theory, it could be that less frequent spikes in muscle protein synthesis but with a higher amount of protein could potentially build more muscle just because of the higher availability of amino acids that could be absorbed much more efficiently.

Eating fewer meals and consuming more than 30 grams of protein in one sitting with intermittent fasting has not been shown to have any negative consequences in terms of lean tissue maintenance. One study done on women who ate their daily protein requirements of 79g of protein in either a single meal or 4 meals saw no difference in terms of protein metabolism and absorption[419].

Several intermittent fasting studies have also shown that eating your entire days' protein in a 4-hour eating window has had no negative effects on muscle preservation[420,421,422,423]. When it comes to body composition and fat loss, then meal timing has been shown to be irrelevant as intermittent fasting doesn't slow down your metabolism or make you lose muscle.

Like I mentioned, eating 4-6 small meals a day may elevate muscle protein synthesis more frequently, but more frequent surges of muscle protein synthesis won't necessarily mean more muscle growth because what matters most is how much protein your body ends up absorbing over the course of the 24-hour period. If you're doing intermittent fasting with 2 meals a day, you can spike muscle protein synthesis twice a day and that's going to be more than enough to force your body to grow. What matters more for muscle growth is the training stimulus and adaptive signal.

In fact, being in a fasted state makes you more protein sparing and anti-catabolic by increasing growth hormone and ketones. Higher levels of growth hormone and IGF-1 can stimulate muscle protein synthesis and it definitely improves the body's sensitivity to protein intake[424]. You'll end up absorbing your food better because it's perceived as more scarce.

How to Increase Protein Absorption

It's not about how much protein you put into your mouth and eat. It's about how much protein you absorb and how many of those amino acids get used for your goals.

The reason some people need to consume ultra-high amounts of protein to build muscle is that they're not able to assimilate it effectively and it's simply a waste. You want to eat as little protein as possible to get as much effect as you can.

Here's what to do to increase protein absorption with intermittent fasting:

- **Exercising before food will also deplete your body's glycogen stores which makes the muscle cells more eager to absorb carbohydrates from food**. Every food that's used for recovering from exercise will be prioritized much higher for both muscle growth and maximal absorption.

- **Before eating you want to create an acidic intestinal environment by stimulating the production of hydrochloric acid (HCl) in the gut.** Foods high in HCl are lemons, celery, olives, and vinegar. Warm lemon water with apple cider vinegar is great for blunting the blood sugar response of eating and helps with digestion. Ginger, in particular, helps with protein absorption.

- **Before eating you also want to put yourself into a parasympathetic state and become relaxed**. If you're stressed out or are feeling anxious, then your adrenals will direct blood flow away from the gut into the extremities. This is going to shut down digestion and promotes constipation. That's why never eat when you're stressed out and always sit down with your meal. Chewing your food is also incredibly important because it's going to make it easier for your gut to absorb it. Don't eat on the run and always be mindful of what you put into your mouth.

- **Pre-biotics are plant fibers that travel through the intestines unchanged and they help with digestion by feeding the bacteria in your gut**. Pre-biotic foods you should eat as the first part of your meal are garlic, onions, asparagus, leeks, artichokes, dandelion, chicory root, and green bananas.

- **Probiotics promote the creation of live bacteria in your gut that are going to help with breaking down the nutrients from your food**. They're microorganisms that are going to influence your mood, your immune system, your cognitive functioning, and body composition. Therefore, you want to maintain a healthy gut flora by eating fermented foods like sauerkraut, pickles, kimchi, olives, miso, raw kefir, and even very dark chocolate.

- **Combining large amounts of protein with starches and carbohydrates are also going to inhibit the digestion of those foods because they require different digestive enzymes**. That's why animal foods are best eaten with just vegetables and fermented foods. If you do eat starch then it should be easily digestible cooked starch like sweet potatoes, white potatoes, white rice, or buckwheat.

- **Slow walking after your meal is also going to lower the blood sugar response from what you ate and helps with digestion**. It has to be a very slow and peaceful stroll because if you become too sympathetic you'll shut down the digestion again.

The over-arching message is that when it comes to protein or any other type of food, you want to eat food that's going to support a specific goal whether that be health, muscle growth, or performance. Just eating large amounts of protein or whatever macronutrient just for the sake of eating it is a waste. That's why you want to absorb more nutrients from eating less instead of simply burning off calories.

Should You Consume Protein Before Working Out

In the case of wanting to maximize the anabolic effects of consuming protein, of course, you'd want to have slightly more frequent episodes of elevated protein synthesis just due to the fact that if you're in an anabolic state you'll build more tissue.

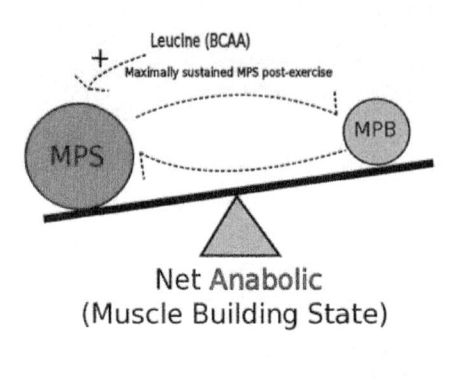

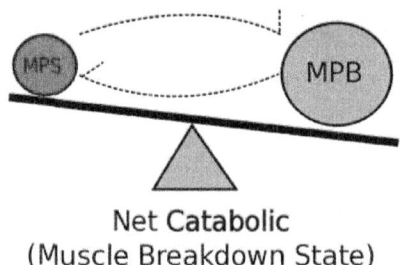

When you're doing resistance training, then you're inducing damage to the muscle cells and tissues. If you consume adequate protein after the workout, you'll be able to heal that damage and hopefully result in sarcoplasmic hypertrophy.

However, when working out fasted you have limited amino acid availability and thus are subject to increased muscle damage. That may not be ideal for someone trying to build more muscle because they may end up with a NET negative muscle homeostasis.

Fasting itself is already a pro-catabolic stressor. Adding training on top of that and then being forced to consume all of your daily protein within a small time frame is indeed like pushing a boulder up a hill.

The amounts of protein you can absorb in one sitting is arbitrary for maintaining lean tissue but it's probably not optimal for positive NET growth. That's why having some amino acids circulating the bloodstream during the workout will not only minimize the muscular damage but will also promote additional muscle protein synthesis after training.

Hoffman et al. (2010) found that 42 grams of protein consumed immediately before and after a strength training session improved recovery in strength trainees compared to a maltodextrin placebo[425]. Unfortunately, energy intake and nutrient timing weren't equated between these 2 groups so it's hard to identify what caused the faster recovery. It might have been the caloric surplus, protein timing or the supplementation itself.

Studies by Brad Schoenfeld and Alan Aragon (2017) haven't seen any significant differences in muscular adaptations when consuming protein either pre- or post-workout[426]. However, these studies aren't probably done on subjects who exercise fasted. If you've eaten something a few hours before training, you'll have plenty of amino acids circulating the blood which will then be used for energy. That's why if you've had a real meal pre-workout it's not necessary to be taking any protein shakes or the like. Intermittent fasting throws a monkey wrench into the spokes of most studies because it puts the person into a different metabolic state.

Despite the catabolic factor, fasting has many physiological benefits if not on muscle growth then at least on longevity and general health. However, there are still some advantages to working out fasted

- **Consuming carbohydrates and raising insulin suppresses lipolysis and fat burning during exercise**[427]. You'd have to burn through those carbs first before getting access to your adipose tissue. Whether or not it matters is subject to context and the goal of that particular training session.
- **Training fasted improves glucose tolerance and improves insulin sensitivity**[428], which doesn't help with muscle growth directly but it'll help you to stay leaner while building muscle.
- **Fasting increases blood flow in abdominal subcutaneous fat, which can help with losing that stubborn belly fat.** A 3-day fast increased abdominal subcutaneous blood flow by 50%[429]. However, this change is minute and probably not relevant.
- **Fasted resistance training causes a bigger anabolic response to eating post-workout than fed training by increasing the p70S6 kinase**[430]. Theoretically, you'll create a bigger super-compensatory effect for muscle hypertrophy by working out while fasting and then refeeding properly.
- **Increased growth hormone from working out fasted may help to preserve lean muscle and protect against excessive catabolism**[431]. This may be worthwhile if you're eating at a caloric deficit but still getting more than enough protein. At other times, it would depend on what macro ratios you eat.

Even Muslim bodybuilders who workout during Ramadan haven't seen any decrease in muscle or body composition[432], despite working out hard and even during dry fasts. They may train differently than normal but it comes to show how more advanced trainees can even workout fasted.

However, in most people, fasted exercise leads to a decrease in performance. It's natural to feel weaker and slower if you haven't eaten for too long. That can cause mental fatigue, lethargy, and lower your workout results. At the end of the day, if you're able to work out harder, then you'll be able to burn more fat and build more muscle, which then helps you achieve a better body composition. If fasting gets in the way of that, then you need to adjust your approach.

Ramadan style fasting has been shown to decrease anaerobic performance and power output[433]. High-Intensity Training modalities like CrossFit, P90X and Tabata intervals can stop muscle growth and repair processes by down-regulating mTOR through AMPK activation. It'll also cause further muscle fiber teardown, which adds an additional catabolic stressor to the body[434]. In that case, increased AMPK and autophagy actually protect against muscle degradation by promoting cellular turnover and lean muscle maintenance[435].

Although working out fasted limits your performance slightly it can only hinder your potential to a certain extent. In reality, the difference will only be like 5-15% which isn't detrimental nor always necessary. You don't need to be performing at 100% of your maximum every workout to see results or to get stronger. In fact, most people will get as good of a physique in spite of that minor limiting factor.

There's this dichotomy between wanting to fast longer during the day and trying to build lean muscle mass with it.

- To maximize the autophagic benefits of fasting, you'd want to fast for as long as possible every day. In most cases, that would entail about a 20-hour fasted window.
- For optimal muscle growth, you'd want to have a smaller amount of amino acids and protein in your system before working out.

- For better body recomposition and nutrient partitioning, you'd benefit more from backloading most of your calories into the post-workout scenario where your body prioritizes recovery, not storage.

This gulf of wanting to build muscle while trying to fast for as long as possible even before training is quite difficult to breach. However, it's possible with something I call Targeted Intermittent Fasting (TIF).

Enter Targeted Intermittent Fasting

You don't need to be consuming any calories all the way up until the point you're about to do resistance training. At that moment, you'd want to have some protein and amino acids to protect you against excessive catabolism.

To minimize time spent in a fed state, the fastest and most efficient solution would be to consume a protein shake with about 20-30 grams of protein immediately before or during the workout. This will promote recovery and muscle growth despite having fasted for that long beforehand.

Here's what the Targeted Intermittent Fasting Protocol looks like:

- Fast for the majority of the day as long as you can before working out.
- Consume only water and zero calorie teas or coffee all the way up until 18-20 hours of fasting.
- When starting to workout at 18-20 hours, consume a protein shake with 20-30 grams protein. It's preferable to drink it during the actual workout and use quality protein powders that don't have artificial sweeteners or other additives.
- While working out, focus on heavier compound movements and hypertrophic exercises to stimulate muscle growth. In between your sets, sip on the protein shake.
- In the post-workout scenario eat the rest of your calories within 2-3 hours or in a single meal. Make sure you still get enough protein after training.

This is what I call the Targeted Intermittent Fasting protocol inspired by the targeted ketogenic diet approach wherein you consume a small amount of carbs during high-intensity training. The idea is that you'll be consuming protein and calories only when your body needs them. At other times you're much better off by fasting because of deepening autophagy and ketosis. This sort of nutrient timing will yield positive effects on muscle growth as well as reaping the benefits of being in a fasted state. It can breach the dichotomy without missing out on potential growth.

However, not all workouts are created equal and they're going to cause different responses in the body. You don't want to be training randomly because you'll be getting random results. Not all types of exercise suit intermittent fasting either. That's why you have to know how to train to meet the nuances of TIF.

Generally, TIF works best with resistance training that focuses on strength and hypertrophy. It's meant to shield your muscles against catabolism while boosting recovery. You don't need any protein or calories if you're doing low-intensity exercise or cardio because you'll be utilizing primarily fat for fuel. Chapter XVIII will talk about the different types of intermittent fasting, including TIF.

Post-Workout Nutrition While Fasting

When working out fasted, it's important to get adequate nutrition and drive protein synthesis post-workout to not cause muscle loss.

In fasting conditions, protein NET balance tends to be negative and resistance training induces further muscle breakdown. That's why it's not a good idea to fast for too long after working out if your goal is muscle growth.

Muscle protein synthesis requires protein and amino acids. Just raising insulin or blood sugar isn't enough to stimulate MPS, which is why you'd need adequate protein intake from food[436].

However, eating immediately afterwards isn't ideal because you may be still under the influence of cortisol and digestive stress, which can promote gut issues and fat gain. Protein synthesis doesn't peak immediately after you finish your last set either as it takes a bit of time for the body to respond and shift from a sympathetic state into a parasympathetic one. The sweet spot for muscle protein synthesis and avoiding catabolism seems to be between 60-195 minutes after training but not before 60 minutes[437].

Will chugging down a protein shake with leucine right after dropping the weights make you build more muscle? Maybe it will if it's going to increase the overall daily nitrogen balance but maybe it won't if your other macros aren't on point. It's definitely not necessary as you have at least 1-2 hours post-workout wherein you won't be losing muscle thanks to increased ketones and growth hormone, but it won't be bad for you either.

Dietary protein supplementation has been shown to significantly increase muscle strength and size during prolonged resistance training in healthy adults[438]. It's another one of those things that's not necessary as long as your other variables are met but it can still be useful for driving up muscle protein synthesis after the workout. Use it only if your diet lacks whole food complete protein sources.

The timing of protein supplementation in the context of anabolic autophagy may be inconvenient or at least problematic. You definitely don't want to be consuming any protein or amino acids during fasting because it will break the fast, blunt growth hormone, and block autophagy. Protein shakes would have the most effect either immediately before or intra-workout ala targeted intermittent fasting. If you wait for 2 hours and eat, then it'd be better to get your amino acids from real food but if you don't have access to those conditions, then the shake is a great alternative.

When taking any fitness supplements, you have to be wary of their ingredients and content. You definitely want to avoid artificial sweeteners like sucralose, saccharin, fructose, and aspartame because they're linked to insulin resistance and other diseases. They're still going to spike insulin and even disrupt the microbiome by promoting the proliferation of certain gut buts that can extract more calories from the food you eat. That's something you can't read from a nutrition label. Instead, you'd want to use natural protein powders with no additional flavorings or sweeteners. The taste might not be as good but to hell with that! It's not supposed to taste like chocolate mint ginger-bread berry blast…it's fuel.

How fast you drive into the anabolic mode post-workout depends on how long you've been fasted for, what kind of a workout you had, what are your physique goals, and how efficiently your body responds to ketosis.

If you're working out on a 20-24 hour fast, then it's more important to get some protein and building blocks into your system faster than if you were to work out between the 12-16 hour mark. In the case of the latter, you have higher energy stores from the previous feeding thus your buffer zone is also larger wherein you can get away with continuing to fast for longer.

In general, workout somewhere between the 16-20 hour mark of fasting and wait at least 60-120 minutes before eating anything. The fear of missing out on this so-called "anabolic window" wherein you'll start building exponentially more muscle is futile. There might be some truth to it and it's relevant in the context of targeted intermittent fasting but the timeframe expands itself for a much longer time than you'd think. Protein synthesis after resistance training can stay elevated for up to 24-48 hours.

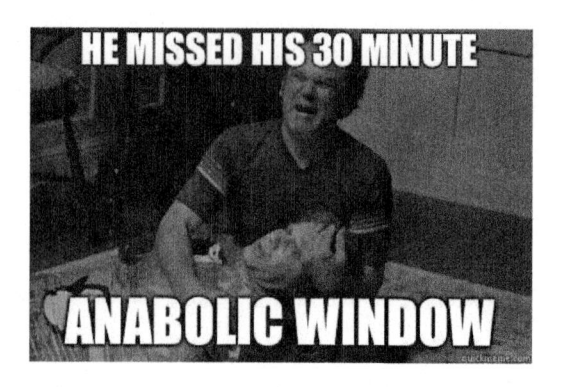

What about other macronutrients like carbs and fats? That would depend on the particular individual, what kind of fuel was used during the workout, overall goals, and general anabolic-catabolic homeostasis.

Re-feeding after fasting, especially on carbohydrates, raises the thermic effect of food, which produces excess body heat and can lead to a positive result in body composition[439]. This is ideal for protocols like carb backloading and the cyclical ketogenic diet wherein you have some days of higher carb intake to promote insulin and muscle growth. How often to do it and when will be discussed in Chapter XIII.

On a daily basis, you'd want to still limit your insulin production and control for blood sugar. That's why the keto template is amazing for maintaining lean muscle while being at a very pro-longevity state of chaperone-mediated autophagy.

Chaperon-Mediated Autophagy (CMA) differs from macroautophagy. CMA responds to increased ketone body production, which will help to salvage fuel for the organs that can't use ketones for energy[440]. Ketone bodies also protect against muscle catabolism and avoid lean tissue breakdown[441]. This adds an additional layer to the autophagic process in which you can still gain a lot of its benefits without having to fast for too long. The best thing about it is that you'll be able to send your body the signal to promote muscle growth without shutting down autophagy completely. Certain ketogenic fat sources like MCT oil and coconut oil have been shown to even promote autophagy in small amounts[442].

Cholesterol has been found to be quite beneficial for muscle growth. In a study by Riechman et al. (2007), 49 elderly people participated in a 12-week strength training program[443]. There was a dose-specific relationship between dietary cholesterol intake and lean body mass increase. The more cholesterol they ate the more muscle they gained. This was confirmed by controlled protein and fat intake as well.

In young healthy adults, a high cholesterol diet (~800 mg/d) has shown 3 times higher MPS rates 22 hours after resistance training compared to a low cholesterol diet (< 200 mg/d) [444].

High cholesterol intake from things like red meat and eggs has been recommended by many bodybuilders from the previous century as well. Vince Gironda from the 1950s became renown for his Steak and Eggs diet where he cycled through 5 days of eating zero carb followed up by a massive carbohydrate refeed on the weekend. According to him, this helped to maintain his leanness year round while still building muscle. Other bodybuilders often go on cycles of consuming a lot of whole eggs and yolks to promote muscle growth with cholesterol. Vince himself said that whole eggs are like natural anabolic steroids. Eggs are also rich in leucine, which will then stimulate MPS further.

Part of the reason why cholesterol helps with muscle growth may have to do with the anti-oxidant properties of cholesterol and the repair mechanisms it triggers. Cholesterol improves membrane stability of cells which enhances their resiliency against muscle damage during exercise. This also controls inflammation during recovery. Cholesterol supports mTOR and IGF-1 signaling by helping with the formation of signaling pathways.

The controversy around eating cholesterol and saturated fat is still controversial and not conclusive. In Chapter X, we'll go through the past and current research regarding their effects on heart disease and how you should eat them.

Whatever the case may be, no doubt that cholesterol and saturated fat, especially from meat and eggs are incredibly potent foods for muscle growth and anabolism. Saturated fat is a building block for cholesterol, which in turn promotes testosterone production as well.

Low saturated fat intake is associated with reduced testosterone production. For instance, men who go from a high saturated fat diet with 40% calories coming from fat to a low saturated fat diet with 25% fat saw their total and free testosterone levels drop[445]. Funny enough, going back to the high saturated fat diet made their testosterone increase again.

In the context of doing resistance training and practicing intermittent fasting, your body needs more anabolic foods to repair the damage it's received. That's why the idea that certain foods give you cancer is completely out of context when doing the Metabolic Autophagy Protocol. If you fast most of the day and lift weights, then you actually want to drive anabolism for recovery.

The upcoming chapters will delve deeper into the topic of what exact foods to eat for anabolism as well as catabolism. Remember that both of them are good in some situations but not all the time.

That's it for protein and fasting. Hopefully, you've now realized that the body is incredibly adaptable in finding out how to produce energy and still build muscle. With smart and strategic nutrient timing you'll make the absurd a reality and push that boulder across the hill. Now let's turn to the general overview of healthy nutrition.

Chapter X
Food Fallacy

Dwight: Through concentration, I can raise and lower my cholesterol at will.
Pam: Why would you want to raise your cholesterol?
Dwight: So I could lower it.

The Office

After the Second World War, the United States went through drastic changes in its dietary landscape. The country was booming economically, influentially, as well as nutritionally. People got access to a lot of commercial privileges they'd been deprived of during the conflict. Instead of sticking to meager rations and eating home-made dishes, a lot of the foods people used to eat were replaced by different kind of inventions of the food industry.

Unfortunately, all did not go well as many in the Western population started to suffer from a devastating tidal wave of deaths caused by cardiovascular disease and its derivatives. In today's day and age, 1 out of 4 deaths in the United States is caused by heart disease, making it one of the biggest dangers to people's health. How the hell did we get here?

Enter the Lipid Hypothesis

In the 1940s, a researcher from the University of Minnesota, Ancel Keys, hypothesized that the sudden rise in heart attacks amongst Americans was caused by their diet and lifestyle.

Keys first did a 15-year long prospective study on Minnesotan business professionals. The study began in 1947 and was followed up in 1963. It confirmed the results of some earlier studies, indicating that the biggest risk factors of heart disease were high blood pressure, cholesterol levels, and smoking[446].

In 1958, Keys collaborated with other researchers from 7 selectively picked countries to try and see a cross-cultural comparison of heart disease risk in the male population. The Seven Countries Study as it was called recruited 12 763 men, 40-59 years of age, living in the United States, Finland, the Netherlands, Greece, Italy, Yugoslavia, and Japan, with a follow-up from 1958-1964[447]. Results showed that the risk of heart attack and stroke were directly correlated with the total level of serum cholesterol, blood pressure, and obesity. See Figure 64.

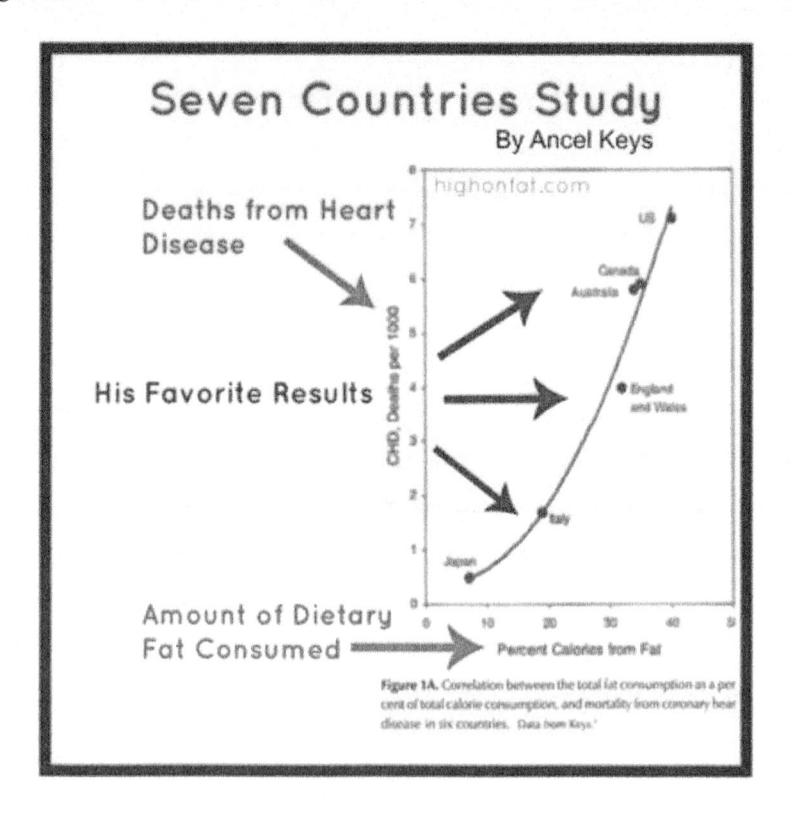

Figure 64 The more dietary fat consumed the more deaths from heart disease

The Lipid Hypothesis postulates that the main cause of heart disease are saturated fat and cholesterol that cause clogging of the arteries or atherosclerosis, which eventually will lead to a stroke or a heart attack. Based on that, lowering blood cholesterol and dietary fat intake decreases the risk of coronary heart disease and is healthier. By the 1980s, the lipid hypothesis or the Diet Heart Hypothesis was widely accepted and considered almost like a medical law[448].

That was a huge red light for the traditional way of eating that included eggs, full-fat dairy, red meat, butter, cream, and other foods high in cholesterol and saturated fat. Instead, inspired by the findings of Keys and other similar researchers, the public dietary guidelines shifted more towards promoting whole wheat grains, fresh fruits and vegetables, low-fat dairy, lean cuts of meat, egg substitutes, juices, and many other processed foods.

Over the course of the following decades, the situation didn't get any better as the obesity epidemic actually got even worse after the introduction of the diet heart hypothesis. There are more people with diabetes, insulin resistance, and cardiovascular events than ever before. Granted the average American still eats a crappy diet with processed carbs and fats but the general population hasn't gotten any healthier despite the change in dietary advice.

The Seven Countries Study has received a lot of criticism since its origin. In the 1950s, other researchers such as John Gofman from Berkeley California agreed that reducing dietary fat might be beneficial for heart disease patients[449]. Others were more skeptical, saying that the cause could be fat and cholesterol or it could also be the elevation of triglycerides from carbohydrates and sugar.

In 1973, Raymond Reiser found several methodological errors in Keys' work, namely misinterpreting the connection of saturated fat and cholesterol with atherosclerosis and confusing it with trans fats[450]. It does pose the question of why ancestral diets with high meat and fat consumption found in Alaskan Eskimos, Tanzanian Maasai, and others don't have such an epidemic of heart disease.

There's also the French Paradox or the Mediterranean Diet, which is relatively high in saturated fat, animal protein, wine, cheeses, and other „heart attack foods". Despite the high amounts of cholesterol and fat in their diet, French have quite low rates of heart disease compared to other Western countries.

Keys had selected those particular 7 countries out of the potential 22 for which he had available data as to fit his narrative. That's why he didn't include France into the study and chose populations with already lower levels of physical activity and poorer lifestyle habits. See Figure 65.

There have been many critiques of the diet heart hypothesis but no serious progress had been made until the 2000s. More data emerged that started to point the blame not on cholesterol or saturated fat but on sugar, processed carbs, vegetable oils, and trans fats.

How Keys Faked It

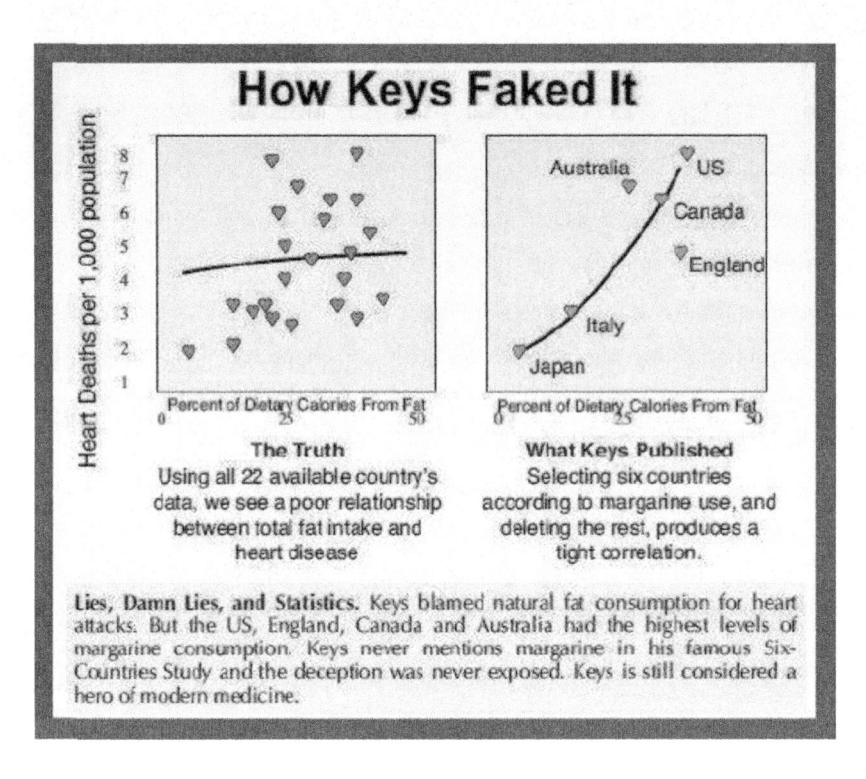

The Truth
Using all 22 available country's data, we see a poor relationship between total fat intake and heart disease

What Keys Published
Selecting six countries according to margarine use, and deleting the rest, produces a tight correlation.

Lies, Damn Lies, and Statistics. Keys blamed natural fat consumption for heart attacks. But the US, England, Canada and Australia had the highest levels of margarine consumption. Keys never mentions margarine in his famous Six-Countries Study and the deception was never exposed. Keys is still considered a hero of modern medicine.

Figure 65 Note how the relationship between fat consumption and heart disease gets skewed when you include all the other 22 countries

On August 2017, True Health Initiative pointed out many of Keys' inaccuracies in the Seven Countries Study[451]. They refuted 4 false ideas: (1) the countries were selected to fit the expected outcome; (2) France was intentionally excluded as to avoid the French Paradox; (3) the data from Greece was taken during Lent, which created discrepancies; (4) sugar or carbohydrates were not considered as a possible contributor to heart disease.

Death By Cholesterol

So, is cholesterol really that bad for you that it'll give you heart disease? Should you be eating foods high in cholesterol and saturated fat or not?

It is true that people who have heart disease tend to have higher cholesterol and blood pressure. But that's not the only thing that matters and it's potentially not the most important variable. In fact, low cholesterol is associated with mortality from heart disease, strokes, and cancer, whereas higher cholesterol has not been seen to be a bigger risk factor[452,453].

Cholesterol is an organic molecule produced by all animal cells. It's an essential component of cell membrane and helps with its functioning. It's also a steroid hormone that promotes hormone, bile, and vitamin D synthesis. Because of its molecular substance, cholesterol doesn't mix with blood and it's carried around the body by lipoproteins. There are different types of them:

- **Very Low Density Lipoprotein (VLDL)** – VLDL delivers triglycerides and cholesterol throughout the body to be stripped off from energy or for storage.

- **Intermediate Density Lipoprotein (IDL)** – IDL helps the transport of cholesterol and fats but its density is between that of LDL and VLDL.

- **Low-Density Lipoprotein (LDL)** – LDL carries energy through the bloodstream and directs nutrients into the cells.

- **High Density Lipoprotein (HDL)** – HDL collects unused cholesterol from the blood and brings it back to the liver for recycling.

HDL is considered the healthy one and LDL the bad one because it prevents the clogging of arteries and accumulation of cholesterol.

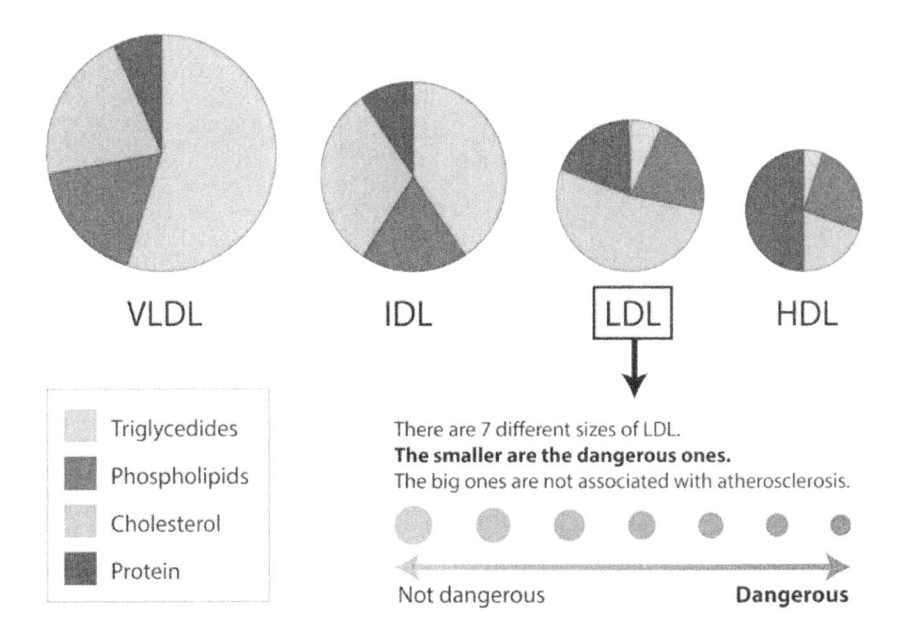

Figure 66 Different types of cholesterol

The amount of cholesterol in your blood is irrelevant to how it affects atherosclerosis and risk of heart disease because the build-up of plaque in the arteries is mostly driven by inflammation. If you suffer from higher inflammation and oxidative stress, then you'll also have more free radicals in the blood which can oxidize LDL. Oxidation of LDL by free radicals is associated with an increased risk of cardiovascular disease[454].

Free radicals and high inflammation damages the arterial lining called the endothelium, which the cholesterol then gets stuck to. Over the long run, this leads to the accumulation of plaques and increased risk of heart disease. Additionally, higher VLDL is considered a better predictor of heart disease risk than LDL[455].

You can get cholesterol from eating animal foods like eggs, cheese, meat, and dairy but your body can also produce its own. An average 150 lb weighing male can synthesize 1000 mg of cholesterol a day. One single egg has 200 mg of cholesterol. The typical US dietary intake of cholesterol is about 307 mg[456]. In that case, about 75% of your body's cholesterol gets produced by the body internally and 25% gets ingested externally[457]. Despite the lower dietary intake of cholesterol from diet in the US people still get heart disease.

Most of the cholesterol you eat gets esterified and is poorly absorbed. In the short term, cholesterol levels will rise as fats are being distributed around[458]. The body also reduces its own cholesterol synthesis when you consume it from food[459]. Therefore, eating more cholesterol will not raise your cholesterol in the long run. When you look at the mechanism of cholesterol transportation, then higher cholesterol in your blood also means that you're burning more fat for fuel as well.

The body uses very low-density lipoproteins (VLDL) to transport fatty acids, triglycerides, cholesterol, and fat-soluble vitamins around the bloodstream to either burn them off for energy or use in other cellular processes. Once the delivery has occurred, VLDL remodels itself into LDL.

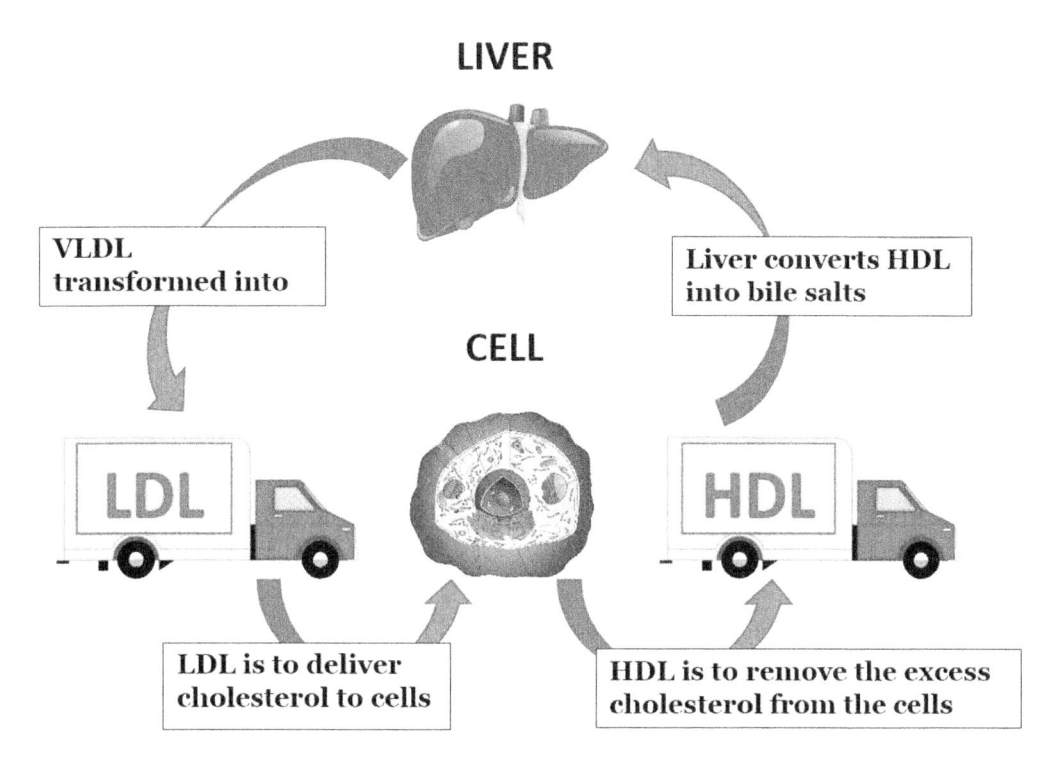

Figure 67 Transport and conversion of cholesterol in the body

If there are more cholesterol and triglycerides in your blood due to diet, then it's still going to be brought back down because of using them for cells.

- On a low carb diet, your body uses triglycerides for energy because there's no glucose around
- Triglycerides are transported to the cells by VLDL, which turns into LDL
- VLDL contains both cholesterol, triglycerides, and fat-soluble vitamins

Basically, if you're using more fat for fuel, then you'll by mechanism have higher fatty acids in the blood. Cholesterol just happens to be along for the ride.

Low carbohydrate ketogenic diets have been found to not lower LDL cholesterol but they do increase LDL particle size from small to large and decrease VLDL, all of which have positive effects on heart health[460].

The medical community should seriously re-evaluate the role of cholesterol in heart disease and what dietary recommendations they prescribe.

DISCLAIMER

This book is not supposed to be used as medical advice or recommendation in regards to atherosclerosis. I'm not a doctor and I'm not claiming to have the right answer. What I'm just sharing here is the physiology and research about how certain nutrients affect the body. Whether or not you should be eating a low-carb-high-fat diet depends on your medical history, current condition, goals, and eagerness to try less conventional approaches. The responsibility for your health and well-being is solely on you not me. Consult your doctor first before making any changes to your diet and lifestyle if needed. Let's carry on…

Here are the current cholesterol ranges in conventional medicine:

- **Total Cholesterol – 200 mg/dl (5 mmol/L) or less**. However, it's mostly irrelevant because of not really giving enough detail about HDL and LDL etc.
- **Triglycerides – 150 mg/dl (1.7 mmol/L) or less**. That's a good guestimate because you would expect triglycerides to go down when eating fewer carbs and burning more fat for energy.
- **HDL Cholesterol – 40 mg/dl (1.0 mmol/L) or more**. More HDL tends to be better and if you're eating a healthy diet you should see HDL to be actually higher at about 50-60 mg/dl.
- **LDL Cholesterol – 100 mg/dl (2.6 mmol/L) or less**. However, based on new research on low carb ketogenic diets, the amount of LDL for optimal health can vary greatly between people and is determined by many other things.
    - If your triglycerides are low and HDL is high, then it means you're using up those fatty acids. A higher LDL, in that case, isn't that relevant. Inflammation or C-Reactive Protein (CRP) would then also have to be low.
    - If your triglycerides are high and you have high CRP, then a higher LDL and total cholesterol aren't good because you'll have more inflammation, which will create more scarring of arteries, which then can promote plaque formation.

For hunter-gatherers, free-living primates, healthy human neonates, and other wild animals, the normal LDL range is between 50-70 mg/dl, which prevents atherosclerosis[461]. Based on current medical guidelines, you can have a slightly bigger buffer zone while staying below 100 mg/dl. More is not definitely better nor protective against heart disease. Too much will still be detrimental, even if you think you're eating a healthy diet.

Some people, around 20%, have a genetic variation that makes them absorb or synthesize so much cholesterol that their diet does influence their blood cholesterol level[462]. Even in these hypper-resonders, however, a high cholesterol diet does not generally negatively influence their cholesterol profile[463]. If total blood cholesterol increases at all during a high cholesterol diet, both 'good' HDL and 'bad' LDL cholesterol generally increase in the same proportion.

A review paper on the heart-related health effects of cholesterol concluded: *"Epidemiological data do not support a link between dietary cholesterol and cardiovascular disease."[464]*

You can make the argument that high cholesterol leads to atherosclerosis because the plaques are created by cholesterol build-up. However, the root cause of the issue is inflammation and arterial scarring in the first place. If you'd have lower C-reactive protein, then cholesterol would simply be transported around the body by VLDL and if it's not needed for nutrition it'd be transited back to the liver by HDL.

Therefore, the key to heart disease prevention is to avoid inflammation and foods that promote arterial scarring. That doesn't go to say that having high cholesterol, especially LDL is optimal or something you want to take for granted...The diet heart hypothesis may still hold true if cholesterol levels are high in the wrong context, such as when people eat high-fat foods with carbs or if they're smoking.

The problem with going on a low-fat low-cholesterol diet is that your body would still keep producing its own cholesterol. Arguably you'd be making more of it because of not getting it from food.

In fact, a lot of people suffer from low levels of HDL cholesterol, which prevents their body from clearing out cholesterol from the blood. Most Americans don't have enough HDL to decrease their risk of cardiovascular disease[465]. Crazy enough, low cholesterol levels are actually associated with increased mortality from stroke and heart disease[466].

Low Carbohydrate ketogenic diets with less than 50 g of carbs a day have been found to be better for long-term cardiovascular risk factor management compared to low-fat diets with less than 30% of calories coming from fat[467]. They can also raise HDL 4x times as much as a low-fat diet[468]. Long-term ketogenic diets reduce body weight, decrease triglycerides, lower LDL and blood glucose, and increase HDL cholesterol without any side-effects[469].

If you're going on a low carb ketogenic diet, then you still have to take responsibility for your own cholesterol levels and other biomarkers. It doesn't matter if you don't think cholesterol causes heart disease if you still end up getting a heart attack. Too high LDL and triglycerides in the blood can still be a risk factor because you can't control all the other invisible variables in your everyday life that promote atherosclerosis. In reality, the cause of heart disease isn't the cholesterol – it's the inflammation and high sugar diet combined with oxidized fats.

The Standard American Diet

Speaking of such diets, The Standard American Diet or SAD as it's called is indeed a recipe for cardiovascular disease, obesity, and general health problems.

It doesn't take a rocket scientist to realize how bad fast food is and what it does to you long-term. Everyone can cognitively grasp the idea that French fries and Happy Meals can't be good for you. Unfortunately, there's still a lot of confusion about what's healthy and people are still eating foods that are quite lethal for them without even knowing it. The SAD diet encompasses more than just your average Big Mac menu because not everyone eats out all the time and they still follow the recommended dietary recommendations. Unfortunately, those guidelines are very misleading and out-dated.

Here are some of the most lethal components of the Standard American Diet that not many people know about:

- Trans Fats and Vegetable Oils
- High Fructose Corn Syrup
- Artificial Sweeteners
- GMO Foods and Pesticides
- Refined Carbohydrates and Sugar
- Processed Grains and Wheat

These are all the staples of the food industry because they're very cheap, easy to manufacture, even easier to over-consume, and to hide it in every packaged food possible. I'll now go through each one of these groups and show why you don't want them in your diet.

Vegetable Oils and Trans Fats

When the diet heart hypothesis became more mainstream, the US dietary recommendations were quick to shift from an animal-based model to a low-fat one that minimized meat, egg, and saturated fat content. Instead, polyunsaturated fatty acids and vegetable oils were considered to be much healthier and protective of heart disease. Unfortunately, they have many negative effects on the body that actually promote heart disease.

Vegetable oils are extracted from various seeds, such as rapeseed (canola oil), soybean, corn, sunflower, peanut, and safflower. To get the oil from these plants, they have to be extracted through processing and heat, which damages their fatty acid composition.

These „heart healthy" vegetable oils are heated at unnaturally high temperatures, which makes the fats oxidize and go rancid. Oxidized fats accelerate aging, promote inflammation and damage the cells of your body when consumed. In addition to that, before the oils get put onto store shelves, they get processed even more with different acids and solvents to improve the composition of the product. As a bonus, they get deodorized and mixed with chemical colorings to mask the horrible residue of the processing process.

If you take it a step further and hydrogenize the vegetable oil, then it will eventually become more solid and intact. Meet margarine and all of the other trans-fatty acid type vegetable spreads that I can't believe aren't butter... All that work to get a cheaper product. But is it actually any healthier?

A Medical Research Council survey showed that men eating butter ran half the risk of developing heart disease as those using margarine[470]. Consumption of trans fats has been linked to obesity, metabolic syndrome, increased oxidative stress, heart disease, cancer, and Alzheimer's[471,472]. *Friends don't let friends eat margarine…*

Paradoxically, after the introduction of heart-healthy vegetable oils and trans fats, the situation of cardiovascular disease didn't get better but actually got worse. What gives?

As we described earlier, it's not the saturated fat or cholesterol that is driving atherosclerosis but it's mainly caused by inflammation and sugar that's making the arteries become inflamed in the first place. Cholesterol is just going there to do its job and tries to heal the injuries but if you keep causing damage to the blood vessels it doesn't matter how little cholesterol you eat from your diet as your body will still manufacture it from within. It's a vicious cycle of trying to fix the symptoms instead of going to the root cause, which is inflammation.

Unfortunately, vegetable oils and trans fats are one of the most inflammatory substances in our diet that are heavily driving this process of arterial clogging.

The body tries to maintain homeostasis in every regard, starting with stable blood sugar, metabolic rate, as well as the balance between essential fatty acids. One of the most important ones for survival, omega-9, omega-6, and omega-3 fats, can't be produced by the body itself, thus they need to be derived from diet.

- **Omega-3 Fatty Acids are an integral part of cell membrane** and they regulate many other hormonal processes. They have great anti-inflammatory benefits that protect against heart disease, eczema, arthritis, and cancer[473]. Omega-3s are polyunsaturated fatty acids, which refers to their multiple unsaturated double bonds. Great sources of omega-3s are salmon, grass-fed beef, sardines, krill oil, algae, and some nuts. There are 3 types of omega-3s:
    - **EPA (EicosaPentaenoic Acid) and DHA (DocosaHexaenoic Acid)** are animal sourced long chain omega-3 fats both essential for the nervous system, brain, and general health. They're found especially in seafood.
    - **ALA (Alpha Linolenic Acid)** is mostly a plant-based short chain omega-3 fatty acid. Most animals, including humans, can't directly use ALA so it gets converted into DHA first. Humans can convert only about 8% of ALA into DHA[474], which is why animal foods like salmon and oysters are much better sources of omega-3s and DHA.
- **Omega-6 Fatty Acids are also essential polyunsaturated fats.** They differ from omega-3s by having 6 carbon atoms at the last double bond instead of 3. Omega-6s are primarily used for energy and they have to be derived from diet. Unfortunately, most people are getting too many omega-6s. The most common

omega-6 is linolenic acid (LA). Another omega-6 is conjugated linoleic acid (CLA) with some health benefits. Vegetable oils, processed foods, salad dressings, and some nuts are the highest sources of omega-6 fatty acids.

- **Omega-9 Fatty acids are monounsaturated fats with a single double bond.** These ones aren't necessarily essential as the body can produce its own. In fact, omega-9s are the most abundant fats in cells. They've been found to lower triglycerides and VLDL[475] as well as improve insulin sensitivity[476]. You can get omega-9s from olive oil, and some nuts.

In general, omega-3s are very healthy for you and there doesn't seem to be many negative side-effects to over-consuming them. However, the body operates best when in homeostasis and that's why you need to balance omega-3s with omega-6s. Omega-9s aren't that important to pay attention to because most omega-3, as well as omega-6 foods, contain some omega-9s.

Anthropological research suggests that our hunter-gatherer ancestors consumed omega-6 and omega-3 fats in a ratio of roughly 1:1[477]**.** That's what we evolved with over the course of hundreds of thousands of years and that's probably optimal for general health. After the industrial revolution about 140 years ago, the ratios between omega-6s and omega-3s shifted heavily towards the consumption of omega-6s[478]. That's the result of food processing and using low-quality ingredients for the sake of increased profit. Today, the average American eats a ratio of 10:1, 20:1, or even as high as 25:1[479], which is definitely not good for you. No wonder there are so many chronic diseases related to inflammation and cardiac events.

The high amounts of omega-6 fatty acids in the Western diet are directly caused by the high amount of vegetable oils and trans fats. These things are everywhere – in pastries, 'healthy vegan restaurants' chips, chocolate, ' low-fat granola', frozen pizza, pre-cooked meals, fast food joints,– basically everything that comes in a package.

In addition to that, most dietary recommendations are skewed towards using these oils for home-cooking as well. In fact, almost 20% of the calories of Americans comes from a single food source - soybean oil. That makes 9% of all their daily calories to come from omega-6 fatty acids alone[480]! I mean, it doesn't matter how much kale or chicken breast you eat if you cook it in a bath of oxidized fat – you'll still damage your health.

Omega 6 fatty acids are unstable because they're made of polyunsaturated fats that get oxidized at high heats as do omega-3s. They're linked to increased risk of cardiovascular disease, cancer, and brain degeneration[481,482], and vegetable oils are very high in omega-6s. That's why you'd want to limit your intake of these fats, especially if they're processed.

So, if there's one food group you should avoid ALWAYS no matter what diet you follow, then let it be the camp of highly processed and oxidized vegetable oils/hydrogenated trans fats. This single change alone may prevent you from getting heart disease altogether.

Here's a list of the different oils and their fatty acid content:

FOOD	OMEGA-6 (g)	OMEGA-3 (g)	RATIO 6:3
FISH			
Salmon (4 oz/113 g)	0.2	2.3	1:12
Mackerel (4 oz/113 g)	0.2	2.2	1:11
Swordfish (4 oz/113 g)	0.3.	1.7	1:6
Sardines (4 oz/113 g)	4.0	1.8	2.2:1
Canned Tuna (4 oz/113 g)	3.0	0.2	15:1
Lobster (4 oz/113 g)	0.006	0.12	1:20
Cod (4 oz/113 g)	0.1	0.6	1:6
VEGETABLES			
Spinach (1 cup/110 g)	30.6	166	1:5.4
Kale (1 cup/110 g)	0.1.	0.1	1:1
Collards (1 cup/110 g)	133	177	1:1.3
Chard (1 cup/110 g)	43.7	5.3	8.2:1
Sauerkraut (1 cup/110 g)	37	36	1:1
Brussels Sprouts (1 cup/110 g)	123	270	1:1.3
NUTS AND SEEDS			
Walnuts (1 oz/28 g)	10.8	2.6	4.2:1
Flaxseeds (1 oz/28 g)	1.6	6.3	1:4
Pecans (1 oz/28 g)	5.7	0.3	21:1

Poppy Seeds (1 oz/28 g)	7.9	0.1	104:1
Pumpkin Seeds (1 oz/28 g)	2.5	0.1	114:1
Sesame Seeds (1 oz/28 g)	6	0.1	57:1
Almonds (1 oz/28 g)	3.3	0.002	1987:1
Cashews (1 oz/28 g)	2.1	0.017	125:1
Chia Seeds (1 oz/28 g)	1.6	4.9	1:3
Pistachios (1 oz/28 g)	3.7	0.071	52:1
Sunflower Seeds (1 oz/28 g)	6.5	0.021	312:1
Lentils (1 oz/28 g)	0.0384	0.0104	3.7:1
OILS AND FATS			
Butter (1 Tbsp)	0.18	0.83	1:1.5
Lard (1 Tbsp)	1.0	0.1	10:1
Cod Liver Oil (1 Tbsp)	2.8	1.3	2.2:1
Grain-Fed Tallow (1 Tbsp)	3.35	0.2	16.8:1
Grass-Fed Tallow (1 Tbsp)	1.2	0.8	1.5:1
Peanut Oil (1 Tbsp)	4.95	Trace	1:0.0
Soybean Oil (1 Tbsp)	7.0	0.9	7.8:1
Canola Oil (1 Tbsp)	2.8	1.3	2.2:1
Walnut Oil (1 Tbsp)	7.2	1.4	5.1:1
Sunflower Oil (1 Tbsp)	6	0.0	6:1
Margarine (1 Tbsp)	2.4	0.04	6:1
Peanut Butter (1 Tbsp)	1.4	0.008	17:1
Almond Butter (1 Tbsp)	1.2	0.04	2.8:1
Flaxseed Oil (1 Tbsp)	2.0	6.9	1:3.5
Olive Oil (1 Tbsp)	1.1	0.1	11:1
MEAT			
Ground Pork (6 oz/170 g)	2.83	0.119	23.8:1
Chicken	2.2	0.16	13.8:1
Grain-Fed Beef	0.73	0.08	9:1
Grass-Fed Beef	0.72	0.15	4.9:1
Domestic Lamb	1.9	0.6	3.3:1
Grass-Fed Lamb	1.7	2.2	0.7:1

| Farmed Salmon | 1.7 | 4.5 | 0.39:1 |
| Wild Salmon | 0.3 | 3.6 | 0.08:1 |

This may come off as alarming because a lot of the dangerous vegetable oils we just talked about don't have nearly as high omega-6 ratios than some of the nuts that are considered healthy. Like said beforehand, the omega-6s themselves aren't detrimental and they need to be kept in balance with the omega-3s. It's the processing and oxidization of the vegetable oils we just talked about that's causing the inflammation and all the other issues.

To keep your omega-3 to omega-6 ratios in balance without obsessing over them, eliminate all processed vegetable oils, avoid high-temperature cooking of oils and fats, eat plenty of grass-fed meat, get pastured eggs, mercury-free fish that's not fed grains and avoid packaged foods.

You'd also want to be careful with supplementing fish oil. It's true that the omega-3s from fish oil supplements can help to lower inflammation but they're also quite high in polyunsaturated fats and thus easily oxidized. Most of the commercial fish oil supplements out there have been exposed to some heat, sunlight or have simply gone rancid. This actually makes them pro-inflammatory if you take them consistently. Instead, you should focus on eliminating the inflammatory vegetable oils and eat some wild oily fish.

So far this chapter has been quite depressive, I know, and it seems things aren't getting better. Fortunately, mainstream medical doctors have picked up on the research about trans fats and are trying to heal the damage, literally. The dietary recommendations haven't changed much in regards to processed vegetable oils but we're slowly making progress.

In 2015, the FDA has determined Partially Hydrogenated Oils (PHOs) to be no longer „Generally Recommended as Safe"[483]. As of June 18, 2018, food companies cannot add PHOs or other trans fats into their products anymore. Took them a while... The compliance date has been extended to January 1st, 2020, after which we can hope to never see these abominations ever again.

But not all vegetable oils are bad. In fact, some of them are one of the healthiest fats in the world. Coconut oil and olive oil that have been used for centuries, however, don't require nearly as much processing. These fats get extracted by pressing whole olives or dried coconut kernels and then they get bottled. This doesn't include processing at high heat or exposure to different chemicals. Olive oil and coconut oil are much more vulnerable to temperature and they can go rancid more easily, which illustrates their natural manufacturing ways. Their fatty acid content and ratios are also more favourable for your health.

The health benefits of olive oil are actually quite legendary by now as they're a staple in the Mediterranean Diet, which even got Ancel Keys' sign of approval. Countries like Italy, Spain, Greece, and France consume a lot of olive oil and other heart-healthy foods like fish and vegetables, which is thought to be one of the reasons they don't have that high rates of cardiovascular disease.

The Mediterranean Diet is definitely misinterpreted and poorly described amongst the folklore of public medicine but it does include a lot of olive oil. Part of the benefit comes from the more balanced omega-3 to omega-6 ratio and low inflammatory foods. Olive oil itself also has a lot of polyphenols and the omega-9 fat oleic acid, which lower inflammation and protect against cancer[484].

Ancel Keys himself wasn't actually a proponent of a diet devoid of all fat. He consumed copious amounts of olive oil and other foods fitting to the Mediterranean Diet template. His main concern and criticism were directed towards animal fat and cholesterol.

So, do consume all the olive oil you want as long as it's not rancid and it doesn't make you gain weight. It has great heart-protective benefits. An additional thing to remember is that because olive oil is monounsaturated fat, it's more vulnerable to oxidation and free radical damage. That's why you don't want to be heating it or use for cooking. This is the entire reason why vegetable oils become oxidized in the first place – high heat processing. Instead, olive oil should be preferentially used as a cold dressing on salads and vegetables.

Bottom Line: When it comes to the highly processed vegetable oils, then avoid them like wild-fire because they'll literally light a forest-fire in your arteries and cells.

Saturated Fat

But what about the other sources of fat that have been deemed dangerous and unhealthy such as butter, lard, tallow, and bacon? Saturated fat is as big of a villain in the Lipid Hypothesis as is cholesterol. Dieticians and doctors have been prescribing low saturated fat diets for decades. Things haven't gotten better as you and I know the real issue has to do with inflammation, driven by sugar and vegetable oils.

A large meta-analysis that included 21 epidemiologic studies and over 340 000 people concluded that there is no clear association between dietary saturated fat intake and increased risk of cardiovascular disease or heart disease[485]. In fact, one large study with almost 60,000 Japanese participants found an inverse association between saturated fat consumption and stroke[486].

Olive oil is also quite high in saturated fat, which paradoxically doesn't cause atherosclerosis. The medical community must've missed that. Even wealthy countries that have quite low heart disease like France and Italy consume quite a significant amount of saturated fat. For the French, it comes from cheese, meat, and olive oil whereas the Italians are also eating quite a lot of pork, sausages, cheeses, and oils. Clearly, something is amiss.

In the Footprints of Weston A. Price

In the 1930s, Dr. Weston A. Price was perplexed why the western population had suffered a serious epidemic of tooth decay and other illnesses of nutrient deficiencies. He traveled to many places around the world that didn't subsist on primarily industrialized food such as Native Americans, Polynesians, Aborigines, and even Switzerland to realize how people in those regions had no cavities, crooked jawlines or other related diseases. He published his book *Nutrition and Degenerative Disease* in 1939 where he concluded that the modern Western Diet rich in flour, sugar, and vegetable oils causes nutritional deficiencies, dental issues, and health problems[487].

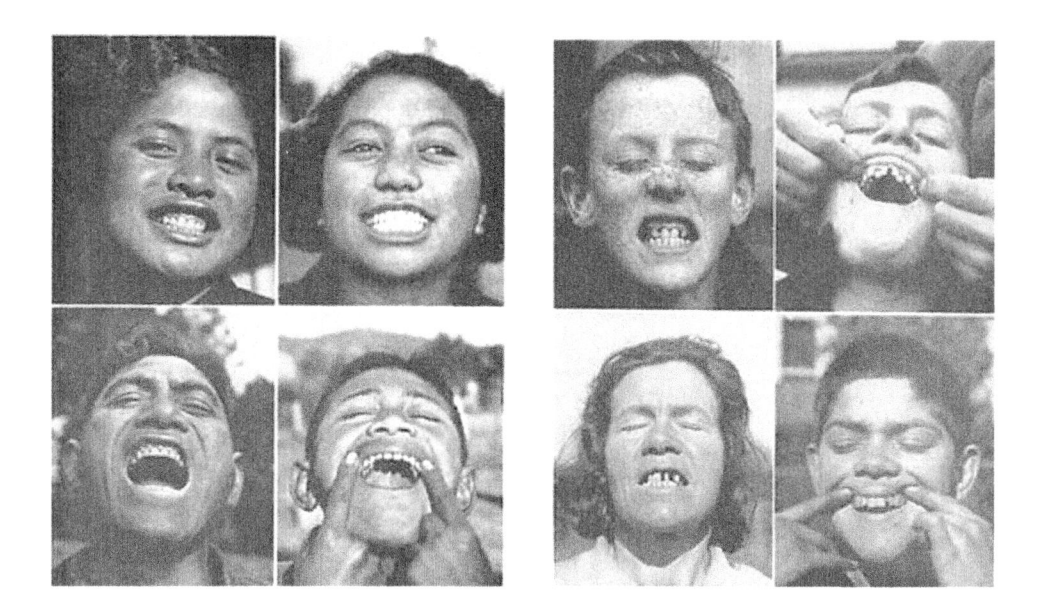

*Figure 68 Price's Pictures of People's Teeth
in the West (Right Picture) and Subsistence Societies (Left Picture)*

Nowadays, inspired by the work of Weston A. Price, ancestral diets such as Paleo, GAPS, and the Weston A. Price Foundation diet include many animal fats, meats, and broths that nourish the gut, promote cellular health, and proper physical development. Part of the reason has to do with the pro-anabolic qualities of these foods that will increase IGF-1, mTOR, and cholesterol. That's not necessarily a bad thing as it's required for proper growth. As a precursor, I'll say that these foods will be a big cornerstone of the Metabolic Autophagy food pyramid as well. It's just that you have to know how, in what amounts, and when you consume them. All of which we'll discuss further down the line.

Dr. Price found that saturated fat, such as butter, cream, lard, tallow, bone marrow etc., was a staple in the diets of these very healthy natives. There's a huge discrepancy with the research done in Western countries where butter is seen so bad that its fat content has to be reduced and swapped out with vegetable oils instead.

Butter actually has many anti-oxidants, such as vitamin A and vitamin E, that protect against the free radical damage that inflames the arteries. It also contains selenium, which is another important anti-oxidant and mineral. The short chain fatty acids of butter like butyrate heal the intestinal lining of your gut that prevents inflammation and autoimmune disorders. Medium-chain, as well as short-chain fatty acids, have strong anti-tumor properties[488].

Healthy fats are essential for your cells to function properly and to carry out many of the other metabolic processes. Most of the important vitamins you need for survival require fats to be absorbed and low-fat diets over the long term can be quite detrimental for your longevity.

When Price was doing his research, he described a "vitamin-like activator" that played a central role in the utilization of minerals, vitamins, growth, and protection against heart disease. He called it *Activator X,* which he found in the butterfat, organs, and fat of animals who consumed green grass, and also in some fish and eggs. Unfortunately, Price died before knowing what this mysterious compound really was. Nowadays we know that Activator X is Vitamin K2, which is a fat-soluble vitamin produced by animal tissues from Vitamin K1.

- Vitamin K1 is mostly found in plant foods and greens. It supports blood clotting and healing.
- Vitamin K2 is divided into many specific forms called menaquinones (MK), ranging from MK-4 and MK-7 etc.
  - MK-4 is found the most in animal foods and it protects tissues against calcium formation and cancer development. It also supports hormones
  - MK-7, MK-8, and MK-9 are mostly found in fermented foods like sauerkraut and miso. They support bone health and hormones much more effectively than K1

K2 is indeed very important for mineral absorption and general health. In the context of atherosclerosis, Vitamin K2 also directs calcium into the right place, namely the bones and teeth, instead of keeping it in the bloodstream to cause plaque formation.

Vitamin K2 works synergistically with two other fat-soluble activators Vitamin A and D. Vitamin A and D signal the cells to produce certain proteins and vitamin K then activates them. Other minerals such as zinc and magnesium as well as dietary fat are needed for the absorption of these fat-soluble vitamins.

The biggest reason why the Western population had such high rates of tooth decay and crooked jawlines had to do with not getting enough Vitamin K2 from their diet. People in non-industrialized countries still ate their ancestral diets and thus got enough vitamin K2 to support their bone development.

Currently, the daily recommended intake of vitamin K is 90 mcg with no official requirements for K2. However, these guidelines originate from 2001 and we've gained more insight into the importance of vitamin K2 since then. Based on current research, you'd want to get at least 100 mcg-s of K2 a day and aim up to 150-200 mcg-s in total.

Although humans can convert some K1 into K2, the biggest effect comes from MK-4 utilization, which is most bioavailable in animal foods. Nevertheless, you'd want to be eating foods high in both to cover your basis. Given you don't need that much K1 from vegetables and K2 is much more difficult to come by, here is a list of foods richest in Vitamin K2, starting with the highest:

- **Natto** is a Japanese fermented soybean dish with a foul smell. It's the richest sources of vitamin K2 I've seen with a whopping 1103.4 micrograms of Vitamin K2 MK-7 in 100 grams.
- **Goose Liver Paste** – a delightful pateé type of cream used a lot in French cooking and similar cultures. It's an easy way of making organ meats more palatable and tasty. MK-4 content 369 mcg/100g.
- **Hard Cheeses** – cheese should be fermented and unpasteurized for the greatest health benefits. A lot of the vitamins and minerals get lost during processing. MK-7 content 76.3 mcg/100g.
- **Soft Cheeses (Brie)** – maybe one of the reasons the French didn't get atherosclerosis had to do with the K2 rich cheeses and pateés that protected them against plaque formation. MK-7 content 56.5 mcg/100g.
- **Egg Yolks** – specifically the cholesterol-rich egg yolks which where the entire egg gets its nutritional value from. MK-4 content 32 mcg/100g.
- **Dark Poultry** – meat that's darker in color such as goose and duck. Much richer in vitamins than white chicken. MK-4 content 31 mcg/100g.
- **Butter** – *I can't believe...*that good old butter with its yellow color. Really good as long as you don't confuse it with margarine. MK-4 content 15 mcg/100g.

- **Liver and Organ Meats** – probably the most nutrient dense foods on the planet are liver, heart, and other organ meats. One of the best sources of dietary vitamin A and D, which are essential for vitamin K utilization. MK-4 content 14 mcg/100g.
- **Sauerkraut and Fermented Foods** – another form of bioavailable vitamin K1 as well as K2. Paradoxically, the K2 and B vitamin content of fermented foods comes from the live bacteria in them, not the cabbage itself. So, sauerkraut is still actually an animal-based food. MK-7 content 5 mcg/100g.
- **Raw Milk and Kefir** – more unprocessed food that's rich in vitamins and some live bacteria. Pasteurized milk just kills all the juice, figuratively speaking. MK-4 content 2 mcg/100g.

As you can see, these foods are again quite high in saturated fat and cholesterol, which have been deemed as dietary killers for so long. Fun fact, consuming trans fats blocks the actions of Vitamin K2, which would make everything even worse regards to arterial health and inflammation.

The reality is that it's not the butter that's causing heart problems – it's the bread that you're spreading it on. Even worse, if you're making fried toast with eggs and oil, then you're literally driving up inflammation through the roof. With a few minor adjustments, such as not using oxidized oils or avoiding the gluten-heavy grains, butter would be one of the healthiest fats in your diet.

Meat Kills?

As we've seen so far, animal foods and fats are one of the most nourishing things for your body and they have many health benefits. However, in the zeitgeist of nutrition, meat, in particular, has received a bad reputation as it's considered not only responsible for heart disease but also cancer.

In a 2017 report, the World Health Organization (WHO) listed bacon, sausage, and other processed meats as "group 1 carcinogens." This puts them into the same category as tobacco, alcohol, and arsenic. Red meat was categorized as „group 2 carcinogen," which makes it „probably carcinogenic" to humans. From WHO:

> *The classification is based on limited evidence from epidemiological studies showing positive associations between eating red meat and developing colorectal cancer as well as strong mechanistic evidence.[489]*

Speculations about animal foods causing cancer have dated as far back as 1975 well into the era of the diet heart hypothesis[490]. Recent reports in the Journal Nature also suggest a possible mechanism linking red meat consumption with heart disease[491]. However, reviews of studies on red meat and cancer have seen inconclusive results[492].

The biggest mistake I see is that processed meats tend to be put into the same category as natural meats. Of course, eating hot dogs and sausages can't be good for you. Eating processed meat such as hot dogs and burgers increases risk of coronary heart disease, stroke, and diabetes[493]. Those meats include many other ingredients such as sugar, chemicals, as well as the gluten-laden wheaty bun it's consumed with. Furthermore, people who eat processed meat ala the SAD diet tend to follow other poor lifestyle habits, such as not exercising enough, over-eating calories, too much stress, not enough sleep, including others.

Eating meat in the context of a healthy diet does not have the same effect as eating meat in the context of an unhealthy diet. A large meta-analysis that included over 1,2 million participants found that the consumption of unprocessed red meat is not associated with an increased risk of coronary heart disease, stroke or diabetes[494].

Most commonly, the idea of eating meat gets juxtaposed with a plant-based vegan diet devoid of all animal foods that are depicted as the epitome of a healthy diet. However, there is no data to prove that. Generally, there are no differences in mortality between vegetarians and omnivores[495].

Those who eat more red meat also have a tendency to smoke, drink, eat fewer vegetables, and engage in other unhealthy behaviors that increase their risk of cancer. This has been proven by many studies[496,497].

People who follow a vegan diet already are simply more mindful of their health if they've taken the dietary path they've chosen. Likewise, a person eating meat and fat can be equally as healthy if they pay attention to these things. Studies that compare vegetarians and omnivores with the general population see both groups living longer than the average person[498].

You can be equally as healthy and equally as diseased on a vegan diet as on a meat-based diet. The determining factor lies in how your body uniquely responds to the food you eat, what other habits you follow, and the general lifestyle. That's why everything regarding nutrition and longevity is context based. Large epidemiological studies that cover large populations are quite arbitrary unless they target a specific way of living that you could identify with.

Meat Myths

Meat and animal foods are thought to increase tumor formation and cancer because of containing certain substances such as Neu5Gc. Eating meat is also thought to be pro-inflammatory and thus dangerous. However, anthropological data in the example of the Masai who eat primarily meat, blood, and milk, don't show these people having inflammatory diseases. It might be that the inflammation is caused not by the meat but the other things such as bread, starch, sugars, or oils. Eliminating those carbohydrates and increasing meat consumption has been shown to actually reduce inflammation markers[499].

Another dangerous chemical that's received some recent attention is TMAO (trimethylamine N-oxide) that's thought to increase the risk of heart disease[500]. However, studies haven't shown that red meat increases TMAO. In fact, a 1999 study found that out of 46 different foods fish was the only one that raised TMAO. Trimethylamine occurs naturally in seafood[501]. Now hold on a minute…if fish is considered the "heart-healthy" alternative of meat for protecting against heart disease, then how come it's high in TMAO that also raises your risk of heart disease? Clearly, the roles of these compounds are misunderstood and taken out of context in attempts to point the finger at certain food groups while being biased with others.

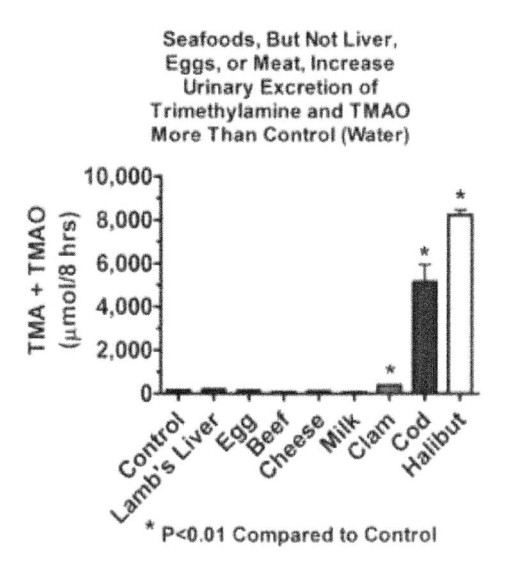

Figure 69 Compared to fish and seafood, meat and dairy are almost devoid or TMAO

Let's look at some more myths about meat consumption...

Meat is high in protein and eating too much protein is thought to cause kidney disease. Kidneys excrete nitrogen by-products from digesting protein, thus too much protein over-taxes the kidneys. High protein diets may be harmful to those who already have kidney disease but there's no evidence they damage healthy individuals[502]. This idea of protein causing kidney disease may be over-exaggerated by diets that combine protein and meat with different inflammatory foods like grains or vegetable oils.

What about nitrates? More chemical compounds that are said to be harmful to us. Usually, nitrates are added to processed foods and meats as to preserve them longer. However, nitrates are found in many other foods and they're produced by our bodies as well. Most of the nitrates you get actually come from vegetables and greens. Did you know that one serving of arugula, two servings of butter lettuce, and four servings of celery or beets all have more nitrites than 467 hot dogs[503]? Obviously, the celery and arugula are healthier than hot dog meat despite their higher nitrate content. Your saliva is also incredibly high in nitrites. So, if you're worried about the nitrites in meat, you should focus your efforts on not swallowing that much.

Neither nitrates or nitrites get accumulated in the body. Nitrates from food get converted into nitrites after coming into contact with our saliva. About 25% of the nitrate we eat gets converted into salivary nitrite, 20% gets converted into nitrite, and the rest gets excreted in the urine within 5 hours of ingestion[504]. Studies show no association between nitrites in diet and stomach cancer[505]. What's more, nitrates may also help boost the immune system and protect against pathogenic bacteria[506][507], which is why arugula is a good ingredient to add to your diet.

Meat itself is actually one of the superfoods of the planet that helped our species to evolve. Red meat is a very rich source of essential vitamins and minerals, especially B vitamins. Vitamin B12 is vital for proper nervous system functioning and general muscle homeostasis. B12 deficiencies can cause accelerated aging, neurological disorders, cardiovascular disease, sarcopenia, and many other ailments. Red meat is also a highly bioavailable source of vitamin D, iron and zinc, which are important for other metabolic processes. It has magnesium, copper, phosphorus, chromium, and selenium. Although some plants such as spinach are high in iron, the body can absorb only about 2-7% of the iron, whereas it can absorb 20% of it from red meat[508].

Aside from all the ethical and environmental implications, meat is an incredibly bioavailable source of most of the vitamins and minerals you need. It also covers the aspect of anabolism and protein, which are needed for maintaining lean tissue as you get older. But as with all things in nature, there are no free lunches and meat isn't that kind of a superfood you could eat in unlimited amounts and stay healthy.

As we've touched upon before, **amino acids and protein, primarily from meat, will stimulate the mTOR pathway and make you anabolic**. This will trigger muscle protein synthesis and makes every cell in your body grow, including cancers and tumors. Overexpression of mTOR or its dysfunction is often related to various cancers and genetic disorders[509]. There's no denying that mTOR is an all-encompassing anabolic switch that makes everything grow – the good and the bad.

You don't want to be under the influence of mTOR because of eating meat breakfast, lunch, and dinner. You definitely don't want to become deficient of it either by fasting too long or eating only low mTOR foods. People who are afraid of IGF-1 and go on fully plant-based diets with low protein tend to deteriorate quite rapidly unless they specifically supplement that.

As is stated throughout Metabolic Autophagy, you want to strike somewhat of a balance between stimulating mTOR with exercise and protein but not too much because it'll accelerate aging. The breach for crossing the chasm again is intermittent fasting and resistance training.

All of the research on eating meat, eggs, saturated fat, cholesterol, and carbohydrates has been done on people who aren't practicing intermittent fasting. There are virtually zero studies on individuals who follow a fasting focused lifestyle and are engaged with other healthy lifestyle habits like resistance training, hormesis, and calorie cycling.

Of course, over-expression of mTOR and protein can be bad for you if you're eating the standard diet with super high meal frequency. However, if you're already fasting ala Metabolic Autophagy style, you're changing the context of the situation completely. When you come from a fasted state or after you've exercised, you want to stimulate mTOR for recovery and muscle growth. In that situation, meat, protein, and mTOR are pro-longevity – they're good for you because otherwise, you'd experience sarcopenia.

If you use mTOR and protein consumption strategically as to support muscle growth and recover from fasting, instead of fat gain and caloric fillage, then that mTOR stimulation is actually helping you to live longer. Shocker, I know.

Chapter XI
The Case Against Sugar (and Fat)

"Any diet can be made healthy or at least healthier,
from vegan to meat-heavy,
if the high-glycemic-index carbohydrates and sugars are removed,
or reduced significantly."

Gary Taubes

Despite the prominence of the diet heart hypothesis throughout the 70s and 80s, there were some contradicting voices in the medical community. In 1972, a physician named Robert C. Atkins wrote a best-selling book called *Dr. Atkins' Diet Revolution.*

Essentially, the Atkins Diet is a low carb diet that prohibits the consumption of carbohydrates and allows unlimited protein and fat. It's said to result in rapid weight loss as long as you stick to those guidelines. Indeed, a lot of people did find success with this way of eating. Because of the higher intake of protein and reduction in carbs, it's very easy to consume fewer calories and still feel satiated.

During its inception, the Atkins Diet was demonized as unhealthy and dangerous due to the high saturated fat content. Even today there are people saying animal fats are bad. However, current data hasn't found any correlation between saturated fat and heart disease[510]. In fact, it's thought that the studies that made those claims in the past suffered from potential biases and selective reporting of results[511].

Nowadays we know that both low-carb, as well as low-fat diets, can be effective for weight loss and improving metabolic health[512]. Which one works better is more a matter of context.

Of course, an Atkins or a ketogenic diet can lead to some serious medical conditions like diabetes and heart disease. Not because of the dietary mechanisms of the diets but because of the individual who's doing it. Some people may indeed react negatively to consuming saturated fat or they may have other poor habits.

It's important to remember that the combination of certain foods may cause a completely different metabolic reaction than eating those same foods alone. For instance, as we remember from the chapter about glucagon and autophagy, consuming protein in a fasted state or on a low carb diet doesn't raise insulin and actually lowers the insulin-glucagon ratio. Combining protein with carbohydrates spikes insulin through the roof and makes the entire meal more insulinogenic. Furthermore, you can indeed promote oxidative stress in the body by consuming oxidized fats and cholesterol or if you end up glycating them with glucose (more on this down the line).

Any diet you follow, whether that be Paleo, vegan, keto, SAD, or omni-lacto-vege-pescetarian, can be equally as unhealthy and dangerous if you fail to understand the underlying metabolic reactions that occur. What foods you eat, what ingredients you combine them with, when you consume them, how much, at what frequency, your methylation status, and overall biomarkers will determine the final nutritional result. It doesn't matter what diet you're on. What matters is how you choose to manipulate nutrient signaling, hormonal profile, and meal timing. That's what Metabolic Autophagy seeks to teach you.

The Carbohydrate-Insulin Hypothesis

Atkins wasn't definitely the first person to promote a low carb high fat diet. It was actually the norm to prescribe it to obese people back in the day. In 1825, Jean Anthelme Brillat-Savarin wrote:

> *The second of the chief causes of obesity is the floury and starchy substances which man makes the prime ingredients of his daily nourishment. As we have said already, all animals that live on farinaceous food grow fat willy-nilly; and man is no exception to the universal law.*[513]

One of the earliest of such dieticians was an English undertaker named William Banting. He was previously obese and failed to lose weight numerous times. In 1863, Banting published a booklet called *Letter on Corpulence, Addressed to the Public,* in which he outlined his diet program[514]. The man ate primarily meat, green vegetables, fruit, and dry wine. He avoided sugars, starches, beer, milk, and butter, which tend to be quite fattening. Nowadays, most low-carb paleo diets follow a similar template with slight differences.

In his books, Arkins made the bold statement that avoiding carbohydrates produces a metabolic advantage because burning fat burns more calories. However, that is not the case as fats and lipids have the lowest thermic effect of food (3-7%), compared to carbs (5-10%) (See Figure 70). The weight loss was probably due to higher protein intake with a 20-30% loss of calories to metabolism. Most diets similar to Atkins, such as The Zone and Protein Power tend to be the same.

Table 1	
Thermic Effect of Food	
• Protein	20-30% of calories ingested
• Carbohydrate	5-10% of calories ingested*
• Fats	3-7% of calories ingested

* When carbs are converted to fat (lipogenesis) it requires ~20% of the calories

Figure 70 The Thermic Effect of Different Macronutrients

In 2002, an American journalist and author Gary Taubes wrote an article in the *New York Times Magazine* called *What if It's All Been a Big Fat Lie?*[515]. He starts off with questioning the US medical establishment and their 30-year long ridicule of Atkins, calling it quackery and fraud. Taubes also makes the claim that the low fat high carb dietary recommendations are the cause of the obesity epidemic in America and that Atkins was right all along. Bold statements indeed.

Taubes does make a good point that despite the introduction of the low-fat guidelines after the 70s, the obesity epidemic took off into even greater heights around the early 80s. Despite smoking less and eating foods devoid of cholesterol, heart disease and metabolic disorders haven't still declined.

Unfortunately, we live in what Kelly Brownell, a Yale psychologist, has called a *"toxic food environment"* of cheap food, empty calories, super-size portions, clever marketing and sedentary lifestyle. That's why the average person in Western countries finds it difficult to lose weight or stick to their diet – their willpower has waned off. However, it's not just that people are eating too much and moving too little. There is another potential metabolic mechanism at play that unfavours the obese and healthy.

In 1920, the Canadian physician Sir William Osler wrote in *The Principles and Practice of Medicine* that most of the doctors in early 1900s considered refined carbs and starches as the primary drivers of obesity[516]. It was common sense back in the 40s and 50s that the most fattening foods were sugars, sweets, and carbohydrates.

The physiology of how insulin promotes weight gain dates back to the 20s when insulin was first discovered. Patients who are underweight tend to be given additional glucose and insulin injections to fatten them up to a healthy range. Type-1 diabetics who take insulin more often will gain more weight[517]. Bodybuilders trying to gain mass also inject themselves with insulin so they could build muscle. Insulin, after all, is a storage hormone that directs nutrients into cells.

The idea that weight loss is simply a matter of calories in vs calories out does not hold water if you put it to serious test. We already showed in the chapter of Anabolic Fasting that it's possible to build muscle and lose fat at the same time with proper nutrient partitioning and meal timing. Body composition is regulated mostly by hormones and the macronutrient ratios of what you eat and when you do it.

Intensive injections of insulin several times a day make diabetics gain weight quite rapidly. In a study by Henry R. R. *et al* (1993), they took 14 diabetics and gave them insulin over a 6 month period. The patients' bodyweight increased on average by 8.7 kg (19 lbs) despite eating 300 fewer calories a day[518]. They were consuming less food but because of the higher levels of insulin, they still gained quite a lot of weight. That's what many yo-yo dieters who eat a low-fat but high carb diet tend to fall into as well. They're keeping themselves in a state of hyperinsulinemia and thus never tapping into their own body fat stores.

On the flip side, medical drugs like metformin and rapamycin can treat insulin resistance and make the person lose weight. In type-1 diabetes, the pancreas doesn't produce enough insulin, which keeps the blood sugar elevated for longer. If the person isn't injecting insulin their insulin levels will drop extremely low. It's very common amongst type-1 diabetics to suffer from severe weight loss. That's why they get injected with insulin – not only to raise their insulin but to also prevent deterioration. The mechanism for insulin-driven weight gain is very well documented.

Insulin is the main regulator of energy storage and body weight homeostasis. It promotes weight gain but also makes you become more hungry and eager to eat. Whenever your glucose drops you want to eat again to prevent hypoglycemia. If a person eats a low-fat high carb diet, then that's going to end up with another rise in insulin, keeping them in a state of chronically elevated insulin (hyperinsulinemia). High meal frequency is as bad as high carb intake both because of hyperinsulinemia and no opportunity to enter autophagy.

Obesity as a disease is primarily a hormonal issue. Fat gain is mostly stimulated by insulin, which will make the body store the food you eat. There are also issues with leptin resistance and chronically elevated cortisol. These hormones signal the person to follow certain behaviours that make them eventually obese, such as binge-eating, become sedentary, or not feel satiated after meals.

Cortisol is the body's main stress hormone that releases adrenaline and breaks down glycogen stores to supply the blood with glucose. Chronically elevated levels of cortisol not only lead to fatigue but can also promote weight gain. Cushing's Disease makes the body overproduce cortisol which results in weight gain. On the flip side, Addison's Disease makes the adrenal glands produce less cortisol, which results in weight loss.

The reason excess cortisol makes you fat has to do with high blood glucose and consequently elevated insulin. If you're stressed out or have entered the *'fight or flight mode'*, you're more prone to store fat because of shutting down digestion and raising insulin. The body wants to supply its muscles to run away from danger and fat loss becomes a secondary goal.

Mechanisms of Cortisol Induced Fat Production

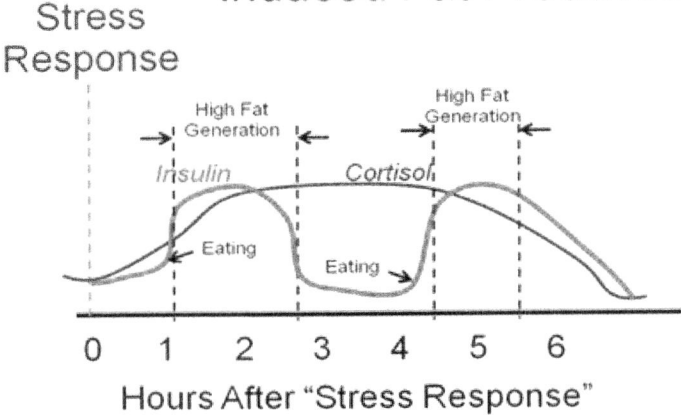

Figure 71 Spiking insulin raises cortisol, which promotes fat gain

Chronic stress promotes chronically high blood sugar and hyperinsulinemia. We also know that insulin resistance walks hand in hand with leptin resistance, that makes the brain desensitized towards satiety signals from food. This is often accompanied by emotional binge-eating, stubborn fat loss plateaus, and less satiety signaling.

Paradoxically, people who are overweight i.e. fat cells are overly full also have higher levels of leptin. That's why they continue to overeat and gain weight despite already having enough energy around – they're brain doesn't get the message.

It's thought that leptin resistance is one of the main drivers of obesity[519]. That's the reason people get obese in the first place and that's why most diets fail. Leptin resistance results partly from hyperleptinemia by downregulating the cellular response to leptin[520]. Basically, too much leptin all the time makes the brain desensitized to the signals that say you're satiated and nourished, thus promoting hunger, cravings, and obesity. It's like the boy who cried wolf...

Leptin resistance walks hand in hand with insulin resistance both of which make you require to produce more leptin and insulin to signal the brain that there's adequate energy.

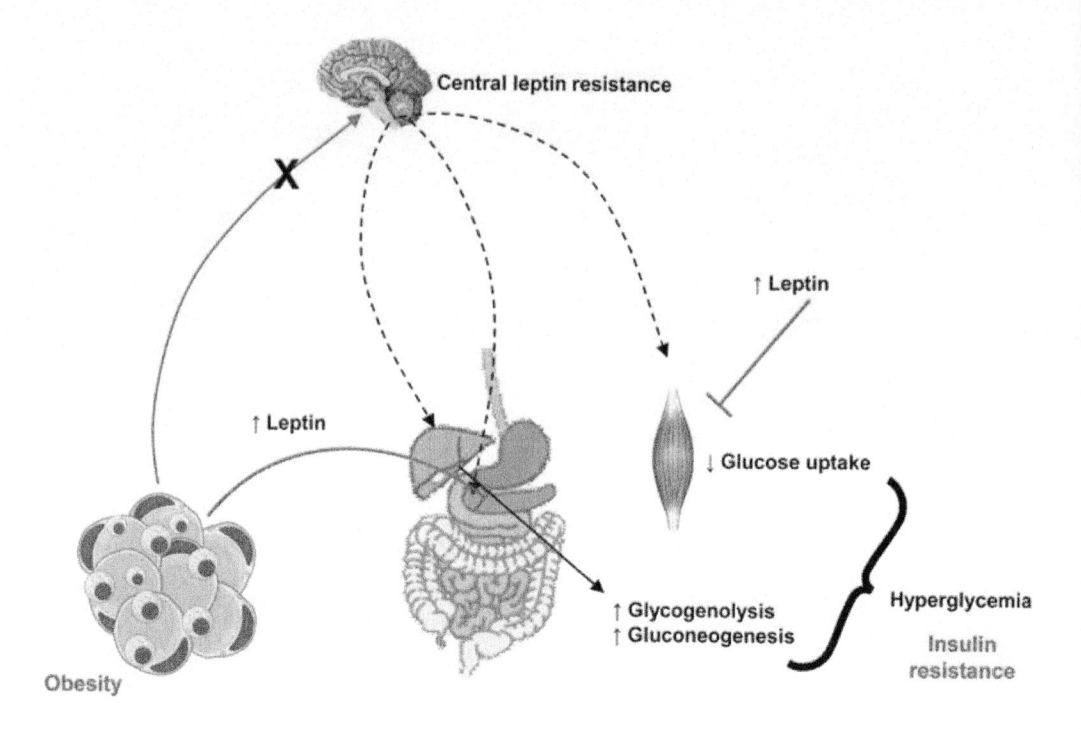

Figure 72 Hyperglycemia causes insulin resistance, which promotes leptin-resistance and obesity

If you look at the physiology and understand the principles of nutrient partitioning you'll realize why the majority of people are obese. They're simply eating too many carbohydrates too often which keeps them in a state of hyperinsulinemia. Add the constantly stressed out lifestyle laden with cortisol and cheap food into the mix and you'll end up with this obesogenic hormonal concoction that's about to explode any moment.

This is called the Carbohydrate-Insulin Hypothesis (CIH), which is based on the idea that insulin makes you fat. However, if you dig a bit deeper you'll see that there's clearly something missing. For instance, the Japanese have very low levels of obesity and heart disease despite eating a diet high in rice and carbs. Other countries with higher carbohydrate intake don't suffer the same way either. Until they come into contact with refined carbs and sugars of the Western food industry that is...

Only after the introduction of processed foods from the West do traditional diets become fattening. This you can see happening in China and India who traditionally eat more carbohydrates but now get access to different indulgences and more added sugars. These countries cover nearly 1/3 of the world's population and there's another epidemic bound to happen. People in Japan have so far remained to be less affected by this global trend thanks to sticking to a more home-made cuisine and avoiding over-eating.

The Innuit have also been devoid of diabetes and obesity throughout their dietary history. Despite getting over 90% of their calories from red meat, seal fat, whale blubber, and fish, they didn't die to heart disease or suffocate to the high amounts of saturated fat they were consuming. Unfortunately, after coming into contact with the white flour and sugar of modern people did they get more of those diseases. *It's not the meat, it's what you eat the meat with...*

These anthropological examples show that the Carbohydrate-Insulin Hypothesis has some truth to it but it's not the entire picture. It's important to understand the bigger milieu in which obesity and disease develop.

Now that this is covered, let's address the deeper cause of the issue, which has been the problem all along – insulin resistance. Shall we begin?

Insulin Resistance

Your body wants to maintain a stable blood sugar level all the time – to be in homeostasis and complete balance. Whenever there's an elevation in blood sugar the pancreas secretes the hormone insulin to store that glucose and stabilize blood sugar again. The cells sense insulin through insulin receptors that will signal the PI3K/Akt/mTOR signaling pathway[521].

Insulin Resistance is a condition where the cells don't respond normally to the elevation of insulin and are resistant to picking up glucose, thus causing high blood sugar. The beta cells in the pancreas continue to produce more insulin but to no avail. This keeps both blood sugar and insulin levels elevated for a longer time.

Hyperinsulinemia is a condition where there's excess circulating insulin in relation to the amount of blood glucose. It's associated with hypertension, diabetes, obesity, and metabolic syndrome[522]. This is one of the underlying issues that's driving cardiovascular disease and other health disorders.

Symptoms of insulin resistance or glucose intolerance include uncontrollable hunger, increased thirst, high blood sugar, high blood pressure, brain fog, lethargy, lightheadedness, easy weight gain around the stomach, stubborn belly fat, elevated triglyceride and cholesterol levels.

In most cases, hyperinsulinemia is both a result and the driver of insulin resistance[523]. The two basically describe the same thing.

A study in *Diabetologia* took 12 non-obese men who got injected with higher and higher doses of insulin over the course of 40 hours[524]. At the end of the experiment, they showed a 15% decreased glucose tolerance, which is another way of saying they became 15% more insulin resistant. Another study took 15 healthy non-diabetic men who were given normal doses of insulin over 96 hours. By the end of it, their insulin sensitivity had dropped by 20-40%[525]. Essentially, frequently high insulin makes the body eventually insulin resistant. High and frequency are relative terms and context-dependent but in most cases, you don't want to keep your insulin elevated at all.

Although chronic insulin resistance is quite harmful, it also has a beneficial adaptive mechanism. Under harsh metabolic conditions, insulin resistance helps to give the brain glucose that would otherwise be taken up by muscles[526]. It's also thought to be a normal physiological response to sustained caloric surplus and overeating, in general, to protect the body against the accumulation of lipids[527]. Animals who eat a lot more calories than needed develop rapid insulin resistance and get obese very fast[528], especially if you feed them a lot of grains and carbs.

However, from the perspective of longevity, it's not a good idea to have continuously elevated levels of blood sugar or insulin. First of all, it promotes fat gain and raises triglycerides, but secondly, it also keeps the body in a continuous state of dysfunctional mTOR signaling. You want to be more insulin sensitive so you'd clear your bloodstream from any excess glucose as fast as possible and get back into ketosis.

Guess what causes hyperinsulinemia and insulin resistance? You can predict that processed carbs and too much sugar play a major role but trans fats are another thing to be blamed for[529]. Fast food that combines a lot of salt, sugar, and fat with sweetened drinks is a perfect recipe for insulin resistance and over-eating because of their high caloric content and palatability[530].

Elevated levels of fatty acids and triglycerides have been found to be associated with insulin resistance as well[531]. However, the prolonged elevation of insulin and triglycerides is most commonly caused by the combination of carbs and fats that keep the blood sugar jacked up for longer. People who eat more fat tend to also eat a lot of carbs. Someone eating fat on a low carb diet will not spike insulin as much.

In Figure 73 you can see that the combination of carbohydrates with fats raises insulin significantly more than carbs alone. Also, note that protein and fats barely cause any difference in blood sugar or insulin. That comes to show why processed foods are so damaging on the metabolic level as well.

You don't want to ever combine high-carb foods with high amounts of fatty acids because it'll not only increase insulin much higher but also promotes more inflammation, oxidative stress, and metabolic disorders. Ironically, that's exactly the combo you can get from a Happy Meal, Starbucks, McDonald's, or any other fast food joint. Unfortunately, you can still make the same error with healthy foods like frying potatoes in oils, making toast with eggs and bacon, or eating chocolate cake.

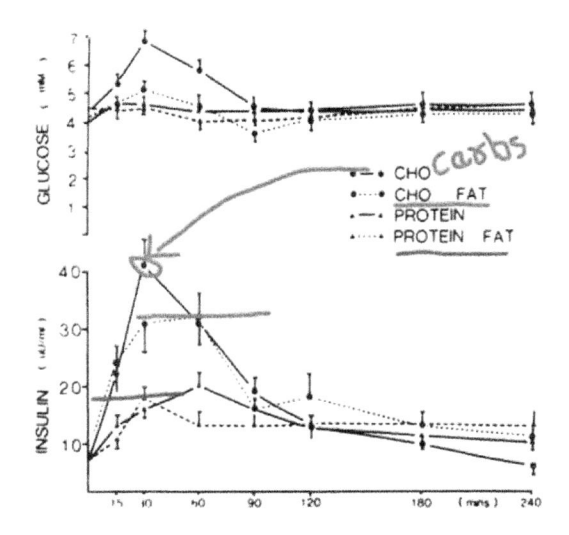

FIG 1. Plasma glucose, insulin, and GIP responses to 50 g carbohydrate ± 50 g fat and 50 g protein ± 50 g fat. Mean ± SEM (n = 8).

Figure 73 Basically, combining carbs with fat will also raise insulin much higher than alone. That's why it's even more dangerous to be combining high fat foods with carbs.

Fructose, which is the sugar molecule of fruit, gets stored in the liver and can stimulate insulin production through a similar mechanism. Excess fructose can be very easily converted into triglycerides that will be rapidly stored as fat. A consistently high intake of carbohydrates and particularly fructose-sweetened beverages contribute to insulin resistance, which has been linked with obesity and weight gain[532]. Natural fruit is slightly less harmful because it contains a bit of fiber and other co-enzymes but it's still very high in fructose sugar that will drive up triglyceride production. For optimal longevity, you wouldn't want to be consuming a lot of fruit every day.

Another contributing factor to insulin resistance is obesity or even just mild adiposity. Visceral fat, which is the type of fat you carry around your belly and organs will continuously release inflammatory cytokines and fatty acids into your system. Abdominal visceral fat is strongly correlated with insulin resistance and type 2 diabetes[533]. That's why consuming excess fructose or carbohydrates drives this entire process – too many triglycerides being stored as visceral fat.

Low carb diets, avoiding processed sugar, and prolonged fasting have been found to be very effective in healing the pancreas and reversing insulin resistance.

- Carbohydrate restriction helps with everything that causes metabolic syndrome, such as high blood sugar, weight gain, elevated insulin, and hypertension[534].
- Diabetics who ate a normal diet were put on a ketogenic one for 2 weeks and they lowered their triglycerides by 35%, dropped total cholesterol by 10%, ate 30% fewer calories, dropped 4 pounds and improved their insulin sensitivity by 75%[535]!
- Compared to a low carb ketogenic diet, a low-fat diet for four weeks has been seen to raise fasting glucose and insulin[536], which is the very opposite to what you want.
- Ketogenic diets have been shown to be very effective at losing body fat in dozens of studies[537].
- Beta-Hydroxybutyrate (BHB) the main ketone body has been found to inhibit histone deacetylases (HDACs) which lowers blood glucose and decreases insulin resistance[538].

If we know that a lot of the modern metabolic diseases are caused by hyperinsulinemia, inflammation and insulin resistance, then it doesn't make sense to purposefully keep our blood sugar and insulin elevated all the time. It also contradicts the information we have about how insulin affects longevity in other species. To live a longer and healthier life, you obviously want to avoid obesity.

However, going very low carb for too long may cause peripheral insulin resistance. Rats fed the ketogenic diet with 95% of calories coming from fat develop hepatic insulin resistance, despite increasing energy expenditure and preventing weight gain[539]. This happens because of the same adaptive response that tries to preserve energy for the brain. In this context, it's not dangerous and not that relevant because the diet itself doesn't require a lot of insulin to clear the bloodstream from glucose. If you're not eating that many carbs, then you don't need extra insulin either. Decreased insulin signaling itself is still a good thing for increased longevity and the insulin resistance induced by carbohydrate restriction is an adaptive mechanism.

After keto-adaptation the body will use primarily ketones and lactate thus decreasing glucose demands but a small amount of glucose is still needed for certain vital organs. If you're already insulin sensitive and lean, then this kind of mild insulin resistance is needed in the context of a ketogenic diet. Without this and gluconeogenesis, your brain wouldn't be able to get the glucose it needs. If you were to be on a ketogenic diet for some time and then took an oral glucose test at the doctor you may potentially score as insulin resistant just because your body wouldn't know how to respond that quickly. You'd have to take a second test after your body has managed to get used to insulin again. To prevent any long-term negative consequences to that, it's still a good idea to occasionally dip in and out of ketosis so your body would maintain its glucose tolerance and become more insulin sensitive again. That's why metabolic flexibility is characterized by both improved insulin sensitivity and glucose tolerance.

Here are more ways of protecting yourself against insulin resistance:

- **Exercise** - Working out and physical activity is one of the biggest determining factors to how insulin sensitive you are[540]. The main cause is thought to be muscle contractions causing the glucose receptor GLUT4 to translocate to the membrane. GLUT4 can improve the uptake of glucose through a distinct mechanism that of insulin[541]. This will lower blood glucose and insulin, preventing hyperinsulinemia.

- **Get Enough Sleep** - Sleep deprivation has been shown to trigger insulin resistance in healthy subjects. Even just a single night of not enough sleep makes you borderline pre-diabetic in the short term. After a bad night's sleep, your glucose tolerance for the next day is going to be drastically lower.
- **Avoid Smoking** - Smoking also induces insulin resistance and causes atherosclerosis[542]. It's causing similar damage to the arteries as does excess glucose in the bloodstream. Eating low carb high fat while still smoking is as bad as eating junk food.
- **Lower Your Stress** - Chronic stress and cortisol are known to raise blood sugar, blood pressure and insulin, which over the long term will definitely lead to insulin resistance[543]. Cortisol literally impairs the uptake of glucose by reducing the translocation of glucose transporters such as GLUT4[544]. It also keeps that stubborn visceral fat around your belly. Mindfulness-based stress-reducing activities like meditation and yoga have been shown to improve insulin resistance[545].
- **Get Enough Sunshine** – Vitamin D deficiencies are associated with insulin resistance[546]. It's an essential steroid hormone that influences every cell in your body, including insulin secretion. Vitamin D is also important for protecting against heart disease. The best way to synthesize vitamin D would be to get it straight from the sun but taking a D3 supplement can also be helpful.
- **Lower Inflammation** - Funny enough, in rats, insulin resistance can be alleviated by fish oil supplementation[547]. This may be partly due to the anti-inflammatory properties of omega-3s that help to fight the inflammation caused by sugar.

At this point, we should move on with the secondary consequence of elevated blood sugar and insulin, which sets this entire party on fire, literally.

Inflammation and Advanced Glycation End Products

Chronically high levels of blood glucose have many detrimental effects on your health. We already know about insulin resistance and the increased risk of metabolic syndrome but hyperinsulinemia also directly damages the body.

Chronic inflammation is connected to most modern diseases, such as insulin resistance, metabolic syndrome, diabetes, and atherosclerosis[548]. In fact, all of those ailments could be umbrellaed under this condition. That's why inflammatory oils, processed food, high fructose corn syrup (HFCS), and high amounts of sugar drive systemic inflammation throughout the entire system.

Advanced Glycation End-Products (AGEs) are these compounds that get formed when sugar molecules react with proteins or fats. AGEs are related to accelerated aging, diabetes, increased inflammation, and mitochondrial dysfunction. AGEs can also interfere with insulin signaling by decreasing insulin secretion, thus promoting insulin resistance[549].

Inflammation from food and AGE formation creates oxidative stress, which increases gut permeability as well. '*Leaky Gut*,' as it's called, allows bacteria, undigested food particles, and unmetabolized toxins to enter the bloodstream and inflame different tissues of the body and lead to obesity[550].

Excess amount of processed carbs and sugars, especially HFCS are linked to inflammation. Even low to moderate sugar-sweetened beverage consumption impairs glucose and lipid metabolism, which promotes inflammation[551]. Better think twice before grabbing for that juice, energy drink, or even diet soda.

It's important to differentiate processed carbs from natural carb foods because they create a different metabolic reaction. In fact, some of the sugars in vegetables and fruit have anti-inflammatory effects[552]. Probably due to their polyphenol and fiber content. However, in the context of insulin and longevity, you don't want to be consuming excess sugar or carbs for nothing.

Even though you may not get diabetes from eating a few apples, the signaling pathways will still be the same with a difference being only in the degree. Additionally, you can gain most of the health benefits of these plant polyphenols from much better sources with fewer carbs, zero insulin, and greater improvement on longevity. If you have to justify eating grapes to get your polyphenols whereas you could get more of them from cruciferous vegetables then you're making a bad trade.

Glycation end products have been shown to promote cancer in animals but not in humans[553][554]. Nevertheless, I think it's still a wise idea to limit your exposure to them because they'll accelerate aging.

When it comes to other foods, then cooking and processing food in general increases the amount of AGEs and other free radicals, such as heterocyclic amines (HAs), and polycyclic aromatic hydrocarbons (PAHs). HAs and PAHs get formed when you cook, grill, fry, or smoke food at high heat. That's why it's never a good idea to char your bacon super crisp or let the vegetables turn too crisp – it causes glycation that's bad for your skin and long-term health.

The highest amount of glycation end products can be found in charred meat cooked over a barbeque[555]. Could it be another reason to limit meat consumption? You definitely don't want to eat BBQ all the time. Fortunately, marinating beef for an hour before cooking reduced its AGE formation by more than half, and marinades can lower HAs in meat by up to 90%[556].

One study found that omnivores tend to have higher dietary AGE intake than vegetarians, but vegetarians actually end up with higher AGE concentrations in their plasma[557]. The authors figured that this was due to the increased fructose intake of vegetarian diets, which induces oxidative stress to the liver. Leafy vegetables also have a lot of PAHs, comparable to the levels in smoked meat even.

AGE formation can be inhibited by the amino acid carnosine, which is found in meat[558]. However, the amounts of carnosine are quite small from natural foods and if you're really worried about it, then you should take a carnosine supplement instead. Either way, AGEs from natural meat aren't that big of an issue, unless you overcook it all the time and expose your food to other free radicals.

The bottom line is this: keep your insulin and blood sugar low most of the time and stimulate mTOR at times you'd benefit from being more anabolic. Then cycle in between periods of being predominantly ketogenic with occasional refeeds and insulin signaling. That's how you'll prevent any dysfunctional insulin resistance or inflammation.

The Case Against Fat

Since Atkins' time, there have been many other low carb proponents trying to make the argument that the key is to avoid all carbohydrates and thus lose weight by entering into ketosis. Insulin and sugar are depicted as the villain because of their nutrient partitioning factors. Whenever you eat carbs insulin goes up and you shut down all fat burning. That may be true to a certain extent but it's not the whole story. We have to keep the context in mind.

There's only a certain amount of glycogen the body can store before it converts carbs into fat. With the right timing of foods and nutrient partitioning, you can cause a completely different metabolic response as the Carbohydrate-Insulin Hypothesis tries to depict. When insulin goes up it stores glucose as glycogen in the liver and muscle cells. Once those deposits get full but insulin remains elevated, insulin will also start promoting the creation of new fat cells. This is called *hepatic de novo lipogenesis* (DNL). Up to that point, the body would benefit from an influx of insulin and carbs. Whether or not it's advisable is a different story but it illustrates the point that not all carbs and insulin will immediately equal fat gain.

Although insulin is the main driver of fat storage, it's not the only one causing weight gain. People who are doing resistance training and intermittent fasting don't have to worry about hyperinsulinemia that much. Despite not eating or being quite glucose tolerant, the body may still end up with a state of insulin resistance or high blood sugar because of cortisol.

Eating any kind of food, whether that be carbs, protein, or fat, will be significantly more obesogenic with high cortisol. Cortisol will inhibit digestion, releases glycogen, raises blood sugar, and spikes insulin. That's why even a „healthy ketogenic low carb high-fat meal" can be damaging to you in the long term. What's more, cholesterol and other dietary fats are more prone to become oxidized if you consume them with high cortisol and insulin. Therefore, the underlying issue isn't as much carbohydrates or fats but more like stress-induced inflammation and hyperinsulinemia.

Furthermore, insulin does not only rise in response to dietary carbs or cortisol but also by non-caloric sweeteners. Since the 1970s, there's been a gradual increase in the consumption of sweeteners and added sugars. Partly this may have been due to the low-fat guidelines, which made food companies increase the palatability of foods by adding extra sweetness to them. Unfortunately, there's evidence to show that artificial sweeteners may promote diabetes, insulin resistance, and obesity. The reason has to do with how the sweet taste stimulates a gut receptor called *Glucose-Dependent Insulinotropic Polypeptide* (GDIP), which is linked with weight gain[559]. That's something to keep in mind the next time you grab a „healthy" low carb meal bar...

Meal timing is as important as choosing what to eat. Of course, keeping blood sugar and insulin elevated throughout the day several times isn't a good idea. It'll promote fat gain and diabetes. However, insulin has an important role in storing energy and replenishing glycogen stores. In some cases, like after a resistance training workout, you'd actually benefit from having more carbs and insulin. That's why you see physically active people being able to get away with eating more food and carbs without getting fat. Of course, they burn a lot more calories throughout the day but they're also more hormonally optimized to tolerate those carbs. Saying that carbs make you fat to a bodybuilder who has to eat over 4000 calories a day to gain weight is falling upon deaf ears. Eating carbs and spiking insulin may be more detrimental at some times than at others.

A good example comes from the circadian rhythms, which are your body's natural biological cycles connected to the light cues from the environment. Essentially, daytime favors better glucose tolerance and insulin sensitivity because the body's metabolic processes are supposed to be more active whereas at nighttime the opposite is true. That's why humans are diurnal creatures, which means we're active during the day and we sleep at night. More about the circadian rhythms in Chapter XXII. Blue light exposure from artificial light sources at night is shown to promote insulin resistance, weight gain, and diabetes[560]. The reason has to do with increased cortisol induced by the highly stimulating wavelengths of most blue light sources.

Other factors that determine insulin release include dietary fiber, protein, fermentation, the addition of vinegar, the thermic effect from spices, gut receptors, consistency and satiety signaling. It's more likely that a person who's overweight already suffers from leptin resistance and insulin resistance. The best treatment I can think of in those situations would be to follow a very low carb ketogenic diet not the low-fat high carb one.

Is Fat Good?

Low carb diets are often depicted as fads in mainstream medicine and the news, promising only short-term weight loss but long-term health consequences. It's true that keto and Paleo have been shown to make the person lose a lot of weight quite fast as long as they follow a well-structured program. However, the idea that higher dietary fat is going to be more detrimental doesn't hold much water either. With that being said...

It's important to remember that studies linking saturated fat and cholesterol with heart disease didn't come out of thin air. The association is primarily created by the combination of high-carb and high-fat foods ala the processed fast food industry.

That's a critical error you must avoid – DON'T COMBINE CARBS WITH FAT! It's going to result in much higher insulin response and AGE formation than if you were to eat that fat or protein in a low carb meal. This is something you mustn't take for granted even if you think you're eating a healthy ketogenic diet. If you end up consuming rancid fats, oxidized cholesterol, glycated proteins, or excessively large amounts of extra calories, then, rest assured, you're not doing your health any favours.

The combination of carbs and fat is also the most appealing yet least satiating one. There's a reason you can't seem to feel satisfied with just eating a single donut. After you take the first bite, you have to eat the god damn box.

As a kid, I sometimes ate white bread sandwiches with butter and table sugar. Not the best thing for children but nevertheless...If I only ate the bread with sugar it didn't taste that good. Once I spread some butter below and then added the sugar my tastebuds were stimulated in a completely different way. The bread combined with butter and sugar just melted in my mouth whereas just the bread with sugar tasted like cardboard.

When you eat either fat or carbs from a whole food source, then this overdrive of insulin and cravings doesn't happen. Not a lot of people find pure lard or syrup appealing by themselves. Only after they get cooked with other ingredients do they become tastier.

Although saturated fat from animals has been recently shown to be not that big of a deal, it's still not something you'd want to make the staple of your diet. Saturated fat is just a source of energy that doesn't have that significant benefit either. It does help with cell membranes and neuronal functioning but you'd gain a whole lot more of that from other healthier fat sources like fish or eggs. There are plenty of populations who get just as much animal fats they need while still thriving and living long. It's only in the Western countries where people tend to fall into the extremes – high carbs all the way, high fat or go home, or even worse, the combination of both.

Some people are also less suited to be consuming saturated fat. This is determined by a specific gene called APOE, which has 3 types (APOE2, APOE3, APOE4). The specific APOE gene we have instructs our body on how to make apolipoprotein E, which combines fatty acids to create lipoproteins. Lipoproteins are used to transport triglycerides and cholesterol around the blood.

If you have pre-dominantly APOE4 genes, then you're going to do worse with increased saturated fat and cholesterol intake. Instead, you'd want to be consuming more monounsaturated fats then. APOE4 carriers are said to be at a 20% increased risk of Alzheimer's disease as well. Having APOE2 makes you more suited for a low carb high-fat diet and APOE3 is suitable for both types of diets. That's why you'd want to get tested before going all in on a single way of eating.

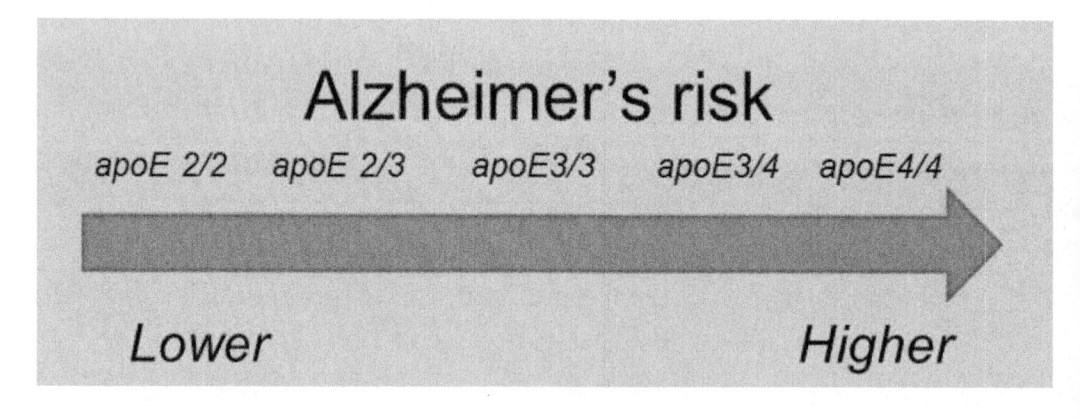

Figure 73 The APOE gene and Alzheimer's risk

I don't know about you but I don't want to be consuming any more fat than my body needs. Any excess can't be good for you in the long-term. Also, we don't know what the future implications might be. It's a much wiser and more evolutionary viable strategy to not be placing all your bets on a single food group or one macronutrient. You can get away with a moderate intake of fat as well as carbs as long as you practice the other habits of Metabolic Autophagy. The context matters a whole lot more.

That's why I'm not advocating a fully high-fat diet. First of all, it's not metabolically suitable for some people. Secondly, it still sustains the potentiality of causing some cardiovascular errors. Thirdly, it's not necessary to be consumed in copious amounts. Fourth, you would benefit your body composition as well as longevity much more by lowering your fat intake slightly. And fifth, the evolutionary trade-off favours the process of deduction. Here are some additional factors to consider that make the case against too much fat:

- **Don't Combine Fats and Carbs** – The single most important thing for your nutrition. Both a low carb as well as a high carb diet can be equally good and bad, depending on the situation. You can be healthy and live long on both diets if you keep the macronutrients separate.
- **Don't Be Stressed Out** – Cortisol will still make you gain weight and raise insulin. In fact, eating a keto diet with chronic stress can be as detrimental as eating fast food because you'll oxidize the fats and cholesterol with elevated blood sugar. I don't have any evidence to prove this but it makes mechanistic sense.
- **Avoid Inflammation** – Whether that be from processed food, charred meat, oxidized oils, trans fats, or the AGE formation of eating carbs and fats together. This will set the entire thing on fire...
- **Eat Whole Foods** – Processed food promotes obesity for a reason – it's easier to overconsume and it's less satiating. You may accidentally end up with eating more calories than you need, which is another driver of insulin resistance and fat storage. More on this in the next chapter.

This book may come off as promoting a low carb high fat ketogenic diet. However, I'm not trying to enforce any food choices on anyone and am just sharing certain principles of best practice when it comes to meal timing, food combinations and such. There are also many ways to do a high carb low fat plant based diet the right way with this program. You just have to adjust the food groups to your preference and take into other considerations when it comes to fasting.

A high carb low fat plant based diet can be effectively used to treat atherogenesis and prevent against coronary heart disease[561]. The reason has to do with lower LDL cholesterol, increased polyphenols from certain plants, and activation of macrophages. Unnecessary cholesterol isn't something to yearn for and you definitely want to avoid it in the context of hyperinsulinemia and inflammation. Fat, cholesterol, and meat aren't going to kill you but it doesn't mean you can eat them in unlimited amounts. There's always some potential trade-offs we don't see right away and usually the body functions best when in homeostasis.

Vegetables, on the other hand, are also incredibly nutritious and healthy. They're probably one of the more safer things you can eat on a daily basis without running into serious health problems. That's why, although a lot of the calories you eat come from animal-based foods, most of what you eat in bulk may still end up coming from plants and veggies. How much and what to actually eat will be covered in Chapter XV but the general idea is that make the majority of your foods come from low carb veggies and leafy greens.

One thing for sure, you shouldn't fall into the extremes on any diet you follow, whether that be paleo, keto, carnivore, or vegan. For optimal health, we have to take the best of everything and apply them in the right context, which more often than not doesn't include consuming an excess of any macronutrient.

The general principles of metabolic autophagy will still apply to any nutrition plan – eat a lot of plants and vegetables for the polyphenols and antioxidants, don't eat too much meat and protein because of the mTOR stimulation, practice daily fasting with minimal eating frequency, stay low carb most of the time and then cycle with carb refeeds.

The simple idea of carbs making you fat is equally as alarming as the one of fat giving you heart disease. Whenever you hear people making bold statements like that without delving deeper into the context you should run. I mean...be very careful with all-encompassing claims about how certain things work in regards to nutrition and physiology.

The human body is such a complex thing that explaining it away with a simple calories in VS calories out model or a black and white idea of mTOR discredits the complexity of reality. Large epidemiological studies are also incomplete in giving you as an individual a lot of practical value. You have to understand the principles of how the body works and then make your adjustments based on the context of the situation.

Chapter XII
WTF Should I Eat

"One cannot think well, love well, sleep well if one has not dined well."

Virginia Woolf

There's so many misconceptions and opinions about nutrition that it's overwhelming. Even though we have a lot of data and different concepts of dieting at our fingertips, most people still find themselves baffled as to *WTF should I eat then?*

Truth be told, there is no one-size fits all answer because everyone's metabolic conditions and energy requirements are different. You won't find a plan you could stick to for the rest of your life because change is inevitable. What you can do instead is teach yourself the core principles of human metabolism and physiology. You'd want to become a true alchemist of anabolism and catabolism, which can then be used according to the situation.

This chapter will give you the main ideas and tenets of what foods you should eat and what not to. It'll give you the 80/20 of diet.

What Humans Evolved to Eat

The first humans, the *Australopithecines and Homo habilis*[562], appeared around 4 million years ago and their diets included primarily plants but also a lot of meat[563]. Basically, they were scavengers who ate fruit, tubers, and small game with occasional remains of large animals.

According to the Expansive Tissue Hypothesis posed by the anthropologists Leslie Aiello and Peter Wheeler, the metabolic requirements of large human brains were offset by a corresponding reduction of the gut[564]. As our stomachs got smaller, our neocortices got larger. This was made possible by getting more calories from less food and not having to spend that much time searching for it. See Figure 74 The Expansive Tissue Hypothesis - Smaller Guts Lead to Larger BrainsFigure 74.

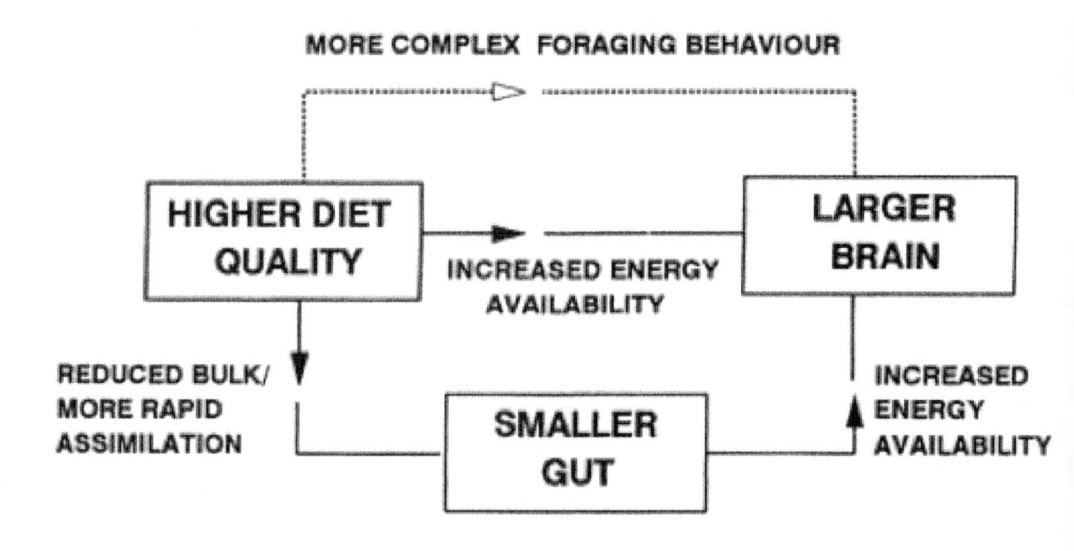

Figure 74 The Expansive Tissue Hypothesis - Smaller Guts Lead to Larger Brains

Before the advent of cooking, it was quite difficult for early hominids to extract that many calories from just tubers and plants. The occasional carcass here and there enabled them to get more of the essential fat-soluble nutrients from bone marrow and ligaments. After we started using fire, we were able to get more nutrition from both cooked starches and meats. The Expansive Tissue Hypothesis doesn't directly indicate that humans got smarter because of hunting *per se*. We were foragers and scavengers before we could hunt.

Around one million years ago, Homo Erectus appeared on the scene and learned the ability to hunt big game. His life was primarily centered around hunting[565], which led to the development of anatomically modern humans about 200 000 years ago. *Cro-Magnon man*, as he was called, was compelled to inhabit many unpopulated regions of the world thanks to a meat-based diet. The disappearance of most large animals such as the mammoth, wild ox etc. wasn't because of climate change but due to people hunting them to extinction[566,567]. That's why you have people like the Inuit living in such barren regions – they followed the animals and survived.

Homo Sapiens has been around for hundreds and thousands of years. Over 90% of that time has been spent hunting and gathering. The agricultural revolution happened about 10 000 years ago and is such a new introduction to our evolutionary lineage.

Before modern agriculture and industrially processed food, we ate primarily a moderate-to-low carb, high fat, high protein diet with a very high nutrient density and plenty of fiber. Even when humans turned to agriculture, a large proportion of the crops was fed to the cattle for rearing their meat, as is today. There hasn't been any 100% vegan aboriginal or even agricultural society because animal foods are much more nutrient dense than just plants. They provide all the essential amino acids, minerals, vitamins, and fats needed for sustaining life. Even vegetarian societies incorporate some dairy or animal fats to cover their essential nutrients.

This would also fluctuate between the seasons and faunal mobility. In the summer and early autumn, foragers would've been exposed to more carbohydrates from fruit, berries, and vegetables. During winter they'd be eating more animal foods like meat, fish, fats, and very little plants. These cycles would replicate the cycling of anabolism and catabolism as seen in nature.

A 2000 publication found that on average the macronutrient ratios of hunting and gathering tribes fall somewhere between 19-35% protein, 22-40% carbs, and 28-58% fat[568]. These numbers may vary hugely because certain populations have access to different types of wild game and vegetables. Naturally, an equatorial society is going to be consuming a lot more fruit and tubers, whereas an arctic one has to primarily focus on fats and meat.

- **The Kitavans of Papua New Guinea eat a very carb-rich diet with tubers, yams, fruit, coconuts, and fish.** No dairy, no grains, no oils, no sugar, no alcohol, or other beverages. In contrast to the Western diet, they have virtually no cases of diabetes, obesity or stroke[569]. There are other high-carb tribes like the Hadza of Tanzania, and the Bantu in Africa who are equally as healthy. More evidence to show that you can't blame it all on the carbs…

- **The Inuit in the arctic climate consume primarily caribou, seal meat, wild salmon, whale blubber, and very few berries.** Their diet is high protein, high fat, and low carb but they're not fully in ketosis because of a genetic disorder.

- **Other animal-based bands are the Masai in Africa, who eat primarily cattle, drink milk and cow's blood.** The Hadza also get a lot of their calories from meat during certain seasons as do the Hiwi in South America.

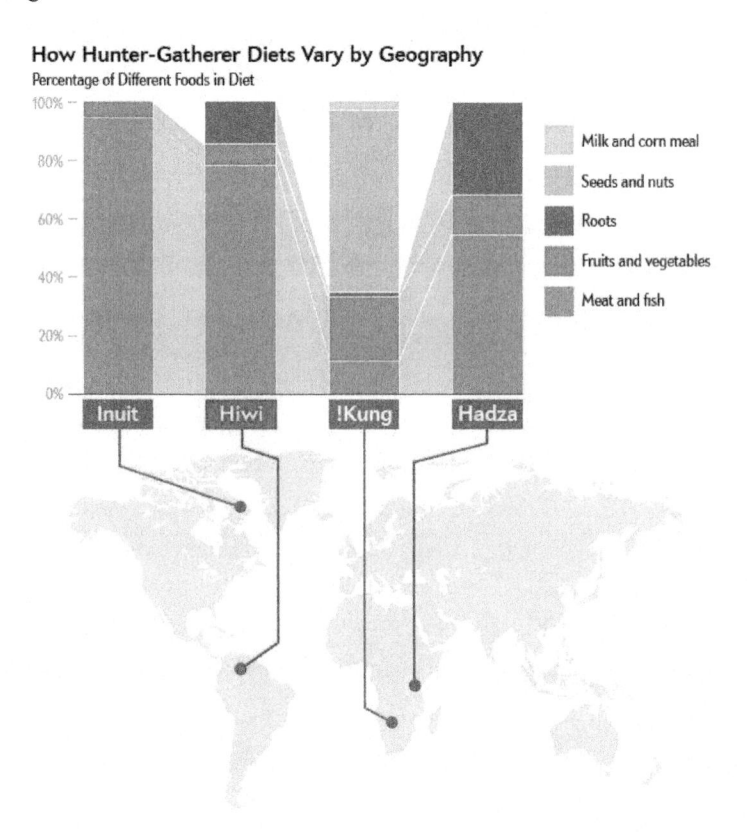

Figure 75 Percentage of different foods in hunter-gatherer diets

Although these hunter-gatherer diets aren't very diverse, they still have enough variety as to cover the essential micronutrients.

The „Natural Diet" Fallacy

However, nature doesn't care about your longevity and well-being. It only wants you to survive long enough until you could reproduce and carry on your genes. What's natural to eat may not always be the best for you. A few examples.

- **Poisonous berries, mushrooms, and leaves are growing in the forest but they're lethal in high doses.** Likewise, certain compounds in grains beans and vegetables actually cause a lot of digestive issues and inflammation when eaten in excess. There are even some plants that other animals can digest but humans can't etc.
- **Some „un-natural" foods such as MCT oil or olive oil can be healthy for you.** You have to change the composition of the fruit from its original form into a processed one but these are one of the healthiest fats for you if you consume them in moderation.
- **Even though most modern humans evolved on a meat-based diet, it doesn't mean that it's most optimal for longevity.** Although hunter-gatherers are devoid of nutritional diseases and degeneration, they don't exhibit exponentially long life-spans either. Partly due to the harsh living conditions of their environment but I believe the aspect of constant mTOR stimulation and not enough deliberate autophagy may play a role in this. In the contemporary setting, high eating frequency combined with a lot of meat probably isn't good for your health.
- **Eating high amounts of fruit and honey during certain seasons as hunter-gatherers do isn't ideal for metabolic health and longevity either.** They deliberately gorge themselves on sugar and carbs as to gain fat for the coming winter. However, that also induces mild-insulin resistance in the short-term. This kind of behavior on a habitual basis isn't probably optimal for increasing life-span because of the insulin signaling. In the modern world, we don't have such selective pressures from the environment and we have access to more food year-round. We're not pressured into excessively gorging ourselves and storing every calorie in sight.

That's why I think we shouldn't romanticize the idea of eating like a hunter-gatherer because it doesn't include the whole picture. If you're so adamant about following the „original" human diet, then you should also incorporate the other practices of foraging, such as scavenging for animal remains, eating nose to tail, sleeping in caves, fighting off predators, exposing yourself to the elements, getting bitten by parasites, tribal warfare, and sometimes starving for days.

Another problem with eating natural is that certain foods can actually do you more harm than good. It turns out even some „healthy" fruits and vegetables can be quite damaging for most people.

The Plant Paradox

The problem is that plants don't want to be eaten. They have their own evolutionary pressures and incentives that make them as self-preserving as any other species. Because plants can't run away from predators the same way animals can, they protect themselves by producing toxic chemicals and compounds.

One of the most common plant anti-nutrients are lectins, which are sticky proteins that attach to other food molecules and cause inflammation. This can lead to leaky gut syndrome, brain fog, autoimmune disorders, weight gain, and other inflammatory conditions. They can even contribute to atherosclerosis due to intestinal permeability and arterial damage.

Lectins can be found in seeds like sunflower, cashews, peanuts, beans, legumes, grains, such as wheat, barley, oats (gluten is a lectin), nightshades, like tomatoes, eggplants, cucumbers, peppers, potatoes. That's why you want to limit your consumption of these foods because they're pro-inflammatory. Even if you're not gluten intolerant or have no symptoms of leaky gut, lectins will still do your body more harm than good. They're simply not worth it because you can get more nutrients from less harmful foods.

Steven Gundry M.D. is a former heart surgeon who wrote a book *The Plant Paradox* where he talks about the dangers of lectins in our modern diet[570]. It's true that after the introduction of lectins into the human food system about 10 000 years ago, people's health has dramatically changed for the worse. Before agriculture, there weren't a lot of the inflammatory diseases we deal with today, such as arthritis, cardiovascular disease, and cancer. Gundry himself lost over 70 pounds, cured his arthritis, and metabolic syndrome by eliminating lectins from his diet.

However, it's not that lectins are inherently bad – it's just the accumulation of too much gastrointestinal stress from lectin-rich foods, other allergens, and the chronic stress everyone's experiencing. That tomato is literally a cherry on top of the cake that sets the entire thing on fire, especially if it's a cherry tomato...

FODMAPs are another food group that may cause digestive issues. It stands for „Fermentable Oligo-, Di-, Mono-saccharides And Polyols," which are short-chain carbohydrates that don't get absorbed by the small intestine that well. Because of that, they can cause bloating, autoimmune conditions, irritable bowel syndrome, and leaky gut. FODMAP foods aren't the cause of these issues but a low-FODMAP diet can help to improve the symptoms. FODMAP foods include wheat, rye, barley, onion, garlic, artichokes, cabbage, beans, lentils, tofu, tempeh, and some fruits like apples, apricots, avocados, cherries, plums, and chicory.

I'm not saying that you should avoid all lectins and plant phytonutrients for the rest of your life because that can create another situation of hypersensitivity to them. Hunter-gatherers would also forage a lot of wild plants and other edibles that had some medicinal benefits despite their toxicity. That's why you do want to eat the other vegetables like cruciferous and then have the questionable ones less often. Different culinary techniques like sprouting, soaking, fermenting, and plain simple cooking can also reduce the amount of lectins in them.

People in the modern world have simply become way too soft and domesticated that they can't handle even just a little bit of digestive strain. The perfect example is antibiotic resistance and lack of microbial diversity in the gut. Children who aren't exposed to bacteria from dirt, animals, grains, and other sources are much more likely to develop autoimmune disorders later in life. In fact, C-section newborns are more prone to suffer from all chronic diseases. Another reason to get your hands dirty every once in a while.

The entire concept of hormesis and stress adaptation applies to both environmental conditions as well as nutrition. You want to be consuming these plants and herbs that have a mild hormetic effect so that your body could get stronger and more resilient. What makes a poison deadly is the dosage.

Mithridates and Hormesis

Mithridates VI of Pontus was the king of Pontus and Armenia Minor at about 120-63 BC. His father was poisoned and he thought that their mother ordered small amounts of poison to be added to the king's food to slowly kill him off. The man fled into the wild where he began to self-ingest non-lethal amounts of toxins and mixing them together into potions as to make himself immune to poisoning. Unfortunately, when he was captured by Roman conquerors, he tried to kill himself with poison but failed because he had built an immunity towards it. Talk about bad luck...

Mithraism is the practice of developing resistance to certain poisons by consuming them in very small non-lethal amounts. You can find examples of toxicity resistance in the ability to handle more alcohol without passing out, being able to digest lactose, gluten, FODMAPS, and not feeling tired after consuming too much food.

You don't want to be eating lectins on a habitual basis because they do indeed cause a lot of gastrointestinal problems and inflammation. It may happen in healthy people without them even knowing it. However, introducing these compounds into your diet every once in a while as hormetic conditioning is a good idea. You don't want to become hypersensitive to these allergens just because of avoiding every potential danger and being too strict.

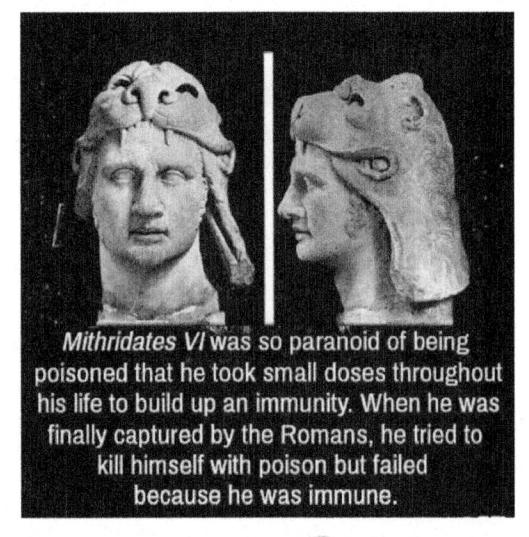

Mithridates VI was so paranoid of being poisoned that he took small doses throughout his life to build up an immunity. When he was finally captured by the Romans, he tried to kill himself with poison but failed because he was immune.

Ya played yourself 💀 ■ ■ Hey guys ✌ I know this isn't scary but I laughed really hard when I saw this so I thought id share it.

The most common allergens include dairy, eggs, fish, nuts, seeds, gluten, grains, soy, mustard, celery, peanuts, shellfish, sulfites, strawberries, pollen, and the common suspects. Ideally, you don't want to be intolerant to any of these foods and have a gut that's capable of handling everything.

Of course, there are a lot of exceptions to the rule. You definitely shouldn't eat raw beans or legumes because they can actually be dangerous. Likewise, there are a lot of lethal berries and plants. Someone who's already suffering from autoimmune disorders or inflammatory disease should also not treat themselves with hormesis.

In nature, foragers are very much subjects of their environment and the food they can find there. Being a part of the technological civilization we can consume many other kinds of nutrients that were previously inaccessible due to geographical issues. That's an amazing thing and it'll allow us to optimize our diets a lot better.

So, the goal of an optimal diet for modern humans living at this day and age shouldn't be to replicate the nutrition of ancient hunter-gatherers. Of course, you should eat a lot of the nutritious and nutrient dense ancestral foods but you also want to incorporate other tricks of the trade related to Metabolic Autophagy.

Essential Nutrients

An essential nutrient is something that the body cannot synthesize itself and thus it needs to be derived from diet. In humans, there are 9 amino acids, 2 fatty acids, 13 vitamins, and 15 minerals that are considered essential nutrients[571].

- **The minimum daily protein requirements** are 0.8g/kg or 0.36g/lb of bodyweight, which for an average adult who weighs around 100-200 pounds is roughly 40-80 grams of protein. However, that's for covering your bare nitrogen balance and I don't think it's optimal. For muscle growth and healthy aging, you definitely need more than that. Higher than minimal protein intake has many benefits such as increased preservation of lean body mass and weight loss. Out of the 20 amino acids, 9 cannot be synthesized by the body itself and thus need to be obtained from diet. They are phenylalanine, valine, threonine, tryptophan, methionine, leucine, isoleucine, lysine, and histidine.
- **Daily dietary fat intake** is suggested to be around 15%, including 2.5% as Linoleic Acid (LA) and 0.5% as Alpha-Linolenic Acid (ALA)[572]. Reference intake value for LA is said to be 10g and for ALA 2g, which on a 2000 calorie diet can be covered with about 20-30 grams of dietary fat. DHA and EPA are conditionally essential for development and growth, which practically makes them essential. Although LA and ALA can be converted into DHA, it's not that

effective of a process. It's recommended to get a minimum of 250-500 mg-s combined EPA and DHA per day[573].

- **Essential vitamins are** Vitamin A, B1, B2, B3, B5, B6, B7, B9, B12, Vitamin C, Vitamin E, Vitamin K, and Choline.
- **Essential minerals for humans are** Calcium, Cobalt, Chloride, Chromium, Copper, Iodine, Iron, Magnesium, Manganese, Molybdenum, Phosphorus, Potassium, Selenium, Sodium, and Zinc.

Conditionally essential nutrients can be synthesized endogenously but they're in most cases insufficient. They become more essential in conditions such as pregnancy, premature birth, malnourishment, childhood growth, healing, and certain diseases. That's why you don't really want to put kids or old people on nutrient deficient diets in hopes of making their body do extra work by converting some foods into essential nutrients.

Non-essential nutrients are not necessary for survival and they can either have a beneficial or a toxic effect.

- **Carbohydrates and glucose aren't essential because the body can shift into ketosis and use ketones instead.** The brain and other vital organs do need a very minuscule amount of glucose for optimal functioning even after becoming keto-adapted. However, as we've mentioned before, the process of gluconeogenesis can create that glucose from dietary fat and protein intake so carbs aren't needed. Nevertheless, this may not be optimal all the time, which is why we'll be using a more cyclical approach.
- **Dietary fiber isn't essential because humans can't digest it.** The emphasis on eating a lot of fiber comes from the idea that it helps with bowel movements, feeds the gut bacteria, and lowers cholesterol. Although fiber isn't essential, it's still advisable to eat some vegetables, especially cruciferous and sprouts for the sulforaphane and anti-cancer benefits. Fiber's not meant to be digested by us but by our microbiome. When the bacteria in our gut eat fiber they produce short-chain fatty acids (SCFA), such as butyrate, which will heal the intestinal lining. However, too much fiber and vegetables can cause digestive issues, bloating, and constipation, which is why you want to aim for a minimal effective dose.

Fortunately, butyrate can be gained from animal fats as well, such as butter, tallow, and meat but to get the other pre-biotic SCFAs you'd want to eat some plants as well.

- **Phytochemicals and phytonutrients are non-nutritional parts of plants.** They're not essential for survival in humans but they help the plants survive harsh conditions and protect against predation. However, those same compounds can have a beneficial hormetic effect by making our own bodies more resilient. Different polyphenols, flavonoids, resveratrol, lignans, and catechins have all been shown to have great benefits on longevity and metabolic disorders. However, what makes a poison deadly is the dose. That's why too much of anything will still be bad.

- **Alcohol is a non-essential non-nutrient that still has calories.** It means the body can't get any nutritional value from alcohol other than the empty calories. Now, the hormetic effect is something you may benefit from as certain spirits can fight off infections and promote ketone production even. However, the dose is probably quite small and you shouldn't be drinking every day. You definitely don't want to get hammered or even seriously intoxicated because you'll do more damage than good. Instead, one shot of vodka or a glass of red wine a few times per week is probably the minimal effective dose. Different kinds of vinegar also have small amounts of acetic acid and alcohol but they're great for blood sugar control and insulin regulation.

Recommended Daily Allowances for the Essential Vitamins and Minerals

Here's a chart for the Recommended Daily Allowances for all the essential vitamins and minerals:

NUTRIENT	Average RDA	Upper Limit
Vitamin A	700-900 mcg	3000 mcg
Vitamin C	75-90 mg	2000 mg
Vitamin D	600-800 IU	4000 IU
Vitamin K	120 mcg	Not Established
Vitamin E	15-22 mg	1000 mg
Vitamin B1 (Thiamin)	1.2 mg	Not Established

Vitamin B2 (Riboflavin)	1.3 mg	Not Established
Vitamin B3 (Niacin)	14-16 mg	35 mg
Vitamin B5 (Pantothenic acid)	5 mg	Not Established
Vitamin B6 (Pyridoxine)	1.3-1.7 mg	100 mg
Vitamin B7 (Biotin)	30 mcg	Not Established
Vitamin B9 (Folate)	400 mcg	1000 mcg
Vitamin B12 (Cyanocobalamin)	2.4 mcg	Not Established
Calcium	1000-1200 mg	2000-2500 mg
Choline	425-550 mg	3500 mg
Chloride	1800-2300 mg	3600 mg
Chromium	35 mcg	Not Established
Copper	900 mcg	10 000 mcg
Fluoride	3-4 mg	10 mg
Iodine	150 mcg	1100 mcg
Iron	8-18 mg	45 mg
Magnesium	300-450 mg	500 mg
Manganese	1.8-2.3 mg	11 mg
Molybdenum	45 mcg	2000 mcg
Phosphorus	700-1250 mg	3000-4000 mg
Potassium	4700 mg	Not Established
Selenium	55 mcg	400 mcg
Sodium	1200-1500 mg	2300 mg
Zinc	8-11 mg	40 mg

Keep in mind that these are your average values for the average person. There are huge variations between populations, ethnic heritage, lifestyle habits, and dietary preferences.

For instance, there's some evidence showing that the need for vitamin C increases only in diets of glucose based metabolism. In fact, ascorbic acid and glucose compete for cellular transport[574]. High levels of blood glucose inhibit the uptake of vitamin C because both of them use the same membrane transport chain and because glucose is a much more prioritized nutrient.

Also, the required intake of sodium will also fluctuate depending on your levels of physical activity, blood pressure levels, hydration levels, and what phase of the diet you're at. Use them as guidelines for what's the bare essentials. In the later chapter, I'll talk about what would be the optimal doses for all the relevant nutrients and supplements.

But now let's carry on with nutrient density, which is one of those topics that doesn't get discussed that often.

Nutrient Density

The modern diet includes a whole lot of calories but not a lot of nutrition. In fact, most of the calories you can eat at a fast food restaurant or the supermarket are empty – they have virtually zero benefits on your long-term health and longevity. They don't even significantly impact muscle growth as the macronutrient ratios are heavily in favor of making you more obese. There's hardly any protein or amino acids but instead a lot of processed carbs, sugar, and fat, which is a recipe for insulin resistance and diabetes.

Diets with low amount of nutrients but high caloric density such as the darlings of the processed food industry don't lead to full satiety because of their poor nutritional value. It doesn't matter how many calories you eat, you'll still be hungry and malnourished if you fail to get enough of the essential nutrients.

On the Standard American Diet with high amounts of processed carbs, grains, and low-fat meat, and dairy, you'd have to be consuming copious amount of calories to get your essential nutrients. Probably over 5000 calories to cover your basis, which kind of explains why people are so ravenous on these low nutrient foods.

It's thought that humans prioritize protein when regulating food intake[575]. This is called The Protein Leverage Hypothesis, which is the idea that satiety is mostly regulated by protein. You eat until you get sufficient protein, which in the majority of cases falls somewhere between 20-30% of total calories.

Most processed foods and ingredients are low in protein, high in carbs, high in fat, and engineered to increase palatability. They're literally designed by scientists to make you overeat because you're not getting enough protein for satiety, no micronutrients, and way too many over-stimulating sugars and other ingredients that make you lose your sanity. Food manufacturers have a financial incentive to replace protein with cheaper forms of calories and manipulate with the other tastes. Can't blame them... high-quality nutrient-dense food, especially meat and protein, is more expensive but it's what your body runs best on.

However, the protein leverage hypothesis has a few flaws because you can eat a lot more protein than your body needs and still stay hungry. The term 'Rabbit Starvation' describes a situation wherein a person in the wild would starve to death if they only ate lean proteins such as rabbits, very lean game, or just chicken breast without enough fat for calories. This would lead the body becoming malnourished and cannibalize itself, hence the starvation.

A diet with low carbs and low fats is a recipe for malnourishment and starvation. That's what a lot of fitness competitors tend to do when dieting – they eat only broccoli, egg whites, and white fish with low carbs, low fat and high protein to really lose that last bit of fat. In the short term, it can work but I can't imagine living in such a condition for long.

A much more reasonable thing to consider would be a protein + nutrient density leverage hypothesis that prioritizes the intake of essential nutrients. Protein's a component but so are fats, vitamins, and minerals.

In general, a good ratio for protein is about 15-35% of your daily calories. The average Western dieter, unfortunately, consumes between 11-18% protein, which leaves them unsatiated as we can see. More about optimal protein intake during the anabolic and catabolic cycles in Chapter XV.

The critical component to eating just the right amount of food for your body's needs is satiety. It's the feeling of fullness and satisfaction – you feel that you've gotten all the necessary nutrients and are satisfied with the meal. This will also keep your body and brain energized for hours, potentially days. An unsatiating meal has empty calories that leaves you wanting for more. It's what the entire processed food industry is based upon.

Getting a bunch of empty calories will leave you malnourished and unsatiated. However, there's also the danger of getting too many nutrients but not enough satiety. For instance, MCT oil, whey protein, and dextrose powders are quite dense in nutrition as well as calories but I think you don't feel very satiated after consuming them. Likewise, a multivitamin that's said to cover all of your RDAs 300% doesn't fill you up either.

In order to maximize nutrient density without over-consuming calories, you'd want to also prioritize the satiety factor of what you eat. So, what are the most satiating foods?

The satiety index of common foods was put to the test in a 1995 The University of Sydney study. Subjects were fed 240 calories (1000 kJ) of 38 different foods[576]. See Figure 76. Their perceived level of hunger was measured every 15 minutes and the final score was based on how much they ate at a buffet 3 hours later.

It turns out that the most satiating foods were potatoes, porridge, fish, red meat, and some fruit. The lowest satiety factor was on hyper-palatable foods like cake, donuts, chocolate, bread, pasta, cookies etc.

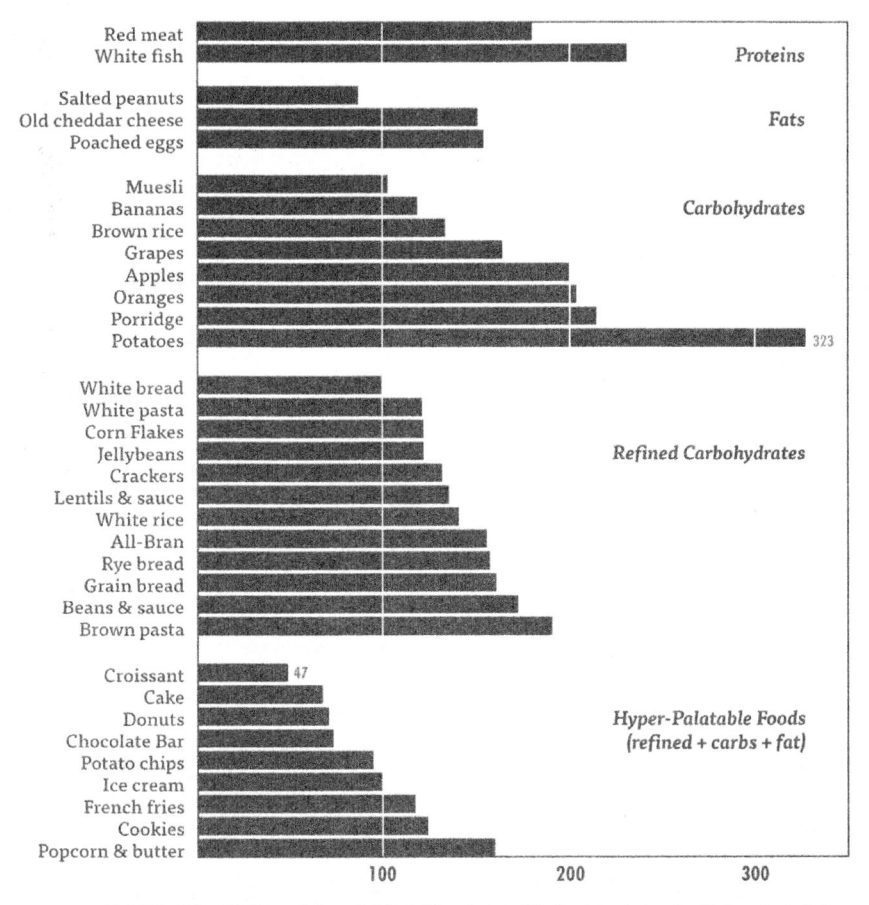

Holt, S.H., Miller, J.C., Petocz, P., Farmakalidis, E. (Department of Biochemistry, University of Sydney, Australia.) "A satiety index of common foods." European Journal of Clinical Nutrition, Volume 49, September 1995, pages 675-690.

Figure 76 The Satiety Index of different foods

As you can see, potatoes are one of the most satiating things but if you cook them in oils and make French fries you'll greatly lower the satiety index. The most satiating food groups are fibrous vegetables and protein that not only fill you up but also satiate you faster. Likewise, a high carb whole food based diet with not a lot of added fats can be quite satiating and healthy because of that same reason. This puts the calories in vs calories out module into a much bigger perspective. It's much easier to over-eat starch and fat because, in nature, you rarely come across hyper-palatable foods that combine both carbs and fats. That's why these primal urges make the person more prone to overconsume these combinations.

Sensory-Specific Satiety describes a hedonic adaptation that occurs when consuming certain types of food. Essentially, if you eat only a single thing with its own taste and consistency, for instance just potatoes, then you're going to get fuller faster than if you were to combine it with other ingredients. The rationale is that if you're only consuming potatoes then you get used to its taste quite fast and that decreases your motivation to continue eating.

If you've ever tried to eat a bunch of any single food, then you can probably relate to how it eventually makes you sick. Even with some junk food – you can eat several bags of chips before you get sick but after a while, you'll still feel disgusted by it. The same principle applies to other foods. Your taste buds tend to tolerate only a certain amount of a single type of sensory stimulation, which then makes you feel satiated. However, if you introduce another kind of taste into the mix while eating, then you'll regain the desire to eat because it provides a novel sensory experience. As the saying goes: *„there's always room for dessert..."* That's so true because even if you eat a bunch of potatoes, you can still eat something with a different flavour profile. Up until that point, you feel like you can't eat anything anymore but after that first bite, you're back in the game. This is a crucial thing to remember when trying to be healthy and not overconsume empty calories.

Whenever you're cooking your meals or planning a diet, you'd want to limit the amount of sensory-specific stimuli in that meal. If you're having steak, then just have the steak with some vegetables or eggs. Don't add a bunch of different sauces and desserts because they'll keep your taste buds constantly stimulated. Your brain will go: *„Oh, this is something new! We better keep eating because it might give me some nutrients for the* *future*". In most cases, you're not actually hungry – your sensory cortex is just getting hijacked by these novel tastes and keeps motivating you to continue eating.

It's never a good idea to combine a bunch of different food groups for the sake of digestion either. In order to digest what you ate, the gut needs to release certain digestive enzymes and a certain amount of stomach acid to break it down. By eating things with conflictive interests, you're at least hindering your full potential for easier assimilation. That's also one of the root causes of autoimmune disorders and gut issues – people eating the wrong foods in the wrong context.

On Metabolic Autophagy, you can eat a wide variety of foods and you're not excluding entire macronutrients completely. Instead of combining them and eating all together you simply have to cycle between the ones you eat. That's why the best diet is a cyclical one – you eat low carb keto most of the time and then on some days you have some carbs.

Principles of Food Combining

- **Don't Combine Fats and Carbs** – Not only will it hijack your sensory satiety but is also more dangerous to your health. Eating calorie-dense foods with high insulin makes it easier to store the food as fat and induce insulin resistance.
- **Combining Starch and Meat** – Starchy foods like potatoes and rice require different enzymes and acidity than meat. Combining them together may cause some conflict of interest in the gut. The body will prioritize one or the other but in both cases, the other food that's left out can begin to ferment in the gut if it stays there for too long. Starch and protein also spike insulin much higher. If you're eating once a day it might not be that big of an issue as there isn't much food in the stomach already. However, for optimal results, you want to eat meat with vegetables and starch with something easier to digest like fish or other plant-based proteins.
- **Don't Combine Fruit With Anything** – The simple sugars of fruit don't require much digestion and they get stored as liver glycogen. If you do eat fruit, then you should do it on an empty stomach with empty glycogen. Eating fruit after a meal makes it easier to store it as fat and become fermented in the gut. Forever alone fructose…

- **Don't Drink Your Calories** – If you're drinking juices, shakes, meal replacements etc. then you're not really feeling that satiated afterwards. It's better to limit liquid calories because it'll bump up your daily calories while limiting satiety. The only exception might be like a green juice powder or some protein shake every once in a while.

- **Get Enough Fat-Soluble Vitamins and Protein** – In order to feel satiated with fewer calories, you should focus on whole foods with fat-soluble vitamins and essential amino acids. This prevents cravings and binging. The best foods for that are meat, fish, eggs, and organ meats.

- **Promote Acidity When Eating Protein** - To digest proteins and meats, you want to have higher acidity in the gut. This will help to break down the food and promotes digestion. That's why combining meat with starch can also cause further issues – one promotes alkalinity and the other wants acidity. To promote hydrochloric acid (HCl) in the gut, drink a little bit of apple cider vinegar before eating, add it to your food, or take some digestive enzymes.

- **Don't Drink While Eating** – Drinking a lot of water before, during, or immediately after the meal can dilute stomach acid and hinder digestion. This can cause undigested food particles to float around for longer and promote inflammation. It'll definitely make you more bloated and constipated. You should wait at least 30 minutes after eating before having something to drink.

- **Limit 'Healthy Desserts'** – Even though some foods can be 'low carb' or 'healthy', you'd still want to avoid snacks and desserts as often as possible. The problem has to do with sensory specific satiety. If you eat a bunch of steaks and then have a keto nutbar, you may end up over-eating on the bar because of the new taste it provides, especially if it has some artificial sweeteners or the like. They should be a treat every once in a while but not the norm.

- **Be Mindful of What You Eat** – Most importantly, don't eat mindlessly and sporadically. If you do eat, then sit down and actually enjoy it. Feel all the flavours and sensory stimuli take you over and be present. This will make you more grateful for what you eat. Eating more slowly will also enable the satiety signals to reach your brain faster without eating unnecessary calories.

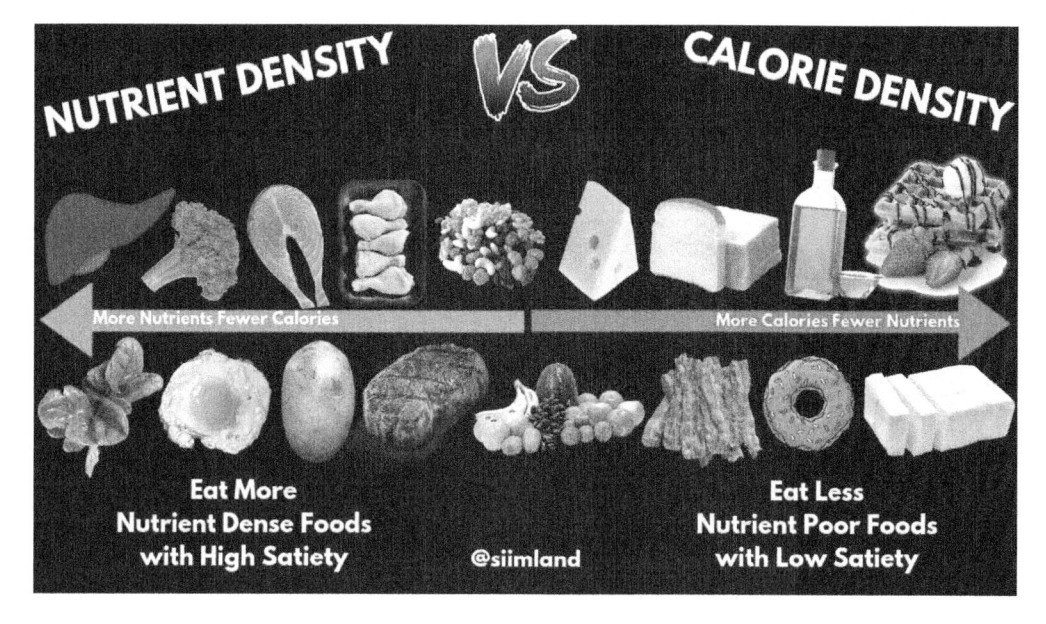

Figure 77 Eat more nutrient dense foods with higher satiety and fewer calories

The goal is to prioritize nutrient-dense foods that promote satiety and cover your essential nutrients. This way you'll feel fuller with fewer calories and you don't have to keep eating to reach satisfaction. What foods are best suited for that? Fibrous vegetables, animal-based proteins, and some healthy fats but not too much.

Eat Less Move More?

Another reason why it's thought that the average modern human is overweight has to do with eating too much and moving too little. It makes sense if you stick to the idea of calories in vs calories out. Our early ancestors who moved around all day and didn't eat a lot, which is why they didn't get obese.

Modern-day hunter-gatherer tribes and subsistence societies don't seem to be burning that many calories despite being lean and active most of the day. In fact, studies have found that the Hadza foragers in Tanzania, Africa have the same average daily energy expenditure than Westerners despite their physical activity level being greatly higher[577]. This may seem like a paradox as the Hadza are lean and free from diabetes. How can these very active foragers subsist on less food while still physically performing at high levels? Part of it has to do with their environment – they're forced to move and stay in motion so they could feed themselves and survive. Another component is nutrient density – they're eating higher quality food with more nutrients that allows them to get more nutrition from less quantity of food.

The Hadza eat quite a limited variety of foods from the wild – primarily meat and tubers in the dry seasons and more berries and honey during the wet seasons. This is enough to cover their average daily caloric intake of 2600 calories for men and 1900 calories for women, which is roughly the same as in Western populations.

Natural selection favors adaptations that make the organism become more efficient with its energy resources. Although physical activity burns calories it's eventually going to make your body adapt to that demand and, after a while, you'll be burning less energy from the same type of activity. This is perfect if you were to live in an environment of nutrient scarcity and dangers. However, it's not that ideal for living next door to supermarkets and unlimited amount of food.

Caloric restriction and causing less oxidative stress on the mitochondria are one of the few known ways of increasing lifespan in many species as we already know. That's why your goal with nutrition and exercise should never be to just burn calories and over-eat afterwards. You may potentially out-exercise a bad diet but it's not optimal for longevity.

Instead of thinking: „*how many calories I can get away with without getting fat,*" you should think: „*how much nutrient density can I get from as few calories as possible to build lean muscle, burn fat, and promote longevity.*" That's a completely different perspective from what mainstream fitness advice tells you. Don't live and exercise to eat but eat to live and exercise.

While you don't have to start living like a Hadza, there's still quite a lot of wisdom to be learned from the diets of hunter-gatherers. Namely that nutrient density plays a key role in the sustenance of any diet. Eating foods with more micronutrients and fewer calories allows them to stay fit and physically strong without suffering from starvation or energy deprivation. This is a good thing as it'll help to maintain a lower caloric intake, which has been shown to be inductive of longevity.

You shouldn't aim for a completely ancestral way of eating either because the hunter-gatherer diet is very opportunistic and doesn't take into account the general goal of life extension. Instead, here are some principles to remember when trying to maximize nutrient density for health and longevity.

- **Maximize Nutrient Density** – You should aim for eating foods with the highest nutrient density that will give you a bunch of essential vitamins and minerals without having to eat too much. The best foods for that are organ meats, whole eggs, fish, vegetables, herbs, and spices (See Figure 77). A more specific overview of which foods to eat in Chapter XV.
- **Avoid Empty Calories** – There's no point in eating foods that don't give you a lot of nutritional value but come with a hefty caloric load. Obviously, processed food, even the „healthy kind" has a much lower nutrient density than whole foods. I'm talking to you low carb paleo bars. The idea is to also keep your overall daily caloric intake relatively moderate as to not tax digestion, mitochondrial functioning, or accelerate aging. Instead, you would want to always stay around your maintenance calories even when trying to gain muscle and strategically cycle your caloric intake.
- **Eat Plenty of Protein** – High-quality protein should be central to every meal because of its satiety factor and benefit on lean muscle. On this program that includes resistance training and intermittent fasting, you don't have to worry about overstimulating mTOR with protein because you're already in an effective metabolic autophagy state. The daily protein requirements are typically quite low and not optimal for muscle or even longevity for that matter. A higher protein diet tends to be better for body composition and predictably on metabolic health as well. In reality, it's only high relative to the extremely low RDA of protein. When doing the more advanced type of intermittent fasting, then you'd need to be consuming slightly more protein as well to trigger more anabolism within a smaller time frame. In general, you would want to aim for 0.7-1.0 g/lb of lean body mass. There's no additional benefit for muscle growth for going beyond that. Sometimes you may find yourself going up to 1.2 g/lb whereas at others you'll be at 0.6 g/lb, depending on which stage of the anabolic cycle you're in.

- **Balance Your Healthy Fats** – Animals who eat their natural diet have a 1:1 omega-6 to omega-3 ratio. The optimal ratio for humans is also around 1:1 or 1:2. If you're experiencing high levels of stress, inflammation, or sickness, then you may benefit from increasing your omega-3 intake a little bit. You can even do well on a 2:1 ratio favoring omega-6s but anything beyond that will make your body more pro-inflammatory. The main idea is to eliminate all vegetable oils, trans fats, processed carbs, added sugars, other artificial ingredients, and eat plenty of wild oily fish, grass-fed meat, and get healthy fats. Additional fish oil supplementation isn't advisable unless you're using a safe source or taking high-quality krill oil or cod liver oil.
- **Eat High Fat But Not Too Much** – Getting the other healthy fats into your diet, such as butter, olive oil, MCT oil, coconut oil, avocados etc. will be okay. However, fat is still a rich source of calories and there's only a certain amount your body needs. More fat won't make you burn more fat or make you metabolically healthier. Although we can say that saturated fat isn't the main cause of heart disease, it's still not a superfood you could eat in unlimited amounts. The potential evolutionary trade-off of excess fat consumption simply isn't a wise move and not optimal for longevity. As said earlier, the minimum daily dietary fat intake is 20-30 grams, which isn't that good either. A healthy fat consumption on non-ketogenic diets should be somewhere between 20-35%, which on a 2000 daily caloric intake would be around 40-80 grams. On a low carb keto diet, it should be slightly higher but you don't need to be eating copious amounts of dietary fat because more won't be always better. Most people can stick between 100-180 grams of fat and be perfectly healthy. There is no metabolic advantage to eating more fat beyond using it for daily caloric maintenance.
- **Get Some Fiber** – The Hadza clog up about 100 grams of fiber a day from tubers and vegetables, whereas the average fiber intake in the Western world is about 15-20 grams. You definitely don't need fiber to survive but it can be helpful in promoting satiety, helping with digestion, clearing constipation or simply bringing in more volume to your meals while keeping the calories low. I wouldn't recommend aiming for any more than 30-40 grams of fiber a day as it can still cause bloating, indigestion, and constipation if you overconsume it. However, if fiber helps you to be less hungry and thus eat fewer calories then it can be useful

for longevity in the long-term. On some days you should be eating more, on some days a little less, and on some days completely nothing.

- **Control for Blood Sugar and Insulin** –Carbohydrate consumption should always be dependent on your body's energy demands – did you work out, what's your general health like, how long you've been fasting for, are you trying to gain or lose weight? Whatever the case, aim for keeping blood sugar and insulin relatively low most of the time and raise insulin only when the body could recover from the spike faster i.e. post-workout.

- **Avoid Inflammation Like Wildfire** – Oxidative stress, free radical damage, and chronic inflammation are the root cause of atherosclerosis, metabolic syndrome, and a myriad of other health conditions. There's acute inflammation whether from lifting weights or eating hormetic compounds that have a beneficial effect but chronic inflammation from being stressed out, consuming too many vegetable oils, eating allergenic things, and environmental toxicity are literally degrading your mitochondria and accelerating aging.

- **Limit Allergens and Phytonutrients** – It's not wise to eat foods you're reacting negatively to, such as gluten, grains, lectins, legumes, nuts, or dairy. You want to heal your gut and metabolism first before trying to play Mithraism. For healthy people, it's okay to occasionally expose yourself to these allergens. In terms of hormesis, you'd want to be eating some beneficial plants and spices, such as wild nettles, green tea, turmeric, cinnamon, berberine, to name a few.

- **Cycle Your Foods** –Hunter-gatherers would go through periods of eating completely different foods throughout the year. The Hadza gut microbiome also changed in between the dry and wet seasons, showing a much bigger diversity than the average Westerner's[578]. This is something we don't really see in modern society. On the Metabolic Autophagy protocol, we'll be cycling between low carb ketogenic periods with brief occasions of higher carbs as well as cycling between the meat and plant foods.

- **Balance Autophagy and mTOR** – Lastly, the thing no fitness guru talks about. Some foods stimulate more mTOR whereas others are more autophagic. I'd say that's a pretty important thing to keep in mind because your longevity is literally determined by these processes. You don't want to have high mTOR all the time nor do you want excessive autophagy. That's why the key to determining what food groups in what amounts you should eat depends on the time of the day, metabolic homeostasis, particular goals of that day, and in what stage of the cycle you're in.

Before we talk about what to eat based on these principles, I want to talk about ketosis and keto-adaptation.

Chapter XIII
The Keto-Adaptation Process

*"The three most harmful addictions are
heroin, carbohydrates, and a monthly salary."*

Nassim Nicholas Taleb

Ketosis and keto-adaptation - you might have heard of these terms being thrown around quite inter-changeably but what do they actually mean?

Ketosis and keto-adaptation aren't mutually inclusive, and they have their differences. Some people might disagree with me on this, which is fine because what matters is that they are distinct conditions. Here's how I define the two.

- **Being in ketosis is the actual metabolic state with the appropriate levels of blood sugar and ketone bodies**. It's said that ketosis begins at 0.5 mmol-s of blood ketones but having 0.3 mmol-s already is quite good. You can be in mild ketosis already after fasting for 24-hours but it doesn't necessarily mean you're successfully using fat and ketones for fuel.

- **The keto-adaptation process makes your body adapt to utilizing fat and ketones as a primary source of energy.** It means you don't have to rely on glucose and can thrive on consuming dietary fat or by burning your own stored body fat.

On the standard ketogenic diet, your macronutrient ratios would be somewhere along the lines of 5-10% carbs, 15-25% protein, and 70-80% fat.

In medical practice or for disease prevention, the ketogenic diet has to be kept quite strict because the purpose is to be in deep therapeutic ketosis with low blood sugar and high ketones between 1.5-3.0 mmol-s. However, people who want to simply reap the benefits of a low carb keto way of eating don't have to be that restrictive and they can safely get what they want by focusing on becoming keto-adapted.

Keto-adaptation results from nutritional ketosis but it's not needed to maintain it. To become keto-adapted, you have to go through a period of being in ketosis where your liver's enzymes and metabolic processes change so you could have the ability to burn fat for fuel, but it's not necessary to be in ketosis all the time to maintain keto-adaptation. You can briefly dip in and out of ketosis for a day or two without fully losing it.

What your own personal macronutrient ratios will look like depend on many things such as genetics, how much muscle you have and what kind of training are you doing. After keto-adaptation, you can be less strict with the macros.

- Carbs should still be quite low around 5-15%
- Protein can be increased for the muscle building benefits up to 25-30%
- Fat will stay around 55-65%, which will cover the essentials and gives extra energy.

This is a good balance between glucose depletion and being reasonable with eating fat.

Keto-Adaptation Symptoms

There isn't a specific point where you can draw a line and go – now I'm completely switched over to a fat-burning engine.

When ketosis is quite binary – you're either in it or you're not – then keto-adaptation is more of like a matter of degree – a wide range of efficiency. In reality, everyone is keto-adapted to some extent.

- If you eat fewer calories than needed, you're going to lose some fat.
- If you eat bacon and eggs, you're going to get some energy from it.
- If you walk or jog, then you're burning some fat.

The problem is that when you're not that well keto-adapted and you're causing metabolic stress to your body through caloric restriction or exercising on an empty glycogen tank, then you're producing some ketones but because your body isn't that efficient at using fat for fuel, you'll also start converting some of your muscle tissue into glucose through gluconeogenesis.

How much fat you're able to burn and how much protein you'll compensate with depends on your level of keto-adaptation.

- Eating the high-carb-low-fat-high-protein diet is making your body quite dependent on glucose and frequent eating. The same applies to a high carb, high fructose diet. You have to eat very often to not go catabolic.

- Eating slightly lower carb, like a paleo approach where 30-50% of your calories come from carbs leaves some room for burning fat but it's still making you burn some glucose because you're eating more of it.

- Eating a strict low carb high fat ketogenic diet is the furthest you can promote keto-adaptation with diet. It'll keep you in a state of nutritional ketosis wherein the body is geared towards using ketones as a primary fuel source.

Therefore, the pursuit of getting into ketosis with a low carb ketogenic diet is going to facilitate keto-adaptation, which will enable you to shift your metabolic engine more towards producing energy from fatty acids.

The goal of the ketogenic diet for most people is not necessarily to be in ketosis but to become keto-adapted. However, in order to make it happen, you still need to go through the keto-adaptation process wherein you get into ketosis and eat a low carb ketogenic diet. Otherwise, you'll never cross the chasm between burning glucose and ketones but will stay in the periphery all the time.

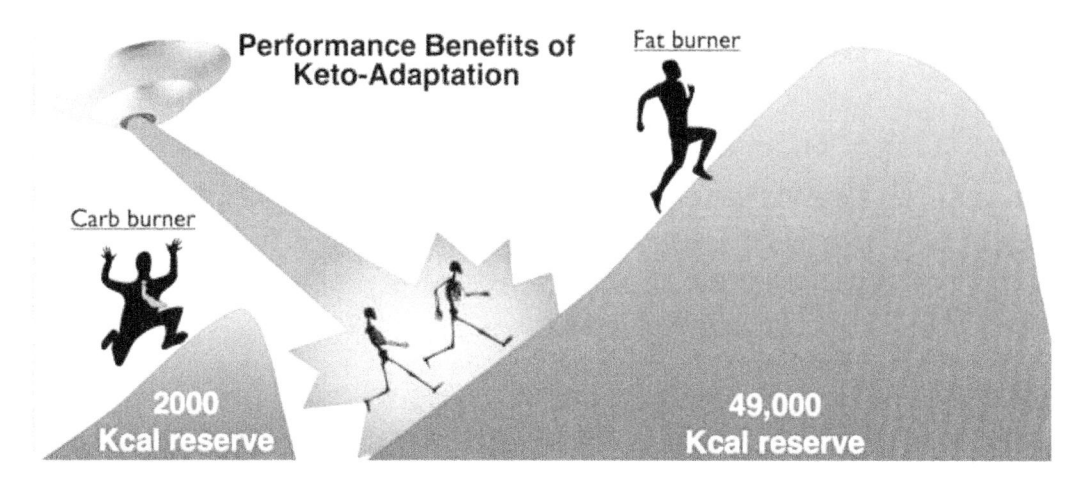

Figure 78 The Difference Between Burning Carbs and Fat In Terms of Energy Quantity

Figure 78 depicts the notion of burning carbs and fat quite well. There's only a limited amount of glycogen our bodies can store compared to the vast fat reserve. In order to access the fatty acid tank, we need to prime keto-adaptation and condition the body to run on ketones in most situations. The chasm in between can be somewhat difficult to cross, which is why some may fail completely

Here's how the keto-adaptation process looks like:

- **Carb Withdrawal** – you go on a low carb ketogenic diet and remove all carbohydrates from your diet. On keto, you eat leafy green vegetables, fatty meat, fish, eggs, and some other fats. The foods on Metabolic Autophagy are slightly different and prioritized differently, which will be the topic of Chapter XIV.
- **Keto Flu Period** – you may experience some fatigue and exhaustion because the brain doesn't know how to use ketones for energy that efficiently yet. This may last from a few days up to several weeks, depending on your sensitivity.
- **Getting Used to Ketones** – you begin to feel better and more energized from eating low carb high-fat foods. The process can be accelerated by implementing intermittent fasting and making sure you're not starving yourself. This may last from 2 weeks up to several months and the longer you do it the better it gets.
- **Fat Burning Mode** – your exercise performance will improve or at least you'll regain the vigor you might have initially lost during keto flu. Here you can begin to see increased time to exhaustion, faster recovery from workouts, less fatigue during the day, mental clarity, and reduced hunger.
- **Keto Adaptation** – you can run very efficiently on dietary fat as well as your own body fat without needing carbohydrates to perform or feel energized. Thanks to burning ketones, you don't get that hungry and whenever you do it's temporary.
- **Metabolic Flexibility** – you can also use carbohydrates for fuel and you're not going to get brain fog from being kicked out of ketosis. This is the ultimate goal of keto-adaptation – to not be dependent on ketones nor carbs and to use both in various situations. More on this shortly.

The process of becoming keto-adapted takes about 2-4 weeks or even up to 3-6 months. How long it's going to end up taking depends on how easily your body begins to accept ketones and fatty acids as a fuel source.

Some of the side-effects you may experience include losing water weight because of low levels of insulin, increased thirst, a slightly metallic and fruity keto breath, slight fatigue, and lack of appetite. Fortunately, these things can be quickly overcome and even avoided. They're definitely not permanent and will pass away shortly.

To get past the initial gauntlet of keto-adaptation you need to have patience and perseverance. The severity of your symptoms depends on how addicted to sugar your body has been before. If you come from the background of the Standard American Diet (SAD, indeed), then it will take you longer than someone who is eating Paleo and already used to less sugar.

That's why the ketogenic diet receives such a bad rep. Because your body is still addicted to sugar, you get tired and lethargic. Your metabolism is geared towards running on glucose and it hasn't been adjusted to burning ketones yet.

Some of the good signs of proper keto-adaptation include no hunger whatsoever, mental clarity, high levels of energy all the time, increased endurance, reduced inflammation, stable blood sugar, and no muscle catabolism.

To know whether or not you're in ketosis, you can measure your blood ketones using Ketostix. Optimal measurements are between 0,5 and 3,0 mMol-s[579](See Figure 80). The same can be done with a glucometer. If you're fasting blood glucose is under 80 mg/dl and you're not feeling hypoglycemic then you're probably in ketosis (Figure 79). Ketoacidosis occurs over 10 mMol-s, which is quite hard to reach.

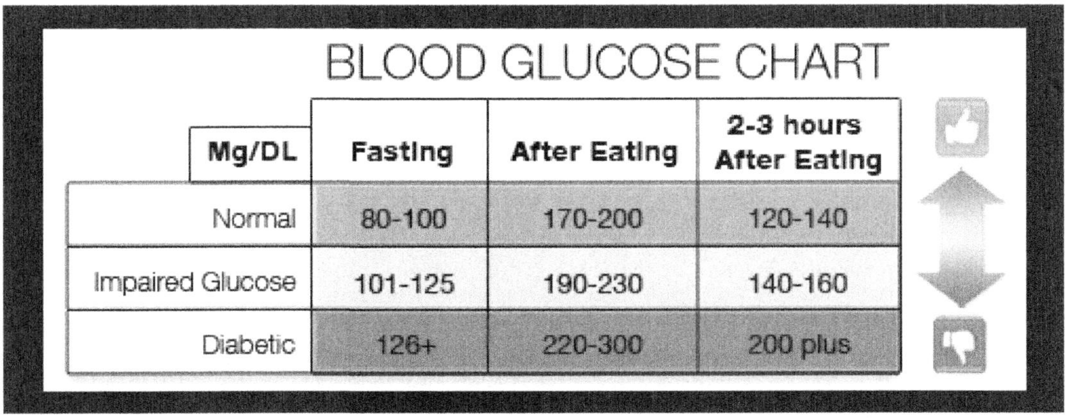

BLOOD GLUCOSE CHART

Mg/DL	Fasting	After Eating	2-3 hours After Eating
Normal	80-100	170-200	120-140
Impaired Glucose	101-125	190-230	140-160
Diabetic	126+	220-300	200 plus

Figure 79 Blood Glucose Chart

What level of Ketosis is optimal?

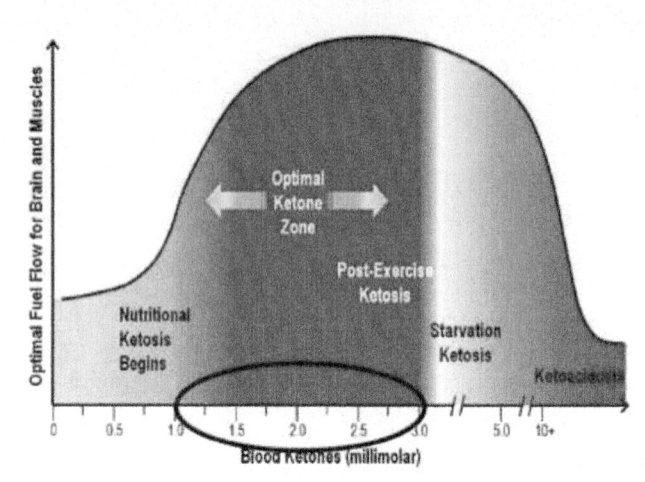

Page 91: The Art and Science of Low Carbohydrate Performance
Jeff S. Volek and Stephen D. Phinney

Figure 80 The ranges of blood ketones

Ketone Breath Meters indicate the amount of acetone in your breath. Acetone gets produced by the breakdown of acetoacetate in the blood. This measurement means that your mitochondria actually take the initial ketone body and then convert it further into an additional source of energy.

When it comes to choosing which ones are the best, then I'd say that for the best results, you'd want to know your blood ketones, blood glucose as well as the amount of acetone in your breath. Urine strips are generally useless because you can have higher amounts of acetoacetate in your urine because of dehydration or nitrogen overload as well but it doesn't really tell you much about how well you're using ketones for energy. The advantage of using breathalyzers is that they're easier, more convenient, and much cheaper to use than using a lot of blood test strips.

Low blood glucose in the context of a non-ketogenic diet generally indicates hypoglycemia, wherein you may feel tired, lethargic, and your brain can't get access to energy.

If your blood glucose is lower than 60 mg/dl or 3.0 mmol/L and you're feeling energized, then it means your brain is getting an alternative fuel from beta-hydroxybutyrate and you're not going to pass out. That's why people on the ketogenic diet have much lower blood glucose than is considered normal. In my opinion, lower blood glucose is also better for your health and longevity. To a certain extent of course.

When glucose goes down, then a metabolically healthy person should see an elevation of ketones as the body shifts into a fat burning mode. If you're not keto-adapted, then you're going to crash and feel exhausted.

The general guideline is that the longer you do the ketogenic diet the easier it gets and the better you'll start performing. But like I said, our bodies are different – what's high-carb for a sedentary person may not be high for an athlete. That's why some people can even be keto-adapted without necessarily being in ketosis all the time. That's where the concept of metabolic flexibility comes into play.

Metabolic Flexibility and Keto Adaptation

Using carbohydrates strategically will not only improve your performance but overall health as well. There are a few reasons why you should occasionally get out of ketosis.

- Some people get hormonal imbalances, like low thyroid or testosterone, if they restrict their carbs for too long or if they fail to adapt fast enough.

- Your energy levels may also suffer from time to time because of overtraining or too much stress. In that case, carbs will speed up recovery and lower cortisol.

- Low mucous production of the ketogenic diet will prevent your body from creating enough mucus that surrounds and moisturizes your gut and eyes. That can be the cause of too low insulin and other growth pathways.

- Some carbohydrate foods can promote a healthy gut by increasing diversity in your microbiome. Changing up your menu will help to reset food intolerances and prevent them from developing in the first place. Too restrictive diets all the time may develop autoimmune disorders.

- Carbs can be used to boost your performance while working out but they can also be used for better sleep. Sometimes being low carb for too long may lead to some serotonin deficiencies and carbs can help to fix that.

- Eating carbs seasonally will fit better with the circadian rhythms and your own individual genetic blueprint. During some seasons it's natural to be eating more carbohydrates and at others less.

- And of course, it's nice to sometimes eat foods that aren't bacon or vegetables. Although I'm the kind of guy who could eat just steak and eggs for the rest of my life I still agree that some variety is not only healthy but also beneficial.

But don't worry, getting kicked out of ketosis doesn't mean you'll lose keto-adaptation. You'll still be able to effectively burn fat for fuel. It's just that you'll gain some of the other benefits of metabolic flexibility.

It's probably not the best idea to be in ketosis all the time from an evolutionary perspective either. The most commonly used anthropological group used to justify a low-carb-high-fat diet are the Inuit who eat almost only animal foods with minimal plant matter year-round. However, you might be surprised to hear that the Inuit actually suffer from a genetic disorder known as '*the Arctic variant of CPT-1a deficiency*'. It inhibits the use of ketones and promotes the risk of hypoglycemia during fasting – low ketones and low glucose. Virtually every native tribe in the Arctic has this and they're asymptomatic. In fact, this defect developed during a specific era in human history that favoured such adaptation.

The Arctic variant was selected by the environment about 6000-23 000 years ago to be an advantage because of some unknown reason, which simply swiped the entire population through positive adaptation to become normal[580]. Maybe, during some time in the past, it was simply not advantageous to be in ketosis and avoid fasting. Because of the cold Nordic regions of the Inuit, it probably had to do with the continuous need for calories and heat in such a harsh climate. Instead of producing ketones, the fats they ate got converted into glucose for the brain and the triglycerides were used up by muscles and heart. Ketones became simply secondary. Furthermore, persistent ketosis in such extreme living conditions was not favourable because of potential ketoacidosis in conjunction with other environmental stressors. Whatever the case might have been, it's simply fascinating to see that even on the evolutionary scale, it's not always the best to be in ketosis. It also says that humans evolved under intermittent ketosis, not a continuous ketogenic diet.

As I said, you don't need to maintain nutritional ketosis 24/7 to be keto-adapted. You're not going to get into ketosis by eating keto for one day, and you're not going to lose your fat-burning metabolism by getting our of ketosis from time to time either. The body is trying to maintain homeostasis and not go through random changes all the time.

Why Bother With Keto?

You're always burning a mixture of different energy molecules whether from glucose, ketones, fatty acids, lactate, and many more. What proportion of each you're currently using depends on the intensity of physical exertion, your overall metabolic condition, and your degree of keto-adaptation.

The simplistic view of fuel utilization is that your body will burn primarily fat at lower intensities of exercise and at higher levels of exertion it starts using more glycogen and glucose[581,582,583]. This is known as the *Crossover Effect,* with a progressively increased contribution of carbs as a fuel source (See Figure 81). It promotes the idea that you can't perform high-intensity sports while eating a low carb ketogenic diet.

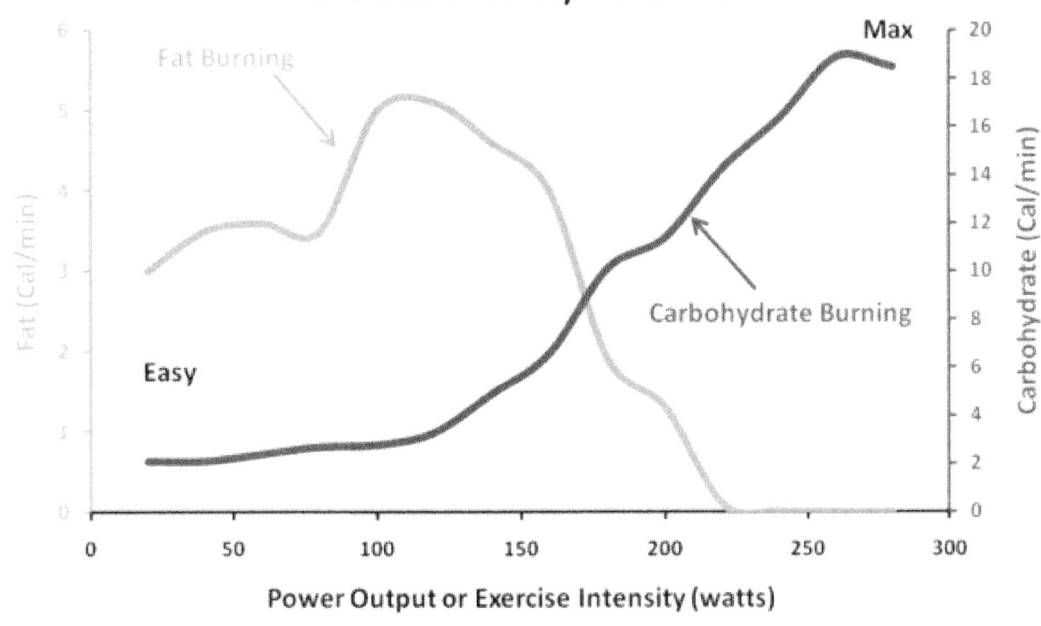

Figure 81 The Crossover Effect

In reality, your body needs glucose from carbohydrates under certain metabolic conditions. At other times, or during periods of starvation, the body will come up with many alternatives that are actually much more suitable in certain situations.

Ketogenic Diets and Endurance Performance

In 2016, Jeff Volek and Steven Phinney did a study called FASTER (FASTER=Fat-Adapted Substrate oxidation in Trained Elite Runners), which showed that ultra-endurance athletes who had keto-adapted for 9-36 months showed extraordinarily higher rates of fat oxidation[584]. Here are the main findings:

- They were using fat for fuel even at intensities of 70-80% of their VO2 max, compared to the 55% of the high carb control group.

- Peak fat oxidation was 2.3-fold higher in the low carb group and it occurred at a higher percentage of VO2 max.

- During submaximal exercise, fat oxidation was 59% higher in the ketogenic group.

- There were no significant differences in resting muscle glycogen and the level of glycogen depletion even after 3 hours of exercise.

Over the course of a 3 hour run at 65% of VO2 max, the low carb group was burning 30% more fat and 30% fewer carbohydrates than those who ate high carb. Some of the subjects on the keto diet actually burned 98% fat and only 2% glucose at 65% intensity. (See Figure 82).

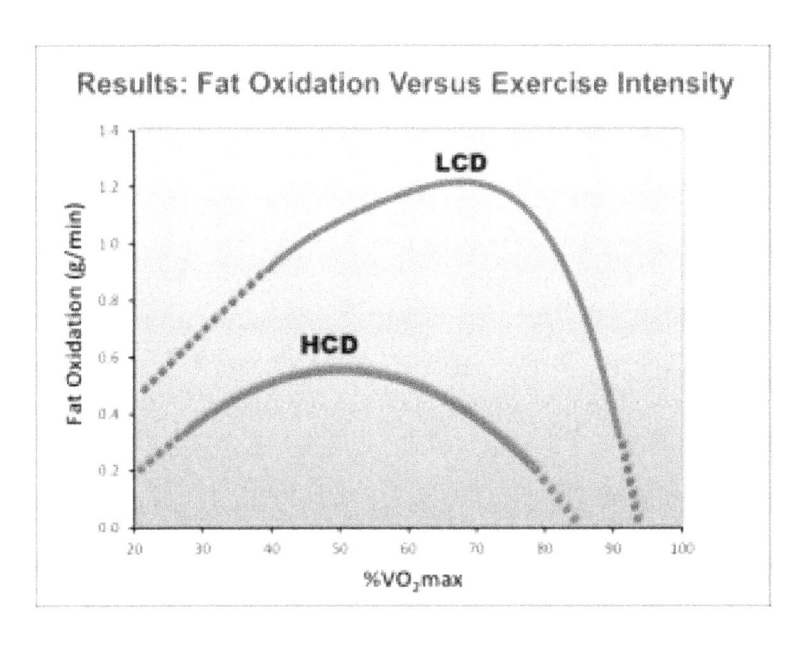

Figure 82 Fat oxidation rates between low carb and high carb athletes

Keep in mind that, fat oxidation doesn't refer to losing body fat directly – it refers to what proportion of fatty acids get burned in the Krebs cycle for energy. In the case of these low carb athletes, they were able to fuel their exercise with primarily fat. Therefore, they could've continued to exercise for several more hours without running into an energy crisis as long as they still had enough body fat with them. The fat from food and the fat on your belly are metabolically very similar.

These results contradict the consensus view of the Crossover Effect that states you can't burn fatty acids above 60% of your VO2 max. They show that after keto-adaptation, the body becomes increasingly more efficient at using fatty acids during higher intensities of exercise, which, in turn, spares muscle glycogen for only near maximum efforts and decreases the need for carbohydrate refeeding for maintaining performance.

Despite glucose being the body's default main fuel source, most of the day you're still using fat for fuel because doing daily chores, walking, or even low-intensity cardio maintains aerobic respiration. You only tap into your glycogen stores whenever you're sprinting, lifting heavy stuff or training hard. Even then the degree of how much glycogen you'll end up burning depends on your level of keto-adaptation because as we've seen ketones can be used at even higher intensities of exercise.

You don't want to burn glycogen for nothing because it can only be stored for a limited amount. With keto-adaptation, you're able to sustain most of your submaximal exertions by tapping into your unlimited adipose tissue stores. Even people with 7% body fat have over 40 000 calories with them at all times. That's why the ketogenic diet is amazing for low-intensity exercise.

In the 1980s already, it was shown that a 4-week ketogenic diet with less than 10g of carbs a day didn't compromise endurance performance in elite cyclists[585]. These athletes were using more than 90% of fat oxidized fuel during exercise at 64% of their VO2 max. They also showed a 3-fold drop in glucose oxidation, a reduction of resting muscle glycogen by half, and they used 4 times less glycogen during exercise.

This kind of muscle glycogen responses in low carb athletes have been shown to share similarities to that of highly trained Alaskan sled dogs[586]. Sled dogs often perform at submaximal intensities for several hours in a row with incredible endurance capacity by eating a high fat low carb diet. In a study, dogs who run 160 km/day for 5 days showed no cumulative depletion in muscle glycogen, despite eating a diet with 15% carbohydrates. A subsequent study on dogs who ran 140 km/day for 4 days reduced muscle glycogen by 66% (similar to the 64% glycogen reduction in the FASTER low carb athletes) and a progressive increase in muscle glycogen over the following days of running[587]. A low carb diet has also been shown to increase the time to exhaustion during prolonged exercise[588].

Keto and Anaerobic Training

It's clear that keto-adaptation is excellent and even optimal for endurance activities that prefer fat for fuel. However, compared to glycolytic pathways, the ketogenic path is less ideal for tapping into the anaerobic threshold at higher intensities.

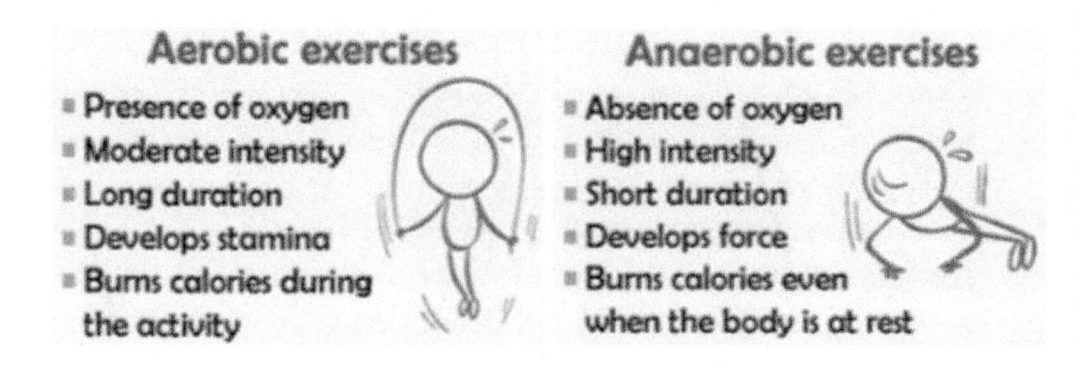

Aerobic exercises
- Presence of oxygen
- Moderate intensity
- Long duration
- Develops stamina
- Burns calories during the activity

Anaerobic exercises
- Absence of oxygen
- High intensity
- Short duration
- Develops force
- Burns calories even when the body is at rest

High-intensity training, such as HIIT cycles, CrossFit, sprints, bodybuilding, and gymnastics are anaerobic by nature and span the creatine-phosphate system which requires you to be using glycogen in the presence of no oxygen. Because of that anaerobic environment, you can't maintain it for any longer than a few seconds. You're only burning glycogen for that specific time length and will revert back to using other fuel sources during rest.

Some reports suggest that muscle can use circulating glycerol for intramuscular triglyceride synthesis[589]. Previous studies have shown that after a short-term low carb high-fat diet, the ability to transport circulating fatty acids into muscle greatly improves[590]. This means that keto-adapted individuals experience higher rates of adipose tissue oxidation, which results in an increased release of both glycerol and fatty acids into circulation, which increases the overall uptake of fatty acids into skeletal muscle.

A study done on recreational CrossFit people showed that 3 weeks of ketogenic dieting didn't lead to a decrease in performance compared to the control group[591]. However, they showed an improvement in body composition and decreased whole-body adiposity. They lost fat mass and thus improved their relative fitness.

Comparative studies between resistance trained individuals who eat a ketogenic diet and a standard diet have also shown that although both of them lead to similar improvements in workout performance and muscle gain, the keto subjects tend to have more lean body mass and less fat mass[592].

Another study was done in 2014 by Jeff Volek, Dominic D'Agostino, and Jacob Wilson compared the effect of a very low carb diet with a traditional diet in resistance trained individuals[593]. The results showed that lean body mass increased to a greater extent in the VLCKD (4.3 +/- 1.7 kgs) as compared to the traditional group (2.2 kg +/- 1.7). Muscle mass increased in the keto group as well (0.4 +/- 25 cm), as opposed to the other one (0.19 +/- 0.26 cm). On top of that, fat reduction followed the same pattern, benefitting the ketogenic individuals (-2.2 kg +/- 1.2 kg) versus the (-1.5 +/- 1.6 kg).

Another study done on elite level artistic gymnasts showed that after going on a low carb diet, the athletes didn't lose strength or muscle but actually lost body fat while increasing lean muscle mass[594]. They only followed the keto protocol for 30 days and already showed great improvements in body composition. These aren't your recreational fitness enthusiasts either - they train at high intensities for several hours a day. If they can do their crazy explosive routines and superhuman feats of strength without carbohydrates, then it's safe to say that the average person will do perfectly fine on a less strenuous workout routine.

Based on these findings so far, we can say that a ketogenic diet hinders exercise performance in neither endurance or anaerobic sports. In fact, with the knowledge we currently have, it's safe to say that it's actually more superior to decreasing fat mass and improving body recomposition, especially if eating at a caloric deficit.

In summary, for the vast majority of people, a ketogenic diet can provide all the necessary fuel sources needed for maintaining daily aerobic activities as well as giving enough energy for short anaerobic exertions.

However, when it comes to high-intensity training, then there are still some implications and constraints that have to be kept in mind.

- **If you're performing high-intensity exercise for longer and with fewer rest intervals, then your body will have troubles resynthesizing its glycogen with just fatty acids.** Examples would include Ironman triathlon, a competitive sports game, a 2-hour high volume bodybuilding workout, a CrossFit game event, or having more than 2 workouts a day.

- **Workouts that aren't as taxing or frequent don't require the addition of carbohydrates although they may still help.** Examples include Olympic weightlifting, powerlifting, gymnastics, endurance, cycling, or short HIIT cardio.

The 2 main ways of adding carbs around workouts are the targeted ketogenic diet where you consume a small number of carbs during high-intensity exercise and the cyclical ketogenic diet where you eat low carb for a week and then have a refeed day.

Do You Need Carbs

This also raises the question, why bother trying to become keto-adapted if you have the possibility to refuel with carbs during exercise?

The answer is quite self-explanatory and obvious – you want to maintain lower levels of blood glucose and insulin the majority of the time as to promote your body's ability to burn its own fat stores and reduce oxidative stress.

Carbohydrate restriction has been shown to have many health benefits, starting from fat loss, neuroprotection of the brain, better biomarkers, stable energy, and ending with mitochondrial density and longevity. Burning fat causes less damage to the mitochondria and it produces more energy per calorie, which has benefits on cellular survival.

Being dependent of carbohydrate refeeding is also quite a fragile position to be – you're always limited by your glycogen stores and have to structure your entire day around eating. After becoming keto-adapted, you can tap into your stored body fat very fast and easily, which keeps you energized even while fasting for 5 days or more.

You don't have to be in strict ketosis all the time to stay keto-adapted but you do need to maintain a semi-ketogenic state of glycogen depletion to build up these fat oxidation pathways into your metabolism.

Whatever the case might be, anyone who is following a ketogenic diet can benefit from some exogenous carbs when performing at higher intensities. If you're training hard and heavy more frequently, then your body won't have enough time to replenish your muscle glycogen stores solely via gluconeogenesis.

Carbohydrates and insulin are not the enemies here. Insulin is a powerful tool that can assist in growth and tissue repair. It governs nutrient partitioning and influences whether or not the calories consumed go into muscle or fat cells. The benefits to this are immense and useful, but only in a specific context – when glycogen stores are depleted and ready to absorb some carbohydrates. That happens after heavy resistance training, but not all of the time. This promotes metabolic flexibility as well.

How to Increase Metabolic Flexibility

The foundation for metabolic flexibility is the ketogenic diet because you need to be able to burn fat as a primary fuel source.

On a high carb diet without keto-adaptation, you're only capable of burning glucose while not being able to use ketones. But in order to avoid bonking after your glycogen runs out and to not get the keto flu whenever you eat some carbs, you need to go through a period of keto-adaptation.

Both aerobic and anaerobic conditioning is also needed. The purpose of your training should be to increase mitochondrial density – your cells' ability to produce energy whether from ketones or carbs.

Metabolic flexibility is a matter of degree like keto-adaptation – someone with greater muscle glycogen stores and improved insulin sensitivity can absorb more glucose than the one with lower energetic demands. If you do more resistance training, then you'll have a bigger buffer zone for eating carbs but you'll also increase your basal metabolic rate.

Here's how you can improve metabolic flexibility and stay keto-adapted while eating carbs.

- You have to establish nutritional ketosis by doing a low carb ketogenic diet for at least 2-4 weeks.
- After the first period of keto-adaptation, you can start tinkering with some carbohydrates to improve your performance.
- The fact of the matter is that you still want to be eating relatively low carb, especially at times when you're not exercising.
- If you're able to go without food for over 24 hours and not experience hypoglycemia or muscle weakness, then that's a good indicator of keto-adaptation.
- At this point, your physical performance at all intensities is generally the same and you don't need carbs to fuel your training. However, you can still use a few hacks that include strategic carbohydrate consumption.
    - The Targeted Ketogenic Diet (TKD) involves consuming a small dose of carbohydrates during your most intense workouts.

- The Cyclical Ketogenic Diet (CKD) involves eating keto for 5-6 days, then having a day of eating more carbohydrates, and then returning back to keto.
- There's also something called Carb Backloading (CBL) where you eat low carb all day, then you go to the gym to have a muscle glycogen depleting workout that makes you more insulin sensitive, and then have dinner with a few extra carbs like a sweet potato, a bit of fruit, or some rice.

The vast majority of people would still do best on a regular ketogenic diet, at least most of the time. Both the CKD and TKD are viable ways of boosting your performance, improving your metabolic flexibility and upregulating the metabolism.

You can use both the CKD and TKD for promoting any physical endeavor, whether that be ironman triathlon, bodybuilding, Olympic weightlifting, or Crossfit. Of course, this doesn't apply to walking, jogging, yoga, disc golf or something else that doesn't really tax your glycogen stores. Therefore, you want to be using these methods only as tools for becoming stronger, faster, more enduring or resilient, not as an excuse to simply eat some carbs. Let's go through some scenarios where you can indeed use these strategies.

The Cyclical Ketogenic Diet

- Training 4 or more times per week with mostly resistance training.
- Examples: bodybuilding, powerlifting, weightlifting, Crossfit, obstacle course racing
- Goals: build muscle, increase strength and power, have your cake and eat it too
- Refeeds either on your harder training days or the night before.
- If you feel like you're feeling sh#t the entire following week because of keto flu then dial down on the amount of carbs you're consuming and have less frequent refeeds.

The Targeted Ketogenic Diet

- Training 4 or more times per week with mostly resistance training or ultra long endurance.

- Examples: bodybuilding, powerlifting, weightlifting, Crossfit, obstacle course racing, Ironman, rowing, marathon running, swimming

- Goals: build muscle and lose fat while improving performance, power, and endurance

- Have small amounts of easily digestible carbohydrates with protein during your workouts, such as a shake or ripe bananas.

- Adjust your carb intake according to your performance requirements and how you feel. Start off with just 5 grams and slowly keep adding an additional 5 grams per 30 minutes of intense physical activity. Hard training athletes can consume up to 30-50 grams of carbs during training while staying in ketosis.

- If you're training twice a day then have a larger shake during the first workout and a smaller one during the second one. Still, eat keto in between training sessions.

- Eat low carb keto when you're not exercising and in the post-workout scenario.

The Standard Ketogenic Diet

- Training 3-4 times per week with either resistance training or cardio.

- Examples: powerlifting, fitness, weightlifting, endurance, jogging, cycling, yoga

- Goals: lose fat, build strength, stay fit, improve health, battle diabetes, reverse insulin resistance

- If you're not feeling tanked or feeble during workouts then you don't need to be consuming carbs. Also, if you're not planning on pushing yourself extra hard at the gym on that particular day, then you shouldn't feel the need to eat carbs either.

- If you're feeling hypoglycemic and are about to pass out during exercise then it's a sign of not being keto-adapted. You're simply in a state of still running on a sugar burning engine and you need to build up your fat burning pathways through diet before trying the TKD or CKD. Adding more electrolytes can also help.

- Eat low carb keto the entire time with enough protein and healthy fats.

Carb Backloading

- Training 4 or more times a week with primarily resistance training and anaerobic exercise.

- Examples: powerlifting, weightlifting, CrossFit, bodybuilding, competitive sports

- Goals: build strength, gain muscle, increase performance, train more frequently at higher intensities, eat carbs more often

- If you don't want to do the targeted ketogenic diet or prefer to workout with low glycogen, then you can safely eat more carbohydrates the night before your heavier training. This will fill up glycogen stores and primes you to perform more intensely the next day.

- If you're having heavier training sessions throughout the week, then it's even better to have 1-2 nights of carb backloading.

- The best time to eat carbs is post-workout with depleted glycogen. Carb backloading is great if you don't want to eat the keto diet all the time.

- Eat low carb keto in the earlier parts of the day before training. After working out have some carbs. Adding fasting is a great idea for sure.

Most of the time you'd still want to be in ketosis because it's going to maintain your keto-adaptation, but it may leave you vulnerable to some foods that aren't keto-proof.

For instance, if you've been in ketosis for months and then you accidentally eat some gluten or even just potatoes, you're going to feel like crap the day afterwards. Of course, the best solution would be to not eat those foods in the first place and stay keto, but it's still going to leave you fragile to these random changes. A much better option would be to have the ability to utilize those carbohydrates for increased performance while still maintaining your keto-adaptation.

That's a general overview of what I call being keto-adapted – running on a fat burning engine while still maintaining the ability to burn carbs as well.

The problem with strict therapeutic ketosis is that it's not necessarily going to ensure keto-adaptation as you can be in ketosis without using those ketones for fuel and you can have very high ketones without being able to perform at your best. More ketosis doesn't equal more keto-adaptation as it has to involve the aspect of mitochondrial density and energy production. It can also neglect some of the performance-enhancing benefits of carbs.

Of course, you'd have to be eating a low carb ketogenic diet the vast majority of time to maintain ketosis and become keto-adapted. However, your goal doesn't have to be ketosis as it's not going to ensure metabolic health or performance.

A keto-adaptation diet would include high-quality low carb foods that build up the person's fat burning engine and then add some occasions of higher carbs as a leverage point for improved metabolic flexibility.

Chapter XIV
The Anabolic/Catabolic Score of Food

„The wise man doesn't give the right answers.
He poses the right questions"

Claude Levi-Strauss

To determine the anabolic/catabolic score (ACS) of these foods, I'll categorize them into 4 subcategories, moving either towards more mTOR stimulation or autophagy. They are:

- **High mTOR (HiTOR)** – stimulates mTOR and insulin significantly
- **Moderate mTOR (ModTOR)** – stimulates mTOR but not a lot of insulin
- **Low mTOR (LowTOR)** – promotes anabolism without significant mTOR
- **Neutral mTOR (nTOR)** – doesn't stimulate mTOR or anabolism
- **Low Autophagy (LowATG)** – supports the activation of autophagy and glucagon
- **High Autophagy (HiATG)** – activates autophagy and lowers insulin significantly

ATG refers to autophagy-related genes and the listed foods will stimulate pathways related to autophagy activation.

The differentiating factor between high and low foods is based on how they affect not only the respective pathways of mTOR and autophagy but also how they generally affect the insulin-glucagon ratio. Glucose and ketones matter only in the larger context of what was eaten, in what amounts, and at what time. That's why certain foods are known for stimulating anabolism or catabolism more so than others.

Now, let's go through them one by one.

High mTOR (HiTOR) Foods

Food that stimulates both the mTOR as well as the insulin/IGF-1 pathway significantly.

These are the most anabolic ingredients that will trigger muscle protein synthesis the most but they may not be ideal for longevity all the time. That's why you want to limit HiTOR activation to only specific times and do it briefly. The best scenario is once or twice a week after a fasted resistance training combined with carb refeeds. Think of spiking insulin quite high for higher anabolism but allowing it to drop again within the next few hours.

Here are the top 5 HiTOR foods:

1. **Whey Protein + Carbs** – Whey protein itself already is a super concentrated form of very bioavailable protein that triggers MPS quite a lot. However, combining protein with carbs raises insulin exponentially higher than protein alone (See Figure). This will also activate mTOR a lot more. That makes all the amino acids you consume with carbs much more anabolic than if taken by themselves. Leucine, which is the main amino acid that stimulates MPS doesn't spike insulin as much if taken alone.

2. **Rice Protein + Carbs** – Rice protein is another highly bioavailable protein that's plant-based. There's not a significant difference between whey and rice in terms of MPS but rice protein may be less inflammatory and with fewer allergens. Compared to whey, rice protein is better for digestion and keeping IGF-1 lower but it may not be that effective for rapid MPS. What matters more is the overall MPS stimulation of the entire feeding window.

3. **Egg Whites + Carbs** – Egg whites are pure protein with even a bit of carbs. They're not that nutrient dense because all the nutrition is in the yolk. Egg whites are just a source of extra protein some bodybuilders use to keep their fat intake low. In fact, on a ketogenic diet, you'd want to do the opposite and eat just the yolks because egg whites may be allergenic. Nevertheless, if you were to consume egg whites alone with not a lot of fat you'd spike your insulin quite high. Add some carbs into the mix and it'll go even higher.

4. **Chicken Breast/White Fish + Carbs** – Chicken breast or white fish are also very lean and mostly protein. The stereotypical chicken and rice meal of bodybuilders holds true in the sense that it'll spike insulin and trigger MPS. If the carbs are low

fiber and high glucose, such as white rice, white potatoes, or pasta, then it'll be even more effective.

5. **Protein Powder + Fruit** – Any protein powder that's low fat, low fiber, high protein and rich in leucine will spike mTOR and MPS. Adding ripe fruit that doesn't have much fiber, such as bananas, pineapples, mangos, dates, or honey will raise insulin in that context as well. That's why you don't want to be eating fruit on a regular basis. It's not ideal for ketosis or your liver health as it can lead to fatty liver disease and promote insulin resistance.

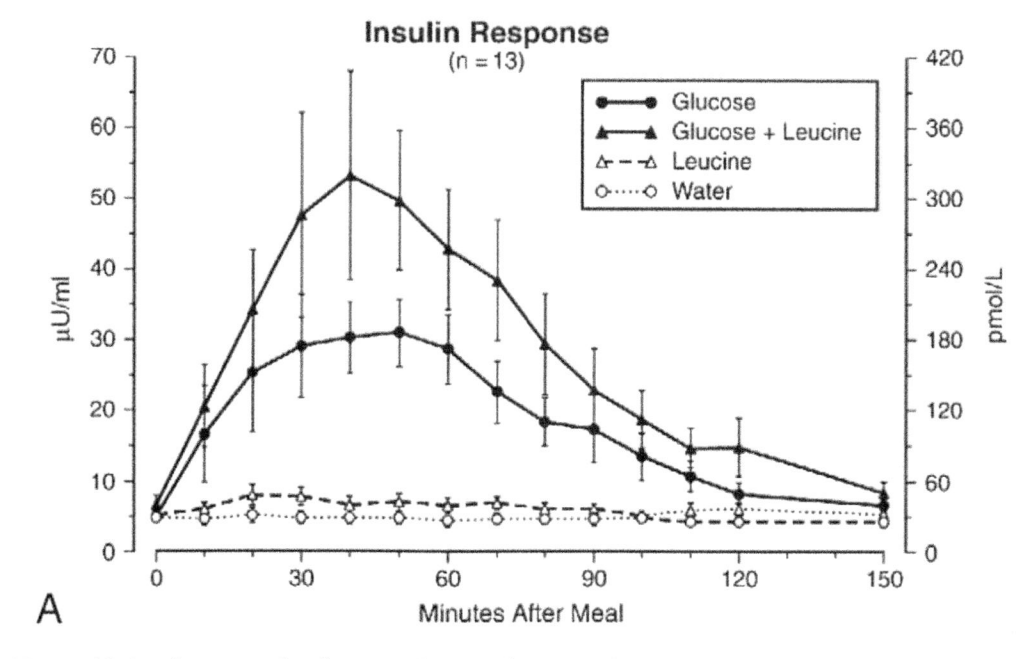

Figure 83 Another example where combining glucose with amino acids spikes insulin very high whereas leucine alone doesn't spike insulin almost at all

Consuming these foods frequently isn't optimal for longevity but they are very beneficial for building muscle mass. How often to have these spikes of High mTOR depends on your training regimen, insulin sensitivity, body composition goals, and the particular stage of the anabolic/catabolic cycle you're in.

Moderate mTOR (ModTOR) Foods

Food that will trigger muscle protein synthesis adequately to promote lean muscle growth but don't raise insulin that much.

These are anabolic ingredients that can maintain ketosis and build muscle at a low carb/low insulin state. If you're working out, then you want to be eating ModTOR foods to not lose lean tissue, especially if you're fasting as well.

For optimal mTOR sensitivity, you'd want to eat ModTOR foods consistently but ideally limit them only to when you're working out.

Here are the top 5 ModTOR foods:

1. **Red Meat** – Meat is a potent stimulator of mTOR and MPS. However, meat by itself in the context of a low carb diet has quite a low insulin to glucagon ratio, thus it's not that anabolic. Red meat is also one of the best sources for most of the essential nutrients you need like protein, fat, B vitamins, iron, etc. For maximum nutrient density, you'd want to eat some organ meats, like liver, heart, kidneys, and the tendons rich in bone marrow. Other great options are beef, pork belly, steak, unprocessed bacon. Ideally, you want to get grass-fed meat or wild game for the better omega-6 to omega-3 balance. Avoid sausages with extra sugar and wheat.
2. **Whole Eggs** –Eggs have all the amino acids you need and they're particularly rich in leucine as well. You can get about 550 mg of leucine from a single egg and it takes 2-3 grams of leucine to trigger MPS. So, the standard 4-5 egg breakfast is quite good for muscle building. Although for optimal longevity you'd want to postpone that breakfast if you know what I mean. The yolk is where most of the nutrition is at so definitely savor them. Cholesterol is incredibly good for building muscle. When cooking, don't over-fry eggs or hard-boil them as it'll damage the nutrients and may oxidize the cholesterol.
3. **Poultry** – Chicken, turkey, and other types of poultry are quite rich in protein. Industrial chicken tends to be quite high in omega-6s and low in other nutrients. Turkey is highest in tryptophan, which can help with serotonin production and thus make you more relaxed. That's why eating poultry may help you to sleep.

Chicken skin and the drumsticks are rich in glycine, which improves skin health and has anti-aging effects.
4. **Oily Fish** – Salmon, sardines, mackerel, trout, flounder, herring, and anchovies are incredible sources of healthy omega-3s, especially DHA and EPA. They're great for cardiovascular health but they also promote muscle growth because of their relatively high protein content. The anti-inflammatory effects will also increase your performance by lowering inflammation, accelerating recovery, and promoting cellular health.
5. **Cheese and Dairy** – Milk is one of the most anabolic foods there is, which is why mammals are breast-fed their mother's milk during the first periods of development. The high fat and protein content is supposed to jump-board the infant's growth and build up the essential parts of the body. However, dairy also raises IGF-1 quite a lot, which will increase inflammation, insulin, skin issues, and accelerates aging. That's why dairy isn't the best thing to consume on a habitual basis. Fermented kefir and cheeses are fine but they're still not something to eat all the time. Compared to something like meat, dairy isn't going to help you build that much muscle but it'll still stimulate the heck out of mTOR and IGF-1, which will lead to a regretful trade-off in longevity.

Although foods like meat and protein are known to stimulate mTOR quite a lot, they're not going to be detrimental for your longevity.

First of all, the low insulin to glucagon ratio will completely blunt the anabolic reaction and the protein will be used primarily for lean tissue maintenance. It's much more difficult to build muscle in a low carb state but the gains you do make are quality and very sustainable.

Secondly, doing intermittent fasting on a low carb diet completely changes the effects of these foods on your body. Of course, eating meat 3 times a day within a 12-hour eating window isn't optimal for long-term health. It's going to over-stimulate mTOR and keep the body out of autophagy. However, eat that same amount of food and protein within 2-4 hours and you'll elicit a completely different response.

Because on the Metabolic Autophagy program, you should be doing prolonged daily fasting as well as resistance training. In that case, you don't have to worry about meat over-stimulating mTOR because you actually want to be more anabolic during the feeding periods. That's the power of time-restricted eating.

Low mTOR (LowTOR) Foods

Foods that can trigger mTOR and MPS but they do it very little compared to HiTOR.

These foods can be anabolic if you consume them in large amounts but their purpose isn't to maximize muscle growth. Think of LowTOR foods as something you'd consume on a rest day as to get the essential nutrients and maintain lean tissue.

If you're following a wider feeding schedule where you're fasting less, then it's also a good idea to keep some of your meals LowTOR as to not have it elevated that frequently. The best time to stimulate mTOR and MPS is still post-workout for recovery.

Here are the top 5 LowTOR foods:

1. **Starches–** Potatoes, sweet potatoes, rice, buckwheat, quinoa, carrots, and beetroot are quite high in carbs and they raise insulin. Of course, the insulin response to white potatoes is a lot different from raw carrots but they're still favouring the insulin to glucagon ratio towards higher blood sugar. It's not just meat or protein that stimulates mTOR – insulin offsets this entire cascade in the first place, which is why your rate of anabolic growth will be much lower if you were to eat protein on a low carb diet. Nevertheless, carbs alone like a rice bowl with vegetables alone won't be that mTOR stimulating because they're low in amino acids. In that situation, it's not worth it to be eating those carbs either because although they'll raise insulin, they won't be that effective for muscle growth. You would've built more muscle and maintained better insulin sensitivity by simply eating a LCHF meal with protein.
2. **Seafood and Algae** – In addition to fish, oysters, shellfish, crabs, lobsters are also quite high in protein. Despite that, they have much more omega-3s and other fats which will lower the mTOR stimulating effect. Also, it's somewhat difficult to over-eat on crab or oysters because of their limited availability and high satiety. Algae like chlorella and spirulina are great plant-based sources of omega-3s and

DHA. Their relatively okay protein content can also help with muscle growth but I wouldn't make my main source of protein algae. Seafood and algae are great for rest days when you want to maintain lower mTOR.

3. **Beans and Legumes** – Azuki beans, kidney beans, lentils, legumes etc. have quite a good amount of plant-based protein but they're also high in fiber, which lowers their insulin response. The Blue Zones are known to eat beans and legumes. However, the longevity effect doesn't come from the beans themselves but because of the overall hormetic lifestyle and caloric restriction. Also, beans are full of phytonutrients and lectins – way too much to eat them every day. You can get much better autophagy activating compounds from other foods that are lower in carbs, cause less digestive issues, and are tastier. Beans and legumes can be eaten sometimes on days you want to limit animal protein consumption and keep mTOR lower.

4. **Nuts and Seeds** – Almonds, pecans, macadamia nuts, walnuts, hazelnuts, chia seeds, pumpkin seeds, and sesame seeds have some good protein that's LowTOR but they also have some phytonutrients. It's not ideal to eat a bunch of nuts every day because (1) they may cause hormonal issues, (2) they may be a potential allergen, (3) they may have become oxidized or exposed to mold, (4) they're easy to overeat, and (5) their nutrient profile isn't something you wouldn't get from other high-quality foods like meat, butter, eggs, or even algae for that matter. A few servings of nuts on some days is acceptable.

5. **Butter and Animal Fats** – Butter is also great for producing short chain fatty acids in the gut, which heals intestinal impermeability. Other animal fats like lard, tallow or goose fat are nTOR if you eat them by themselves or add to your low carb meal. However, excess calories even from fat may still raise insulin and stimulate mTOR because of the increased energy input. That's why use these fats sparingly.

Although some of these foods are low mTOR, they can still stimulate protein synthesis as well as raise insulin. That's why consuming these foods are very context dependent. For instance, potatoes with meat turn a LowTOR food into a HiTOR one because of how carbs and protein interact with each other. Likewise, you can still make yourself anabolic by overeating on fish or algae because of the excess protein intake.

The idea is to simply know how these food combinations will affect the body's biochemistry but in general, on maintenance periods where you haven't worked out or aren't trying to build muscle, it's better to stay on the low-end side of TOR.

mTOR Neutral (nTOR) Foods

Foods that don't really have any significant impact on mTOR and MPS by themselves. They can help you to maintain lean muscle but they won't make you build new ones alone. These ingredients can be combined with either mTOR boosting or autophagy-like foods as to increase their effectiveness. When consumed alone they're quite neutral from the protein kinase perspective but they still have calories so don't think of them as free lunch.

Here are the top 5 nTOR foods:

1. **Olive Oil and Olives** – Rich in polyphenols and healthy fats. However, you have to make sure that your olive oil isn't rancid or oxidized. Use only dark bottled olive oil that hasn't sat on the store shelf for god knows who long. Ideally, freshly pressed extra virgin.

2. **Coconut Oil and Coconuts** – A good source of plant-based saturated fat that has some MCTs. Coconut oil has anti-bacterial properties as well so you can use it for oil pulling your teeth or cleaning the face. MCT oil has even been shown to stimulate Chaperone-mediated autophagy a little bit thanks to the elevation of ketones.

3. **Avocadoes and Avocado Oil** – A low carb high-fat fruit with a lot of monounsaturated fatty acids. They can lower cholesterol and increase potassium intake more than bananas. The small amount of carbs in avocados is quite low which makes it quite neutral on the insulin/TOR scale. However, because of its small protein content, it should be thought of as an additive to your meals not the main source of calories.

4. **Green Leafy Vegetables** – Fibrous vegetables and plants are low in carbs, full of polyphenols, high in fiber and other compounds that help with blood sugar. Broccoli, cauliflower, cabbage, bok choy, and spinach won't really raise mTOR but they can promote autophagy because of their small sulforaphane content.

5. **Fermented Foods** – Sauerkraut, pickles, natto, miso, and kimchi are primarily plant-based but the live bacteria in them actually make these foods animal-based. Sauerkraut won't affect mTOR or insulin but it can help you to build muscle by improving gut health.

Now that we've covered mTOR and anabolism regulating foods, let's turn to the other side of the coin, which is autophagy and cellular catabolism.

Low Autophagy (LowATG) Foods

Foods that can activate AMPK and stimulate autophagy a little bit. They're rich in polyphenols, antioxidants, and other beneficial compounds that will lower blood sugar, improve insulin sensitivity, and trigger a mild hormetic response.

Here are the top 5 LowATG foods:

1. **Coffee** - Coffee induces autophagy and has benefits on cellular metabolism[595]. It can also stabilize blood sugar, enhance fat oxidation, and protect against neurodegeneration, which makes it the perfect drink for fasting. However, too much caffeine will raise cortisol, which can promote inflammation and visceral fat formation around your belly.

2. **Green and Herbal Teas** - Epigallocatechin-3-gallate (EGCG), a green tea polyphenol, stimulates hepatic autophagy[596]. Other herbal teas can also promote with liver cleansing and generally induce a more autophagic state. Bergamot, black tea, chamomile, and ginger tea have polyphenols and other compounds that stimulate autophagy. For biggest effect consume them while fasting. Avoid commercial teas that may have added fruit, sugar, and other carbs.

3. **Apple Cider Vinegar** – Apple cider vinegar lowers blood sugar and suppresses insulin quite a lot. That will indirectly promote ketosis. Whether or not it's going to stop autophagy depends on the type of ACV and when you're taking it. On the Bragg's ACV label it says: *„Contains the amazing Mother of Vinegar which occurs naturally as strand-like enzymes of connected protein molecules."* Raw unfiltered mother contains proteins and bacteria which can technically inhibit autophagy if you take it in a fasted state. The filtered distilled version of ACV without the mother will be better to take while fasting. It'll also lower appetite

and kills of bad bacteria in the gut. Taking vinegar with the mother should be kept around meal time.

4. **Hormetic Herbs and Spices** – Many plants and herbs stimulate autophagy. Curcumin induces autophagy by activating AMPK[597]. Piperine which is a compound found in black pepper induces autophagy[598] and it also boosts the bioavailability of curcumin, which makes it a double whammy! Other similar spices are ginger[599], cinnamon, ginseng, and capsaicin from cayenne pepper[600]. Generally, you'd want to be eating a bunch of herbs like rosemary, thyme, arugula, coriander, parsley, and basil because they're incredibly nutrient dense, virtually zero calories, and with many benefits on blood sugar and cellular turnover.

5. **Polyphenols and Flavonoids** – Certain plant phytochemicals help to protect plants from danger. They have a hormetic beneficial effect on the body. Darker pigments especially are indicative of a high polyphenol count and in folk medicine are said to promote liver health. Maybe because of the autophagy boost, eh?

    a. Phenolic acid includes coffee, teas, grapes, red wine, berries, kiwi, cherries, and plums. For optimal anabolic/catabolic value, you'd want to focus on coffee and tea and have some berries every once in a while. Funny enough, coffee is the No.1 source of polyphenols on the standard western diet. I mean, coffee's great and all but it comes to show how nutrient poor the SAD diet really is.

    b. Stilbenes are associated with resveratrol. Resveratrol from red wine, cranberries, and grape skins is said to have life extension benefits thanks to the polyphenols. Resveratrol also stimulates autophagy and suppresses cancer growth[601]. However, one must be wary of the carbs and alcohol of wine. I wouldn't recommend eating grapes either unless you directly eat just the skins. Red wine's resveratrol content is also quite small and is mainly a marketing hoax. If you want to get some real resveratrol, then consider getting it as a supplement instead.

    c. Lignans are found in legumes, cereal, grains, fruits, algae, flax seeds, and some vegetables. Clearly, it's not ideal to be getting your polyphenols from bread or beans as you can get a lot more of them from other healthier foods.

   d. Low carb berries like bilberries, blueberries, elderberries, seabuckthorn, strawberries, dark cherries as well as dark chocolate and raw cacao are also high in polyphenols and antioxidants. Eat in moderation but not every day.

These LowATG foods can be consumed as part of your daily nutrition even during the limited eating window. Combining these polyphenols with mTOR stimulating foods won't jeopardize the anabolic response that much and will be beneficial for longevity. However, there are some nuances to be kept in mind, such as chasing two rabbits while catching none.

High Autophagy (HiATG) Foods

Foods that will ramp up autophagy quite significantly. These are primarily catabolic ingredients that promote cellular turnover much more so than anything else, especially if you consume them during fasting.

You should eat HiATG foods whenever you want to get into a deeper state of autophagy or liver cleansing. They can indirectly help your body trigger many of the other pathways related to longevity and cellular turnover.

Here are the top 5 HiATG foods:

1. **Berberine** – Berberine (*Berberis vulgaris*) is a compound found in Barberry or other plants with many medicinal benefits, especially in regards to lowering insulin and blood sugar. Berberine has been shown to have a similar effect on lowering blood sugar as metformin, which makes it a great thing to consume after larger meals, and it also activates AMPK[602]. It's better to strategically time your berberine intake after eating HiTOR foods or having carb refeeds because it'll help you to lower the blood sugar and go back into autophagy faster. If you're eating Mod- or LowTOR, then it's not advisable to have berberine all the time because it may drop the blood sugar too much. It's still a poison and the hormetic effect is in the dose.

2. **Medicinal Mushrooms** – Things like Chaga mushroom, Cordyceps, Reishi, turkey tail, lion's mane, and shitake are powerful adaptogens that strengthen the

immune system. They activate the main antioxidant pathway Nrf2 as well as stimulate autophagy[603]. My favourite way to drink coffee is to add a bit of Chaga and Reishi powder to it. It's going to increase the therapeutic effect during the fasted state and prevents the over-stimulation of caffeine.

3. **Caloric Mimetics** – Certain caloric restriction mimetics like Malabar tamarind, rapamycin, metformin, and berberine can activate autophagy[604]. The most effective one is probably berberine but there are others such as bitter melon extract, fenugreek, adiponectin, ursolic acid, spermidine etc. More about these compounds in the chapter about supplements.

4. **Shilajit** – This nutrient-dense mineral has been used in Ayurvedic medicine to energize the body. Modern medicine has shown shilajit to contain fulvic acid and humic acid, which helps with ATP production as well as fighting bacterial infections. It won't stimulate autophagy directly but it can regenerate cells by improving oxygen flow and antioxidant activity. If you've ever tasted shilajit, then you'd know that it may still have an autophagic effect.

5. **Astragalus** – A lot of herbs and plants that belong to the Astragalus family have many medicinal benefits on longevity. Some examples include milkvetch, locoweed, goat's thorn, licorice root, angelica, as well as curcumin, cinnamon, and ginger.

Even if you are eating a higher fat diet, you don't want to roll the dice with your lipid profile and long-term health. That's why it's a good idea to be still consuming a ton of plant polyphenols and other autophagy activating agents that help with lipid clearance and metabolism.

There are no real side-effects to eating a lot of plant polyphenols and vegetables. If you get some sort of digestive strain, then you simply have to focus on healing the gut and establishing a more balanced microbiome. In that case, going through a short elimination protocol that excludes fiber and phytonutrients will be useful. However, these issues should not arise in the first place if you're already practicing intermittent fasting, sleeping enough, keeping stress low, not drinking too much caffeine or suffer from other ailments.

You should strike a balance between the anabolic animal foods and autophagic plant compounds. Most of the nutritional value and calories can come from meat, fish, eggs, and organ meats but the majority of longevity-boosting polyphenols and flavonoids will be derived from vegetables, berries, herbs, and spices.

Whenever you eat, consider the purpose of that particular meal – do you want to promote higher anabolism, maintenance, or stimulate deeper autophagy? With that in mind, choose between these 5 categories and structure your nutrition based on that.

This lays the framework for what kind of foods you'd be actually eating on the Metabolic Autophagy Diet. The next chapter will delve into the full list of potential things you can eat.

Chapter XV
Metabolic Autophagy Foods

„You are what you eat ate"

Michael Pollan

That was an overview of the anabolic/catabolic score and the rationale behind it. I gave you only the top 5 foods of each category but there are many other things we can eat on this protocol.

Keep in mind that the benefit of every food is very context dependent and relative to what you're trying to do in that particular moment in regards to the anabolic-catabolic cycle. That's why for optimal nutrition, it matters what you eat, in what amounts, and at what time. Also, the combinations of certain foods will yield a completely different response as you've already seen.

Now I'm going to list out the principles of every macronutrient from the perspective of optimizing longevity and lean muscle preservation. I'll go through each macronutrient and food group with the pursuit of inducing Metabolic Autophagy.

Protein and Amino Acids

Let's start with probably the most essential macronutrient, which is protein. You'd want to optimize the protein and amino acid profile of your diet because not all protein is the same.

The most abundant proteinogenic or protein creating amino acid is methionine. It has an important role in growing new blood vessels and serves as the initiating amino acid of protein synthesis[605]. Unfortunately, it's been linked to some cancers as well.

Methionine restriction is associated with extended lifespan and reduced IGF levels[606,607,608]. Over-consumption of meat may reduce longevity just because of methionine but we also know that protein reacts differently with different foods.

A 2011 study found that glycine supplementation had the same effects on life-extension and IGF-reduction as methionine restriction[609]. Muscle meat and flesh, in particular, is higher in methionine whereas glycine can be found more in organ meats, tendons, ligaments, drumsticks, and all these ancestral bone broth parts.

You don't need to be going on a low methionine diet just to maybe live a bit longer. The best life-extension still comes from fasting, eating low carb, and training. However, you would maybe want to incorporate more glycine-rich proteins instead to kind of „balance out" the methionine. It's not that natural for our bodies to be eating sirloin steak all the time either. In the past, people would eat the entire animal from „nose to tail".

High-quality protein that optimizes longevity and performance includes all the organ meats, bone marrow, the skin, ligaments, and tendons, not just the steak, filets, and pork chops. These unconventional bits are also more nutrient dense and packed with the other essential vitamins, such as B6, B12, phosphorus, magnesium, etc. Liver and heart, in particular, are probably the No.1 superfoods on Earth.

It's often thought that organ meats and liver aren't healthy for you because they're said to contain toxins. However, the function of the liver isn't to deposit toxins or hold onto them – it simply filters them out and removes potential infiltrations. In fact, the fat tissue would have much more of these toxins. Of course, there's going to be a huge difference of quality in grass-fed and factory grown animals but if the liver being toxic is your concern of not eating liver, then you'd want to avoid all meat, including grain-fed steaks.

The recommended dietary allowance (RDA) for protein is 0.36g/lb of bodyweight which for an average individual weighing between 150-180 pounds would be 55-70 grams of protein per day[610]. However, this is not ideal for the majority of the population and most people actually need more, especially if they're exercising.

In general, the optimal amount of protein tends to be somewhere between 0.7-1.0 g/lb of lean body mass which for the same average individual weighing between 150-180 pounds would be 110-160 grams of protein at a minimum. There are no seeming benefits to eating more than 0.8 g/lb of LBM, even when trying to build muscle.

Here's a List of Proteins to Eat:

Food Source	Calories (kCal)	Fats (g)	Net Carbs (g)	Protein (g)	ACS
MEAT					
Bacon, 1 slice (~ 8g), baked	44	3.5	0	2.9	ModTOR
Beef, Sirloin Steak, 1 oz, 28 g	69	4	0	7.7	ModTOR
Beef, Ground, 5% fat, 1 oz, 28 g	44	1.7	0	6.7	ModTOR
Beef, Ground, 15% fat, 1 oz, 28 g	70	4.3	0	7.2	ModTOR
Beef, Ground, 30% fat, 1 oz, 28 g	77	5.1	0	7.1	ModTOR
Beef, Bottom Round, 1 oz, 28 g	56	2.7	0	7.6	ModTOR
Lamb, ground, 1 oz, 28 g	80	5.6	0	7	ModTOR
Lamb chop, boneless, 1 oz, 28 g	67	3.9	0	7.3	ModTOR
Pork chop, bone-in, 1 oz, 28 g	65	4.1	0	6.7	ModTOR
Pork ribs, 1 oz, 28 g	102	8.3	0	6.2	ModTOR
Ham, smoked, 1 oz, 28 g	50	2.6	0	6.4	ModTOR
Hot dog, beef, 1 oz, 28 g	92	8.5	0.5	3.1	ModTOR
Veal, 1 oz, 28 g	42	1	0	8	ModTOR
Chicken Drumstick, Medium	103	7	0	10	ModTOR
Chicken Wing, Medium	73	5	0	7	ModTOR
Beef Liver, 100	132	3.7	3.7	19.6	ModTOR

Duck, 100 g	337	28	0	19	ModTOR
Wild Boar, 100 g	138	4	0	25	ModTOR
Beef Heart, 100 g	142	7	0	17	ModTOR
Beef Kidney, 100 g	131	6.5	0	16.7	ModTOR
Lamb Brain, 100 g	273	23	0	17	ModTOR
T-Bone Steak, 100 g	247	16	0	24	ModTOR
Rib Eye Steak, 100 g	291	22	0	24	ModTOR
Beef Tenderloin, 100 g	324	25	0	24	ModTOR
FISH					
Cod, 3 oz, 85 g	70	0.6	0	15	ModTOR
Flounder, 3 oz, 85 g	60	1.6	0	11	ModTOR
Sole, 3 oz, 85 g	80	2	0	14	ModTOR
Salmon, 3 oz, 85 g	177	11	0	17	ModTOR
Perch, 3 oz, 85 g	100	1	0	21	ModTOR
Trout, 3 oz, 85 g	120	5	0	17	ModTOR
Shark, 3 oz, 85 g	110	3.8	0	18	ModTOR
Scallops, 3 oz, 85 g	94	0.7	4.6	17	ModTOR
Shrimp, 3 oz, 85 g	90	1.3	1.2	16.5	ModTOR
Tuna, 3 oz, 85 g	110	0.5	0	25	ModTOR
Salmon, 100 g	208	13	0	20	ModTOR
Sardines, 100 g	208	11	0	25	ModTOR
Herring, 100 g	158	9	0	18	ModTOR

Mackerel, 100 g	305	25	0	19	ModTOR
Anchovies, 100 g	138	6	0	20	ModTOR
Sprats, 100 g	150	11	0	10	ModTOR
Oysters, 100 g	199	13	12	9	LowTOR
Mussels, 100 g	172	4.5	7	24	ModTOR
Lobster, 3 oz, 85 g	122	1.6	2.7	22	ModTOR
Crab, 100 g	83	0.7	0	18	ModTOR
EGGS					
Chicken Egg, 1 Large, 50 g	72	4.8	0.4	6.3	HiTOR
Goose Egg, 1 Large, 144 g	267	19	1.9	20	HiTOR
Quail Egg, 1 Small, 9 g	14	1	0	1.2	ModTOR
Duck Egg, 1 Medium, 70 g	130	9.6	1	9	HiTOR
Turkey Egg, 1 Medium, 79 g	135	9.4	1	11	HiTOR
Herring Eggs, 100 g	74	1.9	4.5	9.6	LowTOR
Salmon Roe, 1 tbsp, 14 g	20	1	0.2	3.1	LowTOR

Here's a List of Proteins to Combine With Carbs:

If you want to maximize the anabolic effects of protein and spike your insulin really high, then you can combine the protein with carbs. This can be beneficial every once in a while to really trigger mTOR and promote muscle hypertrophy. For performance goals, this is extremely effective but not for longevity. That's why you don't want to be spiking anabolism all the time with every meal.

Whenever you do choose to spike insulin and mTOR, then it's best to do it with low-fat, high carb, and moderate-to-high protein. This will prevent the fatty acids being directly stored as body fat with elevated levels of insulin. The amino acids and glucose will replenish muscle glycogen and you'll be able to clear the bloodstream faster.

Eating lean protein with no fat on a low carb diet isn't ideal if you're trying to build muscle or maintain it. You need the other fattier chunks of protein for that.

If, however, you're trying to shed some body fat really fast, then you can also dial down your dietary fat intake and get into a much deeper caloric deficit by doing that. Body fat is fuel and if your body is already high fat, then all you need is low carb and high protein. This will work only as a short-term rapid weight loss tool but in the long run, you still want to feed your body enough of the healthy fats and proteins.

Food Source	Calories (kCal)	Fats (g)	Net Carbs (g)	Protein (g)	ACS
Chicken Breast, 100 g	165	3.6	0	31	ModTOR
Tilapia, 100 g	129	2.7	0	26	ModTOR
Whitefish, 100 g	172	8	0	24	ModTOR
Egg Whites, 1 large (33 g)	17	0.1	2	3.6	HiTOR
Cottage Cheese, 1%, 100 g	72	1	3	12	ModTOR
Cottage Cheese, 2%, 100 g	86	2.5	3.7	12	ModTOR
Cottage Cheese, 5%, 100 g	107	4.7	3.7	12	HiTOR
Pork, Tenderloin, 100 g	143	3.5	0	26	ModTOR
Tuna, 100 g	130	0.6	0	29	ModTOR
Rabbit, 3 oz, 85 g	175	7.1	0	26	ModTOR
Ground Beef, 90% Lean, 100g	176	10	0	20	ModTOR
Curd, 5%, 100 g	98	4.3	3.4	11	HiTOR
Cod, 1 fillet, 180 g	189	1.5	0	41	ModTOR
Whey Protein, 1 scoop, 32 g	113	0.5	2	25	HiTOR

Carbohydrates and Vegetables

When it comes to carbohydrates, then the main idea is to eat them according to your metabolic conditions and goals. Carbs are not as anabolic as protein, but they emphasize the mTOR stimulating effect of whatever you eat.

Whenever you're having carb refeeds, you should also focus on nutrient density and the nutritional value of what you eat. It's not worth it to have empty calories that make you over-eat while leaving you unsatisfied. Although cheat days with glorious junk food may sound appealing, it gets boring after a while and totally unnecessary. Trust me, I've experimented with this…

Keep in mind how combining carbs with any other macronutrient, whether that be protein or fat, significantly increases the insulin response (See Figure 73). That's why you shouldn't follow a high-carb-high-fat diet – it's a recipe for hyperinsulinemia and insulin resistance.

Whenever eating foods that are higher in carbohydrates, you should stay relatively lower with the fats, and stick to leaner proteins. That way you'll avoid glycation of fatty acids and limit the formation of AGEs.

It's not advisable to combine cholesterol with carbs or sugars either because of the potential of oxidation. The morning toast with eggs and bacon is literally a ticking time bomb for atherosclerosis. However, take out the bread and go zero-carb and you'll prevent that completely. *It's not the egg, it's the bread…*

Fructose and Fruit

It's also recommended to limit fruit consumption. Fructose can only be metabolized by the liver and can't be used as muscle glycogen. It, therefore, is almost completely useless to the body. In high amounts, it actually becomes toxic because of the liver having to work extra hard.

- **Excess fructose can damage the liver and cause insulin resistance**, which is a precursor to atherosclerosis, diabetes, obesity, and fatty liver disease.

- **Fructose can also cause rapid leptin resistance.** Leptin controls your appetite and metabolism. If you're resistant, then you'll gain weight easily and can't stop gorging yourself.
- **The reaction of fructose with proteins is 7 times higher than with glucose.** Because of that, AGEs get produced at an even greater rate. While your body can't use fructose as energy, the bad bacteria in your gut can and that may cause imbalances in your healthy gut flora.
- **Fructose causes more oxidative stress and inflammation than glucose.** Cancer cells feed upon sugar, especially fructose, and thrive in an oxidized environment.
- **Excess fructose also affects brain functioning**, in terms of appetite regulation and blood sugar. In rats, it impairs memory.
- **Fructose doesn't replenish muscle glycogen and won't promote muscle hypertrophy.** I mean, if you're planning to eat carbs, then why not eat the carbs that will actually make you stronger and more built like potatoes and rice?

I'm not trying to say that fruit is bad – just that excess fructose in the body comes with an array of negative side-effects and that it's not optimal for consumption. There are many people who eat a raw fruit-based diet and seem to be perfectly fine. However, with the principles of Metabolic Autophagy kept in mind, fruit is a weak player.

Fruit isn't simply something you could nor should eat year-round. In nature, you can get fruit during certain seasons and only for a limited amount of time. The function of fruit is to make you rapidly gain extra fat for the coming winter. That's why grizzly bears eat a bunch of berries and salmon in the autumn. Unfortunately, in modern society, the nutritional winter never comes and people actually eat more calories during the holiday season.

Of course, fruit may have some vitamins, minerals, and co-enzymes that help with your overall metabolic process but its nutritional value pales in comparison to other vegetables. For general health, you'd be better off eating cruciferous, and for muscle growth, you'd want to be eating starchy tubers anyway. So, there's no justification to be eating fruit every day. You can have like an apple or two a few times a week but any more than that is pushing it too far.

Whenever you do eat fruit during refeeds, then opt for the ones lower in fructose and higher in glucose. That way you'll keep the liver glycogen more depleted and it'll be easier to get back into autophagic ketosis. The glucose will also be more beneficial for muscle glycogen re-synthesis.

How Many Vegetables to Eat?

In terms of low carb foods like vegetables, tubers, and berries, then how much you can eat depends mostly on your personal preference.

On the ketogenic diet, your daily carb intake would fall somewhere between 30-50 grams of carbs (5-15%). That number can change depending on your level of physical activity, how insulin sensitive you are, the degree of keto-adaptation, and also how long you fast for.

On the Metabolic Autophagy protocol, you'll be very glucose tolerant, meaning you have such a big buffer-zone for eating carbs. Resistance training and fasting both deplete your glycogen quite rapidly and you can get away with a higher carbohydrate consumption without jeopardizing ketosis or autophagy.

However, the purpose of eating carbs isn't to eat as many of them as we can safely handle. Instead, it's about promoting a certain metabolic environment that would make the body more autophagic by default and then use them as a tool for triggering mTOR.

In general, your habitual everyday consumption of carbs would still fall somewhere between 10-30 grams. That's the default you'd want to fall back to. On some harder workout days, you can go up to 50-100 grams from healthy tubers. For anabolic refeeds, it may reach 200 grams or so, depending on your goals.

Figure 84 Metabolic Autophagy Carb Guidelines

Honestly, there isn't a real limit to how much broccoli or cauliflower you should eat because they're an nTOR food and low in calories. For them to kick you out of ketosis you'd also have to be eating like 2 kilos, which would make your gut explode before it could happen.

Some veggies and salads are very low in soluble carbohydrates and they're mostly indigestible fiber. It can bump up the volume of your meals while keeping the calories relatively lower. However, too much fiber and vegetables can cause constipation and digestive issues. So, you have to know where you feel best at.

Short periods of zero-carb carnivore can be quite good for giving the gut some rest but for optimal longevity, you'd want to be eating some veggies packed with the longevity-boosting compounds. It's also probably not ideal for food intolerances or the microbiome to be eating just meat for the rest of your life.

The majority of your food should come from nutrient-dense animal meats, but you do want to get a bunch of vegetables and wild plants into your diet. Therefore, you can eat these healthy veggies I'm about to list here in unlimited amounts (*Or as much as your gut can handle*):

Here's a List of Vegetables You Can Eat:

Food Source	Calories (kCal)	Fats (g)	Net Carbs (g)	Protein (g)	ACS
Asparagus, 100 g	20	0.1	3.9	2.2	nTOR
Artichokes, 100 g	47	0.2	11	3.3	nTOR
Broccoli, 100 g	32	0.4	7	2.8	nTOR
Cauliflower, 100 g	25	0.3	5	1.9	nTOR
Mushrooms, 100 g	22	0.3	3.3	3.1	nTOR
Onion, Green, 100 g	32	0.2	7	1.8	nTOR
Onion, White, 100 g	40	0.1	9	1.1	nTOR
Romaine Lettuce, 100 g	17	0.3	3.3	1.2	nTOR
Butterhead Lettuce, 100 g	13	0.2	2.2	1.4	nTOR
Shallots, Raw, 100 g	72	0.1	17	2.5	nTOR
Beet Greens, 100 g	42	0.2	10	1.6	nTOR
Bok Choy, 100 g	12	0.2	1.8	1.6	nTOR
Spinach, 100 g	23	0.4	3.6	2.9	nTOR
Alfalfa Sprouts, 100 g	29	0.7	3.8	4	nTOR
Swiss Chard, 100 g	19	0.2	3.7	1.8	nTOR
Arugula, 100 g	25	0.7	3.7	2.6	nTOR
Celery, 100 g	14	0.2	3.4	0.7	nTOR
Carrots, 100 g	41	0.2	10	0.9	nTOR
Turnip, 100 g	28	0.1	6	0.9	nTOR

Iceberg Lettuce, 100 g	14	0.1	3	0.9	nTOR
Asparagus, 100 g	20	0.1	4	2.2	nTOR
Eggplant, 100 g	25	0.2	6	1	nTOR
Tomatoes, 100 g	18	0.2	3.9	0.9	nTOR
Green Bell Pepper, 100 g	20	0.2	4.6	0.9	nTOR
Cabbage, 100 g	25	0.1	6	1.3	nTOR
Green Beans, 100 g	31	0.1	7	1.8	nTOR
Brussels Sprouts, 100 g	43	0.3	9	3.4	nTOR
Kale, 100 g	49	0.9	9	4.3	nTOR
Sea Kelp, 100 g	43	0.6	10	1.7	nTOR
Red Algae, 100 g	21	0.1	8.8	1.3	LowATG
Chlorella Powder, 1 oz	115	2.6	6.5	16.4	LowATG
Sea Lettuce, 100 g	130	0.6	41	22	nTOR
Spirulina, 1 tbsp	20	0.5	1.7	4	LowATG
Nori, 100 g	35	0.3	5	6	nTOR
Wakame, 100 g	45	0.6	9.1	3	nTOR

When it comes to cruciferous vegetables and leafy greens, then it's a good idea to cook them slightly as to break down their cell wall a bit for better nutrient absorption. Raw veggies are primarily fiber and very hard to digest by the human digestive tract. Too much rawness can cause digestive issues, bloating, constipation as well as hinder the metabolism directly.

Cruciferous vegetables and leafy greens have different compounds that can damage our thyroid functioning, especially goitrogens and oxalates. Spinach, Swiss chard, and beets should be slightly cooked to reduce their oxalic acid content.

Here's a List of the Yellow Zone Vegetables You

However, not all vegetables are the same. To keep our gut healthy and inflammation low, we'd want to avoid things like grains, gluten, lectins, legumes, beans, and things like that for the vast majority of time. Unless we're doing it as hormetic conditioning.

There are even some seemingly healthy vegetables and tubers that we would want to limit. Whether that be because of their slightly too high carbohydrate content or because eating too much of them may cause some digestive issues. I call them the 'Yellow Zone Carbs' – don't eat them every day but include them into your diet a few times a week.

Yellow Zone Carbs include everything from bell peppers, tomatoes, squash, zucchini, berries, beetroot, carrots, turnips etc.

Food Source	Calories (kCal)	Fats (g)	Net Carbs (g)	Protein (g)	ACS
Carrots, 100 g	41	0.2	10	0.9	nTOR
Turnip, 100 g	28	0.1	6	0.9	nTOR
Beetroot, 100 g	43	0.2	10	1.6	nTOR
Garlic, 1 clove (3 grams)	4	0	1	0.2	nTOR
Green beans, 1 oz	10	0.1	1.3	0.5	nTOR
Bell Pepper, 1 oz	6	0	0.8	0.2	nTOR
Pickles, 100 g	11	0.2	2.3	0.3	nTOR
Peas, 100 g	81	0.4	14	5	nTOR
Spinach, 1 oz	7	0.1	0.4	0.8	nTOR

Squash, 100 g	17	0.3	3.1	1.2	nTOR
Squash, Butternut, 100 g	45	0.1	12	1	nTOR
Squash, Spaghetti, 100 g	31	0.6	7	0.6	nTOR
Tomato, 1 oz	5	0	0.8	0.3	nTOR

Here's a List of the High Carb Foods to Eat on Refeeds:

Food Source	Calories (kCal)	Fats (g)	Net Carbs (g)	Protein (g)	ACS
White Potato, 100 g	77	0.1	17	2	HiTOR
Sweet Potato, 100 g	86	0.1	20	1.6	ModTOR
White Rice, 100 g, Cooked	130	0.3	28	2.7	HiTOR
Brown Rice, 100 g, Cooked	111	0.9	23	2.6	ModTOR
Basmati Rice, 100 g, Uncooked	349	0.6	77	8	ModTOR
Oatmeal, 1 Cup Cooked, 234 g	158	3.2	27	6	nTOR
Grits, 1 Cup Cooked, 234 g	143	0.5	31	3.4	HiTOR
Corn, 100 g	131	1.4	26	3.6	ModTOR
Cream of Wheat, 100 g	369	0.5	78	11	HiTOR
Banana, 1 Medium, 105 g	105	0.4	27	1.3	HiTOR
Orange, 1 Large, 184 g	87	0.2	22	1.7	nTOR
Grapefruit, 1 Large, 246 g	104	0.4	26	2	nTOR
Quinoa, 100 g, Uncooked	368	6.1	64	14	nTOR
Dates, 100 g	282	0.4	75	2.5	HiTOR

Figs, 100 g	74	0.3	19	0.7	HiTOR
Lentils, 100 g, Cooked	116	0.4	20	9	ModTOR
Beans, 100 g, Baked	155	5	22	6	ModTOR
Kiwi, 100 g	61	0.5	15	1.1	nTOR
White Bread, 1 Slice, 25 g	66	0.8	12	2.3	HiTOR
Ezekiel Bread, 1 Slice, 34 g	80	0.5	15	4	ModTOR
Whole Wheat Bread, 1 Slice, 28 g	69	0.9	12	3.6	ModTOR
Pumpkin, 100 g	26	0.1	7	1	nTOR
Apples, 1 Medium, 182 g	95	0.3	25	0.5	nTOR
Peaches, 1 Medium, 150 g	59	0.4	14	1.4	ModTOR
Pineapple, 100 g	50	0.1	13	0.5	ModTOR
Ketchup, 1 tbsp	19	0	4.5	0.2	HiTOR
Pasta, 100 g, Cooked	131	1.1	25	5	HiTOR
Puff Pastry, 100 g	558	39	46	7	HiTOR

I included some processed junk food into the list to give some reference of their macronutrient content. Notice how they're comprised of the worst ratios you'd want – high fat, high carb, low protein. It won't even help with muscle hypertrophy, not to mention longevity.

Of course, you can follow the 80/20 rule and have a few cheat days here and there – I'm not judging. However, you have to take full responsibility for them. Meaning, have them planned out, put higher energetic demands on your body before that, deplete your muscle glycogen, control your food intake, don't get into the uncontrollable binging cycle, and savor it completely.

If you're not able to enjoy what you eat, then it will never satisfy. Not appreciating even healthy food may turn a nutritious dish into empty calories just because you're not fully present.

How often to have higher carb refeeds depends on your workout routine, how much fasting you do, do you want to build muscle, or gain fat, and how many carbs you eat.

In general, it's not advisable to have high mTOR and high insulin refeeds any more than once a week. People who train a lot can get away with 2 times but the average person could even do it only once or twice a year. It's not necessary to have carb refeeds because you can safely build very lean muscle over the long term with a purely ketogenic approach. However, those carbs do make it easier and I find them beneficial for swapping in and out of ketosis as well.

Here's a List of the Fruit and Berries You Can Eat:

Food Source	Calories (kCal)	Fats (g)	Net Carbs (g)	Protein (g)	ACS
Low Carb Berries					
Rhubarb, 100 g	21	0.2	4.5	1	nTOR
Raspberries, 100 g	53	0.7	12	1.2	LowATG
Blueberries, 100 g	57	0.3	14	0.7	LowATG
Strawberries, 100 g	33	0.3	8	0.7	LowATG
Blackberries, 100 g	43	0.5	10	1.5	LowATG
Elderberries, 100 g	73	0.5	18	0.7	LowATG
Cranberries, 100 g	46	0.1	12	0.4	LowATG
Cherries, 100 g	50	0.3	12	1	LowATG
Low Carb Fruit					

Avocado, 1 oz	47	4.4	0.6	0.6	nTOR
Olives, 1 oz	65	7.5	1.9	0	nTOR
Coconut, 1 Cup, 80 g	283	27	12	2.7	nTOR
Watermelon, 100 g	30	0.2	8	0.6	nTOR
Cantaloupe, 100 g	34	0.2	8	0.8	nTOR
Honeydew, 100 g	36	0.1	9	0.5	nTOR

Like I said earlier, fruit doesn't have much value on the nutrient density hierarchsy. Not because it lacks micronutrients and vitamins, but because fructose isn't that useful for overall health nor muscle hypertrophy.

You can get much better results from less effort by eating primarily vegetables and tubers. Low carb berries are an exception because they're more fiber than sugar. Plus, things like elderberries and blueberries have super-concentrated antioxidants and polyphenols that are great.

It's okay to eat like a few servings of fruit a few times a week but it has to be done carefully and with the right food combination.

- **Don't eat fruit with high-fat high cholesterol foods** like eggs, meat, or bacon. It'll oxidize the cholesterol and fatty acids again, creating more inflammation. *Imagine if you'd put apple jam with high fructose corn syrup on your toast and eggs...*
- **Have a few pieces of fruit on carb refeeds.** The low-fat high carb context would make fruit safer to consume. However, too much fruit would fill up the liver glycogen and inhibit ketosis. You would want to eat primarily glucose-rich fruit, such as ripe bananas, dates, and oranges.
- **Eat fruit seasonally.** Despite having access to fruit from the supermarket year-round. You wouldn't want to be eating fructose across all seasons. It's important to have very low carb periods as to keep the body's circadian rhythms optimized.

The best time to consume some fruit is during the harvest season when you could find it growing within your local environment.

When it comes to pesticides and GMOs, then avoiding fruit is also a huge win-win situation. You'll prevent the high fructose load on your liver AND you'll avoid the increased inflammation from all the chemicals and toxins that get sprayed on conventional fruit.

From an evolutionary perspective, not eating fruit in the modern context is an extremely smart and effective strategy. Think about it.

Here's a List of the Herbs and Spices You Should Eat:

In addition to the regular carbohydrates, there are other herbs and spices we would want to add to our diet. There are dozens of herbs that are used as medicine but they're also incredibly potent in boosting longevity.

Food Source	Calories (kCal)	Fats (g)	Net Carbs (g)	Protein (g)	ACS
Rosemary, 1 oz	36	1.6	5.8	0.9	LowATG
Basil, 1 oz	6.4	0.2	0.7	0.9	LowATG
Coriander, 1 oz	83	5	15	3.5	LowATG
Cilantro, 1 oz	6.4	0.1	1	0.6	LowATG
Thyme, 1 oz	28	0.5	6.8	1.6	LowATG
Parsley, 1 oz	10	0.2	1.8	0.8	LowATG
Cardamom, 1 oz	87	1.9	19	3	LowATG
Cumin, 1 oz	105	6.2	12.4	5	LowATG
Turmeric, 1 oz	99	2.8	18.2	2.2	LowATG

Cinnamon, 1 tbsp	19	0.1	6	0.3	LowATG
Ginger, 1 oz	19	0.2	4.3	0.4	LowATG
Ginseng, 1 oz	106	2	23	0	LowATG
Black Pepper, 1 tsp	7	0.1	1.9	0.3	LowATG
Cayenne Pepper, 1 tsp	6	0.3	1	0.2	LowATG

Fats and Lipids

As said earlier, the minimum daily dietary fat intake is 20-30 grams, which isn't that good either.

A healthy fat consumption on non-ketogenic diets should be somewhere between 20-35%, which on a 2000 daily caloric intake would be around 40-80 grams. On a low carb keto diet, it should be slightly higher but you don't need to be eating copious amounts of dietary fat because more won't be always better.

After going through the keto-adaptation process, you don't need to be consuming a ton of extra fat just to meet your daily caloric needs. In fact, it can actually be counter-productive both for body composition as well as performance.

Like I said in Chapter The Case Against Sugar (And Fat), I don't want to be eating any more calories from fat than I need. On a ketogenic diet, fatty acids are used for energy production but there's still a point of diminishing returns. Especially when you add some carb refeeds into your diet, it's not necessary to go all out on fats. Instead, here are the modified macronutrient ratios most people can stick to:

- Carbs should still be quite low most of the time around 5-15%
- Protein can be increased for the muscle building benefits up to 25-30%
- Fat will stay around 55-65%, which will cover the essentials and gives extra energy.

Most people can stick between 100-180 grams of fat and be perfectly healthy. There is no metabolic advantage to eating more fat. Fat should be thought of as caloric leverage, not as a staple.

If you want to raise your blood ketones, then just fast for a bit longer. This is a much smarter way of driving yourself into deeper ketosis as well as autophagy. It'll also help to maintain mild caloric restriction.

Here's a List of the Nuts and Seeds You Can Eat:

Although nuts and seeds can be a good snack, they're somewhat easy to overconsume and with not that high nutrient density. They're definitely high in micronutrients and healthy fats but compared to something like eggs, fish, or meat they're slightly less satiating.

I wouldn't recommend eating nuts when you're trying to lose body fat because they're not that filling in terms of their caloric content.

Eating large amounts of nuts isn't ideal because of their phytate content either. Phytic acid is a compound found in nuts and seeds that tries to protect the nuts from being eaten. In a small hormetic dose, they're great but not something you'd want to make a staple in your diet.

In terms of the nutritional value of nuts and seeds, then they caloric content of nuts and seeds isn't justifiable either. Although things like Chia seeds and flaxseeds can be a good source of low mTOR plant-based protein, their fatty acid content and amino acid profile isn't that bioavailable.

Like mentioned earlier, humans can convert only about 8% of ALA into DHA and you definitely want to be getting more DHA rather than ALA. From longevity as well as a performance perspective, you're better off spending your calories and money on high-quality fish like salmon or oysters. It'll be better for your brain, muscles, cellular membrane, heart, as well as physical output.

Definitely limit your peanut consumption because they're high in omega-6s, phytonutrients, lectins, and tend to get oxidized on shelves. They're not even nuts – they're legumes that kind of look like nuts.

Food Source	Calories (kCal)	Fats (g)	Net Carbs (g)	Protein (g)	ACS
Almonds, 1 oz	170	15	3	6	LowTOR
Brazil Nuts, 1 oz	186	19	1	4	LowTOR
Cashews, 1 oz	160	13	7	5	LowTOR
Chestnuts, 1 oz	55	0	13	0	LowTOR
Chia Seeds, 1 oz	131	10	0	7	LowTOR
Coconut, dried, 1 oz	65	6	2	1	LowTOR
Flax Seeds, 1 oz	131	10	0	7	LowTOR
Hazelnuts, 1 oz	176	17	2	4	LowTOR
Madadamia Nuts, 1 oz	203	21	2	2	LowTOR
Peanuts, 1 oz	157	13	3	7	LowTOR
Pecans, 1 oz	190	20	1	3	LowTOR
Pine Nuts, 1 oz	189	20	3	4	LowTOR
Pistachios, 1 oz	158	13	5	6	LowTOR
Pumpkin Seeds, 1 oz	159	14	1	8	LowTOR
Sesame Seeds, 1 oz	160	14	4	5	LowTOR
Sunflower Seeds, 1 oz	150	11	4	3	LowTOR
Walnuts, 1 oz	185	18	2	4	LowTOR
Almond Butter, 2 tbsp	196	18	6	7	LowTOR
Peanut Butter, 2 tbsp	188	16	6	8	LowTOR

Cashew Butter, 2 tbsp	188	16	9	5	LowTOR
Macadamia Nut Butter, 2 tbsp	194	20	4	4	LowTOR

Here are the Dairy and Milk Foods You Can Eat:

Like said earlier, dairy is one of the most natural anabolic foods that support growth in growing organisms. However, it's not optimal later in life because dairy also raises IGF-1 quite a lot, which will increase inflammation, insulin, skin issues, and accelerated aging.

Dairy can be subtly damaging for most people. Its main inflammatory insult comes from lactose, which can cause autoimmunity and allergies. There are 2 proteins in dairy – casein and whey. Whey gets digested quite rapidly, which is why it raises insulin and IGF-1. Casein is absorbed much slower, which is why it can promote more inflammation and things like skin problems.

To avoid additional inflammation from dairy, you should opt for organic, grass-fed, unpasteurized products. Commercial cattle get injected with antibiotics and steroids, which jeopardizes the quality of their meat and milk. It's also higher in omega-6s but lower in the fat-soluble vitamins like K, A, and D. Pasteurizing dairy also denatures the casein proteins, which makes it harder to digest and more inflammatory.

You can make the argument that milk and dairy are „natural" and very healthy because of its potent protein and fat content. However, Metabolic Autophagy isn't about doing what's „natural" – it's about optimizing human nutrition for longevity and performance.

Fortunately, things like butter and ghee are both virtually zero lactose and casein, which makes them relatively safe to consume. Very few people react negatively to these pure fats. Ghee is even clarified and devoid of allergens. It's also valued in Ayurvedic traditions, which practices a very vegetarian heavy diet. Despite that, they still proponed the use of these healthy animal fats to some extent because of their nutritional value. Fat is essential.

It's not recommended to be consuming dairy on a daily basis because you'd gain bigger benefits for both muscle hypertrophy as well as general health from other foods. If you do choose to consume it, then stick to 2-3 servings a week from raw unpasteurized sources.

Food Source	Calories (kCal)	Fats (g)	Net Carbs (g)	Protein (g)	ACS
Buttermilk, 1 oz	18	0.9	1.4	0.9	HiTOR
Blue Cheese, 1 oz	100	8.2	0.7	6.1	ModTOR
Brie Cheese, 1 oz	95	7.9	0.1	5.9	ModTOR
Cheddar Cheese, 1 oz	114	9.4	0.4	7.1	ModTOR
Colby Cheese, 1 oz	110	9	0.7	6.7	ModTOR
Cottage Cheese, 2%, 1 oz	24	0.7	1	3.3	ModTOR
Cream Cheese, 1 oz	97	9.7	1.1	1.7	ModTOR
Feta Cheese, 1 oz	75	6	1.2	4	ModTOR
Monterey Jack Cheese, 1 oz	106	8.6	0.2	7	ModTOR
Mozarella Cheese, 1 oz	85	6.3	0.6	6.3	ModTOR
Parmesan Cheese, 1 oz	111	7.3	0.9	10.1	ModTOR
Swiss Cheese, 1 oz	108	7.9	1.5	7.6	ModTOR
Marscapone Cheese, 1 oz	130	13	1	1	ModTOR
Cream, Half n Half, 1 oz	39	3.5	1.3	0.9	HiTOR
Heavy Cream, 1 oz	103	11	0.8	0.6	HiTOR
Sour Cream, Full Fat, 1 oz	55	5.6	0.8	0.6	ModTOR
Whole Milk, 1 oz	19	1	1.5	1	HiTOR
Milk, 2%, 1 oz	15	0.6	1.5	1	HiTOR
Skim Milk, 1 oz	10	0	1.5	1	HiTOR

Here are the Fats You Can Eat:

In addition to nuts and dairy, there are many other healthy fats we should be eating. Meat, fish, and eggs also have plenty of fats on them, but sometimes it may not be enough.

Although we can say that saturated fat isn't the main cause of heart disease, it's still not a superfood you could eat in unlimited amounts. The potential evolutionary trade-off of excess fat consumption simply isn't a wise move and not optimal for longevity.

Instead, you should aim for eating more fattier pieces of meat that come with additional protein so you'd be getting more fat from whole foods. Adding the other healthy fats should be thought of as extra caloric fillers. *And to bathe your vegetables in...delish.*

Food Source	Calories (kCal)	Fats (g)	Net Carbs (g)	Protein (g)	ACS
Butter, 1 oz	204	24	0	0	LowTOR
Ghee, 1 oz	225	25	0	0	LowTOR
Lard, 1 oz	255	28	0	0	LowTOR
Beef Tallow, 1 oz	253	28	0	0	LowTOR
Avocado Oil, 1 oz	248	28	0	0	nTOR
Cocoa Butter, 1 oz	248	28	0	0	nTOR
Coconut Oil, 1 oz	241	28	0	0	nTOR
Flaxseed Oil, 1 oz	248	28	0	0	nTOR
Macadamia Oil, 1 oz	248	28	0	0	nTOR
MCT Oil, 1 oz	248	28	0	0	nTOR
Olive Oil, 1 oz	241	28	0	0	nTOR

Red Palm Oil, 1 oz	260	28	0	0	nTOR
Coconut Cream, 1 tbsp	49	5	1	0.5	LowTOR
Coconut Milk, 1 Cup, 240 g	552	57	13	5	LowTOR

Fats on Refeeds

As was stated already, you want to keep your fat intake as low as possible during refeeds. That's why there's no list of fatty foods. However, if you do choose to opt in for some junk food, you'll have to stay mindful of your intake.

The fats in processed foods are the worst kind – vegetable oils and trans fats – which are bad for your health. Consuming them one day won't do you much harm though, especially if you stay diligent on your diet for the rest of the week. Nevertheless...

You should eat your main course meals as clean as possible. I.E. you have a dish of curcumin-curry-chicken with some lemon juice and ketchup sauce on the side. It's a low-fat mouthful and perfect for getting the necessary glycogen re-synthesis.

Afterward, you can opt-in for some less conventional health foods that your average guru highly warns against, such as low-fat ice cream or yogurt. You add in some mashed bananas, cottage cheese, grapefruit or even plain kid's cereal. Including a little bit of gluten is a good idea because it reminds your body how to digest it safely. Yet again, a low-fat meal, but there's still some fat in it.

The danger with fats is that they tend to oxidize if used improperly. That's why heating some of them is out of the question. To not cause inflammation we need to be very wise with how we use our fats. Here's a chart of the smoking points for different fats:

Fat Source	Smoke Point °C/F	Omega-6: Omega-3 Ratio
Unrefined Flaxseed Oil	107°C / 225 F	1:4
Unrefined Safflower Oil	107°C / 225°F	133:1

Unrefined Sunflower Oil	107°C / 225°F	40:1
Unrefined Corn Oil	160°C / 320°F	83:1
Extra Virgin Olive Oil	160°C / 320°F	73% monounsaturated, high in Omega 9
Unrefined Peanut Oil	160°C / 320°F	32:1
Semirefined Safflower Oil	160°C / 320°F	133:1, (75% Omega 9)
Unrefined Soy Oil	160°C / 320°F	8:1 (most are GMO)
Unrefined Walnut Oil	160°C / 320°F	5:1
Hemp Seed Oil	165°C / 330°F	3:1
Butter	177°C / 350°F	9:1, Mostly saturated & monosaturated
Coconut Oil	177°C / 350°F	86% healthy saturated, lauric acid (has antibacterial, antioxidant, and antiviral properties). Contains 66% medium chain triglycerides (MCTs).
Unrefined Sesame Oil	177°C / 350°F	138:1
Lard	182°C / 370°F	11:1 high in saturated

Macadamia Nut Oil	199°C / 390°F	1:1, 80% monounsaturated, (83% Omega-9)
Refined Canola Oil	204°C / 400°F	3:1, 80% of Canola in the US in GMO
Semirefined Walnut Oil	204°C / 400°F	5:1
Sesame Oil	210°C / 410°F	42:1
Cottonseed Oil	216°C / 420°F	54:1
Grapeseed Oil	216°C / 420°F	676:1, (12% saturated, 17% monounsaturated)
Virgin Olive Oil	216°C / 420°F	13:1, 74% monosaturated (71.3% Omega 9)
Almond Oil	216°C / 420°F	Omega-6 only
Hazelnut Oil	221°C / 430°F	75% monosaturated (no Omega 3, 78% Omega 9)
Peanut Oil	227°C / 440°F	32:1
Sunflower Oil	227°C / 440°F	40:1
Refined Corn Oil	232°C / 450°F	83:1

Palm Oil	232°C / 450°F	46:1, mostly saturated and monosaturated
Palm Kernel Oil	232°C / 450°F	82% saturated (No Omega 3)
Ghee (Clarified Butter)	252°C / 485°F	0:0, 62% saturated fat
Rice Bran Oil	254°C / 490°F	21:1, Good source of vitamin E & antioxidants
Refined Safflower Oil	266°C / 510°F	133:1 (74% Omega 9)
Avocado Oil	271°C / 520°F	12:1, 70% monosaturated, (68% Omega-9 fatty acids) High in vitamin E.

Drinks and Beverages

The only drink you could ever need is still water, preferably spring water with more minerals. Unfortunately, that's not possible all the time.

Water is one of the most essential things for life so you should get a high-quality source. Don't drink fluorinated tap water that may have residues of plastics and god knows what else in it. It also contains chlorine at high enough levels to be carcinogenic. Plus there's the potential of heavy metal contamination from plumbing, especially in the city.

- The cheapest and easiest thing you can do is get a countertop filter pitcher like Brita or Longlast. It's not 100% ideal but the least you should do. (<$50)
- Attach a carbon filter to your faucet ($50)
- Get the reverse osmosis water filtration system for the entire house. It's the golden standard for water filtration ($300)
- Install a whole house carbon block filter system ($400-500)

Fortunately, on a fat-burning engine, you can create your own water endogenously during beta-oxidation. That water is deuterium depleted as well, which makes it lighter and healthier for the body. It's amazing to think how being keto-adapted makes you so self-sustainable. I mean, you don't have to eat food and can fast for days AND you don't even have to worry about drinking water that frequently either because of creating it from your stored body fat. That's a true life-hack for situations where you don't have access to quality food or water.

In general, it's not advisable to drink your calories – to have beverages with a bunch of extra sugars or fats. It's just not ideal for satiety and leptin signaling. Plus, I think saving those calories for some fattier steak or vegetables is probably a much better option for both muscle growth and health.

Nevertheless, there are many zero calorie beverages or close-to-zero calorie drinks that are okay for consumption. How much or how often you choose to consume them has to depend on your own preference and individual choice.

Here's a List of the Drinks You Can Consume:

Food Source	Calories (kCal)	Fats (g)	Net Carbs (g)	Protein (g)	ACS
Coffee, 1 Cup	1	0	0	0.3	LowATG
All Teas (Green, Black, Etc), 1 Cup	2	0	0.4	0.2	LowATG
Almond Milk, 1 Cup, 8 oz	60	2.5	8	1	nTOR
Almond Water, 8 oz	55	0	15	1	nTOR
Coconut Water, 1 Cup	80	0.5	20	0	nTOR
Coconut Milk, 100 g	230	24	6	2.3	nTOR
Apple Cider Vinegar, 1 tbsp	3	0	0.1	0	LowATG
Distilled Vinegar (30%), 1 tbsp	3	0	0.1	0	LowATG
Kombucha, 1 Cup, 247 g	34	0.6	5.8	2.9	nTOR
PROTEIN SHAKES					
Whey Protein Shake with Water, 1 Scoop, 32 g	98	0	4.9	20	HiTOR
Rice Protein Shake with Water, 1 Scoop, 32 g	103	0.6	3.4	22	HiTOR
Hemp Protein Shake with Water, 1 Scoop, 32 g	113	3.4	7	12.6	ModTOR
Pea Protein Shake with Water, 1 Scoop, 32 g	113	0.5	2	25	ModTOR

Micro Greens Blend with Water, 1 Scoop, 32 g	20	1	21.5		nTOR
Alcohol					
Beer, 350 ml	154	0	13	1.6	nTOR
Red Wine, 5 oz	120	0	3.7	0	LowATG
White Wine, 5 oz	120	0	3.7	0	LowATG
Champagne, 5 oz	96	0	1.5	0	nTOR
Bud Light, 12 oz	110	0	6.6	1.1	nTOR
Gin, 1.5 oz	98	0	0	0	LowATG
Vermouth, 1.5 oz	64	0	0	0	LowATG
Rum, 1.5 oz	77	0	8	0	nTOR
Vodka, 1.5 oz	104	0	0	0	LowATG
Cognac, 1.5 oz	104	0	3	0	nTOR
Tequila, 1.5 oz	104	0	8	0	nTOR
Chocolate Liquor, 1.5 oz	155	0	17	0	nTOR
Mint Liquor, 1.5 oz	155	0	17	0	nTOR

Drink Timing Dissected

When you're fasting, then the only drinks you can consume are water, sparkling water, minerals, salted water, black coffee, teas, and apple cider vinegar.

During meal times, it's not advisable to drink a lot of liquids because it will hinder digestion. If you drink a lot of water with food, then it can lower stomach acid and can cause indigestion, bloating, or even intestinal permeability. The same post-meal.

Don't drink anything at least 15-30 minutes before eating, minimize liquids during the food, and wait at least 30 minutes to an hour after you eat before drinking again.

In terms of alcohol, you'd want to avoid it for the vast majority of time. I don't care what longevity „gurus" tell you, it's not good for everyday consumption. You can get the hormetic response of ethanol from apple cider vinegar and fermented foods already and there's no need to try and macro-dose resveratrol with drinking wine. Alcohol can be beneficial more as a relaxation agent or social lubricant. *Crazy word, I know.* It will help you relax and wind down, which is why some people in the Blue Zones have a glass of wine at dinner. For optimal longevity and performance, it's recommended to not drink every night and stick to 2-3 drinks a week at max.

What NOT to Eat

It's important to know what NOT to eat. In fact, avoiding the bad is much more critical than adding the good because you'll jeopardize all of your efforts.

By subtracting from your life you will actually add more to it. This applies especially to nutrition because it's very easy to eat just a little bit of the wrong kind of food and then suffer immensely. It's very difficult to out-exercise a bad diet as well.

A much wiser and more strategic approach from an evolutionary perspective even would be to apply certain positive restrictions and limit your options. By focusing on only the highest quality foods and then fasting at other times you'll greatly improve your health as well as mental well-being.

- **Refined Carbohydrates** – These foods are processed in a way that lowers their nutrient density but increases their glycemic load and inflammation. Processed carbs like breads, candy, pastries, cakes etc can promote serious health issues.
- **Sugar** – Clearly, table sugar is quite bad. Unfortunately, it's added to almost every packaged food imaginable. Even „healthy meal replacements" or „low carb bars" may have hidden sugars.
- **Trans Fats and Vegetable Oils** – Highly inflammatory and oxidized fats high in omega-6s and other chemicals. It's regrettable they were considered to be healthy at one point whereas they're actually the complete opposite.
- **Artificial Sweeteners** – Despite their non-caloric content, they still raise insulin and promote diabetes. There are even some studies on how they cause brain cancer and tumorigenesis in rats. By all means, avoid aspartame, sucralose, saccharin,

Acesulphame-K, and everything else you can't recognize. Certain sweeteners like xylitol and stevia are okay in small amounts. Sugar alcohols like erythritol are also fine if you limit them to 1-2 times a week. However, sweeteners may disrupt the microbiome and create bacterial overgrowth. They can also make you crave more sweetness.

- **Grains and Legumes** – They can cause gastrointestinal stress and gut issues in most people. If not the gluten from wheat, then the other phytates and anti-nutrients in beans can promote leaky gut syndrome, headaches, brain fog, autoimmune disorders, and even block the absorption of other minerals. These foods aren't even that nutrient dense compared to something like steak or even healthy starches like potatoes. It's not worth it to be eating grains and legumes because (1) they're not that healthy, (2) they cause issues down the line, (3) they don't even taste good, (4) it's a hassle to make them more easily digestible. Avoid wheat, barley, rye, bread, pastries, cookies, cakes, beans, peanuts, lentils 90% of the time. Introduce them only a few times a year as hormetic conditioning – a birthday cake.
- **GMO Foods and Pesticides** – They are not only poor in nutrients but also cause digestive problems. Commercial fruit and vegetables at the supermarket are laden with invisible chemicals that you don't see. Even washing them thoroughly won't help because the pesticides have penetrated into the plant. Unless you're buying organic, avoid the Dirty Dozen and stick to the Clean 15 list of veggies. Then again, you don't need to be eating 'the entire color of the rainbow' either and can safely stick to high-quality animal foods.
- **Commercial Dairy** – Ideally, you'd want to avoid all dairy because of several reasons. It can cause inflammation and allergic reactions in most people. Compared to other sources of protein like meat or fish, it's not that good for building muscle either and it's too anabolic for the effect you're getting. Fermented cheese, raw milk, kefir, and cottage cheese are okay in some amounts.
- **Alcohol** – Although a glass or two of red wine or a few heavy spirits can have a beneficial hormetic effect, you'd want to limit your alcohol consumption 95% of the time. Definitely avoid the high carb and sugary alcohols like ciders, beer, long drinks, cocktails, margaritas, pina coladas and everything else that's super rich in

calories. Those drinks are basically the equivalent of processed fruit juices with high fructose corn syrup and ethanol.

You should apply the principle of addition through subtraction into the other areas of your life as well, such as food choices, exercise regimes, the books you read, who you spend time with, what projects you pick up, how you structure your daily routines and so on.

Bruce Lee said it correctly: *„It's not the daily increase but the daily decrease. Hack away at the unessential,"* which perfectly describes the process of autophagy as well.

Instead of adding more material belongings, more calories, more carbohydrates, more micronutrients, and more responsibilities into our life, we should take a step back to focus on the vital few. That kind of a minimalistic way of thinking may seem restrictive and boring but it'll give you the ultimate freedom to always choose your own way.

The Metabolic Autophagy Food Pyramid

Figure 85 Metabolic Autophagy Food Pyramid

The idea of the Metabolic Autophagy Food Pyramid is to eat more nutrient dense foods that have a bunch of vitamins, minerals, essential amino acids, fats, and the other longevity-boosting compounds while limiting the consumption of empty calories. It will keep the body running optimally while maintaining mild caloric restriction.

The Dirty Dozen and Clean 15

To increase agricultural output, companies are using a lot of chemicals and pesticides that would protect against parasites and insects. Unfortunately, the insectum and bacteria are much quicker to adapt to environmental changes than mammals. Most bugs become resistant to pesticides quite quickly and thus farmers are forced to use even more lethal poisons. It's a non-stop arms race and we're usually on the losing team.

Genetically Modified Organisms (GMOs) and crops have been shown to be quite dangerous to other animals and they're linked to many health conditions in humans. There is no direct association but I think we know it's there. That's why you'd want to ruthlessly eliminate all GMOs and pesticides from your diet.

By following the principle of deduction, you'll already avoid the majority of foods that aren't even healthy for you. I mean, you don't need to eat bell peppers to survive, yet they're one of the most common vegetables that get sprayed with chemicals.

The US Department of Agriculture's Pesticide Data Program publishes an annual report on the most pesticide-rich foods. They divide it into The Dirty Dozen and The Clean 15. Here's a list for the year 2018:

- **The Dirty Dozen (Buy Organic and avoid conventional)** – strawberries, spinach, nectarines, apples, grapes, peaches, cherries, pears, tomatoes, celery, potatoes, sweet bell peppers.
- **The Clean 15 (Safer to buy but still aim for organic)** – avocados, sweet corn, pineapples, cabbages, onions, sweet peas, papayas, asparagus, mangoes, eggplants, honeydews, kiwis, cantaloupes, cauliflower, broccoli

Truth be told, most vegetables and plants still get exposed to chemicals and pesticides to some degree. That's the unfortunate side-effect of modern agriculture.

The best option would be to grow your own food or at least buy it straight from local farmers who you know and trust. That way you'll avoid the potential of pesticide contamination as much as possible. In the long run, it's worth it to invest a bit more money into getting high-quality ingredients because your health and longevity literally depend upon it.

Another option is to eat primarily low carb and spend your money on more nutrient-dense foods instead. I mean, high-quality meat probably costs just a little bit more than organic kelp but you don't need to eat a bunch of kelp and you'd be better off eating the meat. Kale is very over-priced in terms of its actual nutritional value. Grass-fed liver would give you much more nutrition.

Buying expensive vegetables and fruit won't give you nearly as much nutrient density than high-quality animal foods like eggs, fish, and meat. In total, you'd end up spending much less money but getting more nutritional value.

If you're eating a specific food a lot, then you better ensure it comes from a high-quality source. Think about it – the things you get exposed to the most will impact your health the most. Starting from the thoughts you have, the light you're exposed to, the people you're surrounded by, the coffee you drink, and especially the foods you put into your mouth.

If you were to invest a bit more money into making your meals higher quality, then you don't need to detox that much or go to the doctors. Prevention is the best form of healthcare as you'll avoid all the potential problems that may come in the future. As Hippocrates said: *"Let food be thy medicine and medicine be thy food."*

Chapter XVI
Supplementation

"You have to let it all go, Neo.
Fear, doubt, and disbelief.
Free your mind."

Morpheus, The Matrix

This wouldn't be a nutrition book without a chapter about supplementation. Bodybuilders, health enthusiasts, fitness gurus, and biohackers all alike are known to take different tablets, pills, and powders for improving their physical and mental performance. So it is with Metabolic Autophagy.

In this chapter, I'm going to walk you through many of the supplements we could take. It'll include the essential nutrients we should supplement because of common deficiencies, the ones that will promote longevity as well as anabolism, and a bunch of others that give you the extra edge.

Keep in mind that nothing replaces good food and it doesn't matter how many supplements you're taking if your diet is imperfect. Supplementation should be considered as counter-balancing the deficiencies you may come across due to poor nutritional values of certain foods or not getting access to some ingredients.

Before taking any supplements or meal replacements, you have to do your own research as to understand the potential side-effects and abnormalities that may or may not occur. The responsibility yet again is solely on you.

Additionally, it's advisable to take blood tests to see what your deficiencies are before adding random supplements to your diet.

The 3 Main Supplements

Let's start off with the 3 most important supplements that most people need and what will give you the 80/20.

- **Omega 3s** - The more omega-6 fatty acids you consume, the more omega-3s you may need. A healthy dose of omega-3s is 1000-3000 mg/day. Research shows that more than 5000 mg doesn't seem to have any added benefits. For EPA and DHA, you should aim for a minimum of 250 mg and a maximum of 3000 mg/day in a combined dose. Eat wild fatty fish a few times a week. Taking a fish oil supplement is also optional. Krill oil might simply be a more potent and bioavailable source. Cod liver oil will be even better. Make sure to use wild-caught sources to avoid mercury poisoning. Plant-based supplements for omega 3s include hempseed oil, algae omega, and wheat germ.
- **Vitamin D-3** - It's not actually a vitamin but gets synthesized into one inside the body. Vitamin D-3 governs almost every function within us starting with DNA repair and metabolic processes. It fights cardiovascular, autoimmune and infective diseases. Of course, it would be best to get it straight from the Sun but that is not always possible because of seasonality and location. An average adult should take at least 2000 IUs of vitamin D but it would also depend on how much exposure you get to natural sunlight. The upper limit for adults is 4000-5000 IU/day. Vitamin D may become toxic with high levels of calcium or if you take 10 000-40 000 IUs/day consistently. It can be consumed as oil or in a capsule.
- **Magnesium** - It comprises 99% of the body's mineral content and governs almost all physiological processes. Magnesium helps to build bones, enables nerves to function and is essential for the production of energy from food. Deficiencies can drive cardiovascular disease, depression, and headaches[611]. RDA is 400 mg/day. If you're physically active, then pay especially close attention to this because you may get muscle cramps and other problems.

Now, these 3 are the main supplements, I would add to any diet. It's said that up to 80% of people don't meet their RDA for magnesium. Seasonal Affective Disorder and depression due to vitamin D deficiencies are also very common. Not to mention the already pro-omega-6 diet.

Essential Nutrients

There are other micronutrients that are very important for optimal health and energy. Like potassium, vitamin K, zinc and b vitamins.

However, you would want to take supplement vitamins and minerals when you're actually deficient in them. Consuming too many of certain nutrients will hinder other metabolic functions and mineral absorptions. So, taking blood tests a few times a year to assess your nutrient status should be mandatory.

Here are the Essential Minerals to Cover:

- **Calcium** – RDA is 1000-1200 mg. Calcium deficiencies are common in older people or those who don't consume a lot of dairy. Before supplementing, you should know whether or not you're actually deficient because too much calcium promotes atherosclerosis and plaque formation. Especially if you're not getting enough vitamin K2. Calcium and Magnesium absorption compete with each other in doses higher than 250 mg-s so you shouldn't supplement them together. Consuming more dairy and calcium isn't healthier and won't strengthen your bones. Regions with the highest dairy consumption also have the highest rates of bone fractures and osteoporosis because they're not getting enough vitamin K. It's not recommended to supplement calcium if you're eating meat and veggies.

- **Choline** – RDA is 425-550 mg-s. Choline is a precursor to acetylcholine – a neurotransmitter responsible for cognitive functioning and attention. It's also vital for cell membrane, methyl metabolism, and cholesterol transportation. Foods rich in choline are eggs, meat, and fish. If you eat these foods, then you don't have to supplement with choline. On a plant-based diet, it may be a good idea to take choline and inositol.

- **Iron** – RDA is 8-18 mg with most people getting about 10-15 mg/day from food and other supplements. Iron is essential for hemoglobin transportation, which helps to transfer oxygen to muscles and cells. Overdosing iron can be toxic so consult your doctor first before supplementation. Iron deficiencies are more common in diets with little or no meat.

- **Iodine** – RDA is 150 mcg-s but a lot of people are still deficient. Iodine is important for thyroid functioning and the metabolism. If you're not eating a lot of seafood, like oysters, salmon, algae, sea kelp, and lobster, then you may want to supplement iodine. Taking about 300-400 mcg-s can be good for fixing symptoms of low thyroid. Raw vegetables will also inhibit iodine absorption so if you feel like having hypothyroidism, then make sure you cook your veggies or replace them with starchy tubers.

- **Potassium** - Estimated daily minimum for potassium is 2000 mg/day and the RDA 4700 mg/day. You shouldn't worry about eating too much potassium unless you're taking supplements. If you're not eating a lot of green leafy vegetables and avocados, then you may not be getting enough. Using potassium chloride salts with reduced sodium like NuSalt or taking potassium gluconate can be useful.

- **Selenium** – RDA is 55 mcg-s but optimal doses are somewhere between 100-300 mcg-s. Selenium is important for hormones and energy production, especially testosterone. Over 400 mcg-s of selenium, however, can be toxic and cause nausea. The daily requirements for selenium can be met with eating only 2-3 Brazil nuts. Other foods include seafood, meat, organs, chicken, nuts, seeds, and carrots.

- **Zinc** - RDA for zinc is 8-12mg/day. Zinc is an essential mineral involved in cell growth, protein synthesis and protecting the immune system. The upper limit for zinc a day should be under 100 mg because you may get nausea, vomiting and reduced immune functioning. Oysters are the most abundant sources of zinc with a massive 74 mg per serving. Other sources are beef, poultry, and some nuts. If you're a male, then you should pay close attention to your zinc consumption because it's one of the crucial minerals for testosterone production. But if you're eating a lot of seafood or red meat then you don't need to supplement it either.

Here are the Essential Vitamins to Cover:

- **Vitamin A** – RDA is 700-900 mcg with an upper limit of 3000 mcg-s. Vitamin A or retinol is important for nerve functioning, growth development, building new cells and improving eyesight. The best sources of vitamin A are organ meats with liver giving you about 5000-7000 mcg-s from just 100 grams compared to the

700-800 of carrots. That's why it's better to eat liver only a few times per week. Higher doses of vitamin A like 12 000 mcg-s can become toxic and cause drowsiness and coma. The Inuit are known for developing hypervitaminosis A because of eating polar bear liver. Because polar bears feed exclusively on seals and fish, their liver contains extremely high amounts of vitamin A. Even just a mouthful has nearly 9000 mcg-s, which is why you'd probably die if you ate polar bear liver. If you eat meat and some organ meats, then you don't need to supplement vitamin A.

- **B vitamins are also essential and they can be found in animal foods**. If you're already eating a whole foods based diet that includes some meat then you really don't need to supplement this. Vegans, however, are commonly deficient in B-vitamins so you'd have to look into taking a B-complex supplement. As an omnivore, supplementing can be counter-productive because you may get the wrong ratios of vitamins. Instead, focus on eating some red meat and fermented foods consistently. The optimal doses for B-vitamins are also slightly higher than the RDA. Here's what you should aim for.

    - Vitamin B1 (Thiamin) —1.5 mg/day
    - Vitamin B2 (Riboflavin) —1.7 mg/day
    - Vitamin B3 (Niacin) —20 mg/day
    - Vitamin B5 (Pantothenic Acid) —10 mg/day
    - Vitamin B6 (Pyridoxine) —2 mg/day
    - Vitamin B7 (Biotin) —300 mcg/day
    - Vitamin B9 (Folic Acid) —400 mcg/day
    - Vitamin B12 (Cobalamin) — 10 mcg/day

- **Vitamin C** – RDA is about 75-90 mg-s with an upper limit of 2000 mg-s. The function of vitamin C or ascorbic acid is to reduce oxidative stress by increasing antioxidants. In the 18th century, sailors who went on long sea explorations developed ulcers, rotten teeth, and hair loss. This *"plague of the seas"* was caused by a medical condition called scurvy, which is a deficiency in many vitamins, mostly vitamin C, and B vitamins. The conventional treatment was eating lemons, oranges, sauerkraut, malt, marmalade, and lemon juice. If you're eating fresh cruciferous vegetables and cabbage, then you probably don't have to worry about getting scurvy. Not an excuse to eat a bunch of fruit either. Also, by keeping inflammation low, your needs for vitamin C will decrease. It's not recommended to macro-dose vitamin C or other antioxidants in hopes of boosting your immune system because it may actually make you weaker. The daily recommended dose would be 300-400 mg-s and perhaps 1000 mg-s during sickness but exceeding that can have a negative effect on your body's capacity to deal with sickness naturally.

- **Vitamin E** – RDA for Alpha-Tocopherol is 15 mcg-s with a 1000 mcg upper limit. It's a potent antioxidant and a fat-soluble vitamin. Vitamin E deficiencies are quite rare as it's found in vegetables, fish, and nuts. Instead of taking dietary vitamin E supplements, you can use vitamin E oils on your face and skin to reduce wrinkles, lighten dark spots, and promote anti-aging.

- **Vitamin K** - The RDA for vitamin K is roughly 60-120 mcg, and the optimal level is about 200 mcg. This optimal level is mostly the same for both vitamin K1 and K2. It should be noted that while many sources may claim to hit the RDA for vitamin K, they have poor bioavailability - your body is unable to extract the full amount of said foods. That's why you should eat a lot of organ meats, fermented foods, a bunch of cruciferous vegetables, and a bit of cheese. Supplementing vitamin K should be secondary.

What about a multivitamin? - There are definitely a lot of vitamins to be covered for our body to not only be healthy but function at its peak. It would be unreasonable to take too many tablets or pills while neglecting the importance of real food. Plus there's the potential of interfering with vitamin absorption if you get it all together.

However, taking a multivitamin that has a lot of beneficial minerals all combined into one bottle is very effective and will most definitely be useful. If you haven't taken a blood test to assess your deficiencies, then a multivitamin can be somewhat counterproductive.

Longevity Supplements

Moving on with supplements and compounds that can specifically promote longevity in one way or another. When supplementing vitamins and minerals won't directly make you live longer as long as your diet is optimized for nutrient density, then there are certain compounds that can. These supplements can extend lifespan by either stimulating autophagy, controlling insulin, eliminating pathogens or boosting mitochondrial functioning.

Here are Some Plant Compounds and Medicinal Mushrooms for Longevity:

- **Chaga mushroom.** - Chaga is a mushroom that grows on birch trees. It's extremely beneficial for supporting the immune system, has anti-oxidative and soothing properties, lowers blood pressure and cholesterol, stimulates autophagy and fights cancer. Chaga will promote the health and integrity of the adrenal glands. This mushroom can be consumed as powder, made into tinctures, or boiled into teas. You can harvest and grind it yourself. I myself consume about 1-3 teaspoons of Chaga throughout the day and love to add it to my coffee.
- **Lion's Mane** – A white mushroom that looks like a lion's mane. It's incredible for growing new brain cells and preventing cognitive decline[612]. Supplementing just 3 grams a day has been shown to improve mental functioning of people with cognitive impairment[613]. There are no known side-effects to all of these medicinal mushrooms. However, some people may be allergic to them. Generally, take 1-3 teaspoons a day.
- **Reishi Mushroom** – Reishi or Lingzhi mushroom is a fungus that grows in humid regions. It improves the immune system and red blood cell functioning[614], which makes the body more capable at fighting disease. In fact, a study of over 4000 breast cancer survivors found that 59% of them were consuming reishi[615]. This mushroom has a more relaxing feeling to it and is best taken for relaxation and stress reduction.
- **Rhodiola Rosea** – It's an adaptogenic herb that grows in mountainous regions. The root has many compounds known to reduce stress, fatigue, and anxiety. For men, it can also boost testosterone and increase virility.

- **Shitake Mushroom** – a Dark brown fungus that grows on decaying trees. It contains polysaccharides, terpenoids, and sterols that boost the immune system, lower cholesterol and fight cancer[616].
- **Turkey Tail** – Looks like a turkey tail but doesn't act like one – it's a mushroom, for god's sake! Cancer patients are sometimes given turkey tail extract to recover from chemotherapy and strengthen immunity. It's another adaptogen that lowers stress and makes the body stronger.
- **Gingko Biloba** – Maidenhair, as it's called, is native to China. Not only is it a powerful antioxidant but it also improves blood circulation by increasing nitric oxide[617]. There are other benefits on brain and eye health as well. Gingko contains alkylphenols that may cause nausea, allergic reactions, headaches, and rashes. Taking about 120-240 mg in several doses throughout the day seems to be enough. Any more than 600 mg is probably not a good idea.
- **EGCG** - Epigallocatechin gallate (EGCG) is the main polyphenol in green tea with many health benefits. Green tea, in particular, is probably the healthiest drink in the world after mineral water. Too much green tea, however, may cause anxiety and heart palpitations because of the high caffeine content, which makes using green tea extracts or EGCG supplements a more convenient way to add extra polyphenols to your diet. Doses above 500 mg may become problematic.
- **Astralagus** – It's considered a superherb from Chinese medicine that's been used for thousands of years as an adaptogen. Astralagus boosts immunity, strengthens the heart, and promotes the flow of energy throughout the body. You can take about 250-500 mg-s as a supplement, ½ tsp as a tincture, or 1-2 tsp of dried root powder.

Here are Some Additional Longevity Supplements to Consider:

- **Resveratrol** – We all know the anti-aging benefits of resveratrol by now. The red wine special...Unfortunately, 1-2 glasses of red wine wouldn't give you a significant longevity boost and drinking more than that isn't optimal. Taking resveratrol supplements with medications may cause unwanted side-effects like blood clotting and enzyme blocking. Most supplements have 250-500 mg-s per serving but studies say that to get the benefits of resveratrol you'd have to consume about 2000 mg-s a day. To get that amount, you're going to have to take a high-quality resveratrol supplement.

- **Sulforaphane** – The cruciferous special...It's a powerful antioxidant that turns on the Nrf2 pathway with many anti-cancer properties. Cooking broccoli and cabbage triples their sulforaphane content[618]. Unfortunately, frozen veggies deactivate myrosinase, which is an enzyme that creates sulforaphane. Broccoli sprouts, in particular, contain dozens of times more vitamin K and sulforaphane. If you're not eating a lot of cruciferous or sprouts, then you can take 10 mgs of sulforaphane as a supplement.

- **Carnosine** – It's a combination of the amino acids beta-alanine and histidine with many anti-aging benefits. Carnosine is most known for protecting against free radicals and AGEs. This will keep the cells healthy and prevents aging of the skin. Naturally found in red meat and animal foods, supplementing carnosine has no side-effects. 7 ounces of beef has about 250 mg-s of carnosine but for optimal longevity, you'd want to aim for about 1000 mg-s a day.

- **Astaxanthin** – Wild salmon's flesh is slightly pink and reddish – that's astaxanthin. It's an incredibly powerful antioxidant and mitochondrial supporter, which is why freshwater fish like trout and salmon are capable of surviving such harsh conditions. Astaxanthin supplementation is great for anti-aging and maintaining muscle functioning. Doses of 4-40 mg-s a day have been shown to be safe. Too much astaxanthin may cause an upset stomach.

- **Alpha Lipoic Acid (ALA)** – Lipoic Acid has anti-inflammatory and anti-oxidant effects on the brain, and other tissue. It helps with fat oxidation, blood sugar regulation, and cardiovascular function. ALA is found in every cell of your body and it can be obtained from vegetables and meat. Therapeutic dosages of ALA range from 600-1800 mg/day with doses above 1200 mg-s causing nausea and itching.

- **C60** – Buckminsterfullerene, or buckyballs, or C60 is quite a new and unstudied compound. It helps to eliminate superoxide dismutase, which gets created as a by-product of cellular metabolism[619]. This fights reactive oxygen species and promotes longevity[620]. C60 oil should be dissolved in oil with a centrifuge. Pure C60 can be toxic[621]. Usually, as a supplement, C60 comes in either olive oil, avocado oil, or coconut oil. One teaspoon a day is probably enough for experimentation.

- **Glutathione** – The most important antioxidant in the body that's made of glutamine, glycine, and cysteine. Naturally, glutathione is found in sulfur-rich foods like beef, fish, poultry, and vegetables. Glutathione decreases with age so it's a good idea to supplement it. Increasing vitamin C and selenium may help co-factor the production of glutathione. Milk thistle and curcumin can also increase glutathione. Glutathione supplement doses range from 50-600 mg/day.

- **Apigenin** – It's a compound found in plants and vegetables like parsley, onions, fruit etc. with anti-carcinogenic and anti-inflammatory properties. Apigenin is neuroprotective and fights cancer. However, it can be toxic with 100 mg/kg causing liver toxicity in mice[622]. Parsley is 45 mg/g apigenin. It can also be found in olive leaf and artichoke extracts. No need to take additional apigenin supplements.

- **Quercetin** – Flavonoids are amazing anti-aging compounds and quercetin is one of them. In fact, quercetin is the most consumed flavonoids in the human diet[623]. It protects against free radicals and DNA damage. Quercetin is found in elderberries, red onions, garlic, cranberries, kale, hot peppers, kale, blueberries and the skin of apples. Supplementation is generally safe but not very effective because of poor bioavailability.

- **Melatonin** – The main sleep hormone melatonin is also a powerful antioxidant. You don't want to rely on melatonin supplementation as it may hinder your natural ability to produce melatonin. However, using about 0,3-1 grams on some nights can be useful. Liquid melatonin is absorbed much better and gives a more sustained release.

- **Spermidine** – Originally found from semen, spermidine is a polyamine compound associated with anti-aging and other metabolic benefits. It can also boost autophagy, longevity and assist with circadian rhythm regulation[624]. Foods higher in spermidine are aged cheese, natto, and miso but also beef, mushrooms, salmon roe, wheat germ, and chicken.

Here are Some Supplements for Mitochondrial Support:

- **PQQ -** Pyrroloquinoline Quinone is a non-vitamin growth factor that supports mitochondrial function. This will have a compounding effect on everything else you do. Humans can make about 100-400 nanograms of PQQ a day[625], mainly from dietary sources. Consuming 0.3 mg/kg PQQ is safe but 500-1000 mg/kg can cause death in rats[626]. Foods high in PQQ are raw cacao, green tea, fermented foods, and organ meats. Taking about 20 mg of PQQ as a supplement is the optimal dose for an average weighing individual.
- **CoQ10** – Co-Enzyme Q10 is another mitochondrial supporter and antioxidant. It's important for energy production and tissue development. Found in fish, red meat, especially organ meats, and fermented foods. CoQ10 comes in two different forms — ubiquinol and ubiquinone. The CoQ10 in your blood consists of 90% ubiquinol and it's more absorbable. Therefore, ubiquinol CoQ10 supplements are better. Daily dosage ranges from 90-200 mg. Doses over 500 mg-s are also safe.
- **Nicotinamide Riboside** – B vitamins play an important role in energy and nerve functioning. Nicotinamide Riboside (NR) is a form of B3 that gets easily converted into NAD+ and can activate sirtuins. It's mostly found in cow's milk, whey protein, and brewer's yeast. If you're not eating a lot of animal products or are deficient in B3, then you nicotinamide supplements can work. Doses of 5000 mg/kg haven't shown increased risk of death or anything like that. Nicotinamide Riboside can increase NAD+ as does Nicotinamide Mononucleotide (NMN).
- **Pterostilbene** – It is a polyphenol that's chemically similar to resveratrol that can also act as a precursor to NAD. The benefits include improved insulin sensitivity, reduced cholesterol, increased cognition, and antioxidant capacity[627]. High doses of pterostilbene can raise LDL cholesterol but there are no other known side-effects.

Here are Some Synthetic Medication Linked to Longevity:

- **Metformin** – Used for primarily diabetes, a lot of anti-aging enthusiasts are also interested in using metformin. Its main effect is in lowering blood sugar and suppressing insulin. Metformin also inhibits the liver's glucagon production, which prevents weight gain and blood sugar rises. The most common side-effects include nausea, vomiting, and diarrhea. It can also cause lactic acidosis by decreasing the breakdown of lactate into glucose. Hypoglycemia is another potential issue as well as cognitive impairments. Other than that, it's deemed safe. I don't see any practical value in taking additional metformin to lower your blood sugar if you're already doing intermittent fasting, exercising, and eating low carb.
- **Rapamycin** – The mTOR pathway was discovered by a group of proteins being targeted by the compound called rapamycin, hence the name 'mammalian target of rapamycin'. Rapamycin blocks mTOR and thus has been shown to increase longevity. It's also known as sirolimus and it works as an immunosuppressant that inhibits T-cells and B-cells. There are a few side-effects, such as diabetes-like symptoms, weakened immune system, increased cancer risk, and impaired wound healing. That makes sense because TOR is necessary for tissue growth and maintenance. Blocking mTOR all the time isn't optimal for longevity because it makes the person less insulin sensitive and weaker. Although rapamycin may be promising as a short-term treatment of diabetes and cancer, I don't see it having any practical value for someone already doing the Metabolic Autophagy Protocol.

A lot of the supplements that promote longevity do it by upregulating AMPK, lowering blood sugar, or strengthening the immune system. The natural way to activate these pathways would be to engage in the right type of exercise, exposure to hermetic stressors and caloric restriction.

However, some of these supplements I've listed here are much more potent than just fasting or working out. That's why it's still a good idea to take what nature has to offer and use different compounds as therapeutic additions to your diet. I personally love medicinal mushrooms and mitochondrial supporters.

Anabolic Supplements

In addition to longevity-boosting supplements, there are countless other things that don't directly extend lifespan but will promote more anabolic muscle hypertrophy, which can indirectly help you live longer.

The mainstream fitness and supplement industry has been trying to sell people dozens of different drinks and powders that supposedly make you build massive amounts of muscle and burn fat at the same time. Of course, you can't prescribe the pill of fasting because it's free and there's not much money in that.

However, because of the severity and complexity of the Metabolic Autophagy protocol, we want to make sure we're not wasting our time, money, or energy for nothing. That's why I'm going to list out some of the supplements that can help you to increase physical performance and support muscle growth.

These aren't your anabolic steroid-like substances that lower your natural testosterone, give you Gynecomastia, or make your hair fall out. They're actually quite healthy for you if taken at the right time.

Here are Some Muscle Building Supplements:

- **Creatine Monohydrate** - Creatine is an organic acid produced in the liver that helps to supply energy to cells all over the body, especially muscles. It enhances ATP production and allows for muscle fibers to contract faster, quicker, and makes them overall stronger. This means increased physical performance with explosive and strength-based movements and sprinting. However, it doesn't end there. Creatine has been found to improve cognitive functioning, as it's a nootropic as well, improving mental acuity and memory, especially in vegetarian diets. Naturally, it can be found mostly in red meat. It's dirt cheap and easy to consume, as only 5 grams per day will do wonders and doing so won't make a person big nor bulky. You don't have to load with 30 grams of creatine a day or some other nonsense. Just take 3-5 grams a day, preferably with food.

- **Branched Chain Amino Acids.** L-Leucine, L-Isoleucine, and L-Valine are grouped together and called BCAAs because of their unique chemical structure. They're essential and have to be derived from diet. Supplementing will increase performance, muscle recovery and protein synthesis. There is no solid evidence to show any significant benefit to BCAAs. However, they can be very useful to take before fasted workouts to reduce muscle catabolism. It will protect against muscle catabolism and can even promote ketone body production. A full guide to taking BCAAs in this chapter.

- **Whey protein** - On a standard ketogenic diet, you would want to avoid protein shakes because they spike your insulin. If you're doing CKD or TKD you would benefit from having an easily digestible source of protein. Before you break your fast and begin your carb refeed, make a quick shake to get the juices flowing. You can also use whey protein during fasted workouts with targeted intermitted fasting. This again will prevent muscle catabolism and will increase performance.

- **Phosphatidic Acid** – Phosphatidic Acid (PA) can regulate mTOR and promote muscle growth. It's a unique lipid molecule that turns on MPS in response to resistance training. PA can be found in foods some foods but in extremely low quantities. Vegetables like cabbage contain 0.5 mg-s of PA per gram. That's minute compared to the 250-750 mg doses in studies. More research about the effectiveness of PA supplementation in humans is needed but no long-term side-effects have been noted. Daily intake of 450 mg-s is optional.

- **Dextrose** - It's basically powdered glucose and very high on the glycemic index. You want to avoid it on SKD, but on CKD or TKD it's very useful for a post-workout shake with protein. It's dirt cheap and you'd want to take only 3-5 grams at once. Use it ONLY when doing the TKD or CKD because under other circumstances you're not doing your health a service. Dextrose is pure glucose and it's processed so it's definitely not optimal for autophagy or longevity. Most people don't need it and you may feel better without it but it's just an option to keep at the back of your head.

- **D-Ribose** – It's a simple carbohydrate molecule that's involved in energy production. D-Ribose can be found in all living cells as it's the structural basis of DNA and RNA. The other health benefits include reduced fatigue, improved heart health, better workouts, and kidney protection. Long-term D-Ribose supplementation may promote AGE production so you only want to use it for some hardcore workouts. In total, you can take about 5-10 grams of D-Ribose with pre-workout protein.

- **Cordyceps** – Cordyceps aren't actually mushrooms but a family of parasitic fungi that grow on the larvae of insects. They grow inside their victims, usually ants, and grow stems outside of the host's corpse. These 'zombie-parasites' have been shown to promote ATP production, reduce time to fatigue, increase oxygen uptake, and improve exercise performance[628]. They're not necessarily anabolic or pro-longevity but they will improve your health and lifespan.

- **Peptides** – They're a combination of two or more amino acids in which a carboxyl group of one is united with an amino group of another. Basically, very small protein molecules with anabolic benefits. Peptides are digested more easily and rapidly. There are some common peptides like glutamine, creatine, and collagen, but some less conventional ones have a much stronger effect on human growth hormone production. For instance, IGF-1, GHRP-6, and Ipamorelin are very anabolic and promote muscle growth. Most of these peptides are not available for commercial purposes, only research and clinical situations.

- **Deer Antler Velvet Spray** – IGF-1 promotes cellular growth and anabolism. Deer antler velvet sprays contain growth hormones that can make muscles grow. Although the evidence and bioavailability of such products is questionable, it can still raise IGF-1 a little bit. Use deer antler sprays only after heavy training the same way you'd stimulate mTOR.

- **Colostrum** – It's sometimes called 'liquid gold' because of the yellowish colour. Colostrum is the precursor to breast milk and it's rich in immunity-boosting compounds and growth factors. As an anti-aging supplement, it may prevent tissue degeneration and skin aging. Not something I'd recommend taking every day but on workout days it can be used for muscle hypertrophy.

- **Collagen protein.** Collagen provides the fastest possible healthy tissue repair, bone renewal, and recovery after exercise. It can also boost mental clarity, reduce inflammation, clear your skin, promote joint integrity, reduces aging and builds muscle. Naturally, it's found in tendons and ligaments, that can be consumed by eating meat. As a supplement, it can be used as protein powder or as gelatin capsules.

- **HMB** - β-Hydroxy β-Methylbutyrate is a by-product of leucine, which is an amino acid that stimulates protein synthesis. It's been shown to reduce exercise-induced muscle damage and improve recovery[629]. HMB can cause acute muscle anabolism and MPS independent of insulin[630], thus it will maintain a semi-fasted state. You can take it with the intra-workout protein shake to minimize muscle catabolism. Use pure HMB powder instead of the ones with artificial sweeteners.

- **Beta-Alanine** – An amino acid that reduces fatigue and increases physical performance. It's the main ingredient of many pre-workout drinks and thus it can help you to push yourself during workouts, especially if you're training fasted. Generic pre-workout drinks are way too stimulating and high in caffeine. Instead, take pure beta-alanine with your protein shake.

These supplements that I've just listed aren't necessary and they're not even advisable for most people who workout 3-4 times a week. They're simply anabolic agents that will promote muscle hypertrophy more so than anything else. It's a good idea to know how certain ingredients affect the body and then implement these strategies based on your situation.

I myself stick to primarily creatine, whey protein, cordyceps, and maybe some BCAAs every once in a while. By following a whole foods based diet, I'm already getting enough amino acids and the supplements are just to optimize my fasted workouts.

The Extra Edge

We've covered the longevity as well as muscle building supplements. Now it's time to go into the ones that aren't mandatory but can give you the extra edge.

As a disclaimer, they're not some miraculous alchemical substances that will instantly melt off 10 pounds of fat and increase life expectancy by 5 years. Instead, they're supplements that work in some situations.

- **Pro- and prebiotics** - Naturally, food is full of living organisms. Sauerkraut, raw milk, yogurt, unprocessed meat all have good bacteria in them. Probiotics are live microorganisms in a pill that transport these good bacteria into our gut for improved digestion and immune system. Prebiotics are different, they're not alive, but plant fiber that feeds the bacteria. They're indigestible parts of the vegetable that go through our digestive tract into our gut where the bacteria then eat them. If you're coming off antibiotics or suffer from a gut dysbiosis, then it can be useful to add some probiotics into your diet. However, to know what strands and species to take specifically you'd have to take a gut test.

- **MACA Root** - Another superfood comes from the Peruvian mountains and is the root of ginseng. It has numerous amounts of vitamins and minerals in it, such as magnesium zinc, copper etc. Also, it promotes hormone functioning for both men and women, as well as increases our energy production just like creatine does. It can either be powdered or made into a tablet. You shouldn't take maca every day because of its potency. Optimally, you'd want to take maybe a teaspoon every other day.

- **GABA** - Called gamma-aminobutyric acid, it's the main inhibitory neurotransmitter and regulates the nerve impulses in the human body. Therefore, it is important for both physical and mental performance, as both of them are connected to the nervous system. Also, GABA is to an extent responsible for causing relaxation and calmness, helping to produce BDNF.

- **L-Theanine** – It's an amino acid found in tea leaves, especially green tea. L-theanine has an alertness boosting effect but it's not as stimulating as coffee. The release of caffeine from L-theanine is more subtle and long-lasting. That makes it a great addition to your morning coffee if you want to prevent the crash. It's generally safe and there isn't a lethal dose.

- **Alpha GPC** - Alpha-Glycerophosphocholine or α-GPC is a cholinergic compound that improves cognitive functioning and brain health. High doses of 1200 mg-s appear to be effective in treating Alzheimer's and cognitive decline. Some athletes take 600 mg-s before exercise to hone their mental focus.

- **5-HTP** – 5-Hydroxytryptophan is a precursor to serotonin, which has anti-depressant and relaxing effects. It can help with sleep, weight loss, and anxiety. 200-300 mg doses are safe but higher ones may have side-effects.

- **Bacopa Monnieri** – Also known as Waterhyssop is a nootropic herb that's used in traditional medicine for cognition and longevity. It can improve memory and relieve stress. The standard dosage is 300 mg-s a day with the upper limit being around 750-1200 mg-s.

- **MCT oil** - Medium chain triglycerides are fat molecules that can be digested more rapidly than normal fat ones, which are usually long chain triglycerides. Naturally, it's extracted from coconut oil and is an enhanced liquidized version of it. Additionally, I also eat raw coconut flakes, which have MCTs in them.

- **Exogenous Ketones** – Supplemental ketone bodies that raise your blood ketones for a short period of time. They come in the form of either beta-hydroxybutyrate salts or ketone esters. BHB salts are marketed as a fat loss tool for getting into ketosis. However, the seeming fat loss effect comes from appetite suppression and adherence. They still have calories and they'll actually shut down your liver's own endogenous production of ketone bodies. The best time to take exogenous ketones would be as a pre-workout while fasting. At other times they're not worth it. I sometimes would take some BHB salts to increase my electrolytes but not on a habitual basis.

It's true that most of these supplements aren't as effective as they claim to be with a few exceptions. First of all, processing lowers their nutritional value but they're also less bioavailable. Then there's the issue of not knowing what effects are real and which ones are simply a placebo.

Honestly, it doesn't even matter if certain supplements don't work as long as you believe them to do so. The human mind is incredibly powerful in conjuring up certain stories and images that are as good as the real deal. That's why you have to be optimistic about everything that you do whether that be fasting, working out, starting a business, writing a book, eating food or even taking a pill. It won't reduce the therapeutic value of the supplement but it will keep your brain from making up false narratives of disbelief: *"This will never work"* (even if it does). You may end up negating the positive effects by increasing your stress and anxiety levels.

This is how Neo from the movie The Matrix gained all his powers as well – he started to believe in himself and realized that all the potential was already within him. When one of his mentors Morpheus posed him the question do you want to find out what the Matrix is, Neo had to choose between the Blue Pill and the Red Pill – to either take his chances and see what's possible or fall back to sleep. Morpheus said:

> *This is your last chance. After this, there is no turning back. You take the Blue Pill - the story ends, you wake up in your bed and believe whatever you want to believe. You take the Red Pill - you stay in Wonderland and I show you how deep the rabbit-hole goes.*

Of course, Neo took the Red Pill and went on a crazy adventure that taxed his body and expanded his mind. The Red Pill did have a real physiological effect as it plugged his body out of the computer system but metaphorically it was a placebo. At first, Neo didn't show the traits of being a superhero or anything the like and he even died. After he overcame his own limiting beliefs and chose to believe in himself he was reborn and reached his true potential.

So, the Placebo Pill is probably the most important supplement you have to take – to believe in what you're doing because that belief will facilitate an actual psychosomatic response in your physiology.

When to Take Supplements

Like I said in the beginning, you should focus on getting your nutrition on point first before taking any supplements. And secondly, you don't need to take all the supplements, only the ones you're most deficient in.

Micronutrients actually have a much greater impact on how you feel and how you perform than you think. Your decision making, your cognitive processes, and emotions are very much affected by them. That's why it's important to pay close attention to these things.

Take your supplements with food. At the first meal, take vitamin-D, fish oil, sea vegetable powders. In the evening, GABA, creatine, magnesium, to promote sleep. I like to think of Brazil nuts as supplements as well and take 2-4 of them daily. Don't take any more because you'll get selenium poisoning.

P.S. If you don't know which products and brands to buy, then I've listed out some safe supplements in my Amazon Influencer's Shop (http://www.amazon.com/shop/siimland).

It includes the ones I mentioned here but there are others that will give you even more energy.

Taking Supplements While Fasting

It's true that your body needs a certain amount of essential vitamins and minerals for survival. However, it doesn't mean you need to be getting them all the time every day. It definitely doesn't mean that if you skip your multivitamin in the morning, you're going to become severely deficient and need to be put under an IV. *#hitmeupwiththejuicer*

In fact, most of the minerals in your body are stored in your bones, fat tissue, and other storage locations. The problem is that if you're eating all the time several meals a day, then you're preventing yourself from using those micronutrients. Your body wants to store as much backup energy as it can and therefore an abundant supply of nutrients coming in around the clock actually reduces your ability to utilize those minerals that are currently stored in your body.

Moral of the story is: you don't need to be taking a lot of supplements as long as you're eating a wide variety of foods, even if you're eating just once a day. There are of course some supplements you can benefit from if you're deficient in some of the micronutrients, but in general, unless you're fasting for 3-5 days, you don't have to „get your daily vitamins and minerals" from some sort of a pill that you've been sold to.

It's also easy to get hypnotized by the fancy numbers on labels – *just one tablet gives you all the essential amino acids and vitamins, and minerals your body needs...IN 10x TIMES!!!* Just taking a pill with over 9000% of your daily RDAs doesn't mean you'll end up absorbing all of it.

The thing is that while the body catabolizes itself, you mobilize a lot of the nutrients that are already there. Micronutrient deficiencies are mostly caused by poor dietary choices overall not by inadequate supplementation.

There's also the fear that if you miss out on your daily recommended dose of vitamins for a day you'll turn into a raisin and die. This kind of thinking is already based on a false understanding of how the human physiology works. Skipping certain vitamins and minerals for a short period of time can actually promote the absorption rate of those nutrients if you follow it up with adequate nutrition. It's another example of a hormetic response. For instance, if you're slightly deficient in a specific nutrient, like vitamin K2 or B12, then your body will absorb them more effectively because it's craving for them. Likewise, insulin sensitivity and protein sensitivity increase if you're doing intermittent fasting. Why would mineral absorption be any different?

Taking a multivitamin while fasting is also not necessary if you're fasting for less than 10 days as most of the minerals are stored in your bones and the demand for other vitamins decreases thanks to the hormetic response of fasting. The longest recorded fast lasted for 382 days and the guy survived on just a multivitamin and some brewer's yeast. His plasma electrolyte concentrations all remained relatively the same and he lost 276 pounds

You definitely don't have to worry about micronutrient deficiencies while fasting for just 16-24 hours. Honestly, anything less than 24 hours isn't actual fasting because you're not triggering the deepest metabolic adaptations.

Iron and electrolyte deficiencies happen usually because of dehydration and excessive excretion of your body's salts. The most common reason is coffee and tea consumption. You can drink coffee while fasting but hot liquids and beverages may make you absorb less of the micronutrients. The tannins and caffeine in coffee and teas can lower the absorption rate of your supplements. They can also make you excrete more of the other electrolytes and minerals through urine, so you have to be careful with not taking your supplements together with these drinks. Otherwise, you're simply pissing them out and wasting your money.

But what about specific medications while fasting? Should you take prescription drugs during an extended fast? There are so many prescription drugs out there that it all depends on what particular disease you have and what kind of medicine it requires.

- If you're fasting for just 16-24 hours, then take your medication with food to make sure you absorb it. That short period of abstinence won't make you sicker because most of the drugs are already bullsh#t.

- If you're fasting for several days, then consult your doctor and ask what are the absorption implications and what ingredients the drug has. Most medications have additional filler ingredients like corn starch, dextrose, and other compounds that may potentially inhibit the autophagy effects of a fast.

But honestly…Doing strict fasting for several days is much more effective and healthier for you than taking medicine. The majority of diseases can already be fixed with fasting.

Your body already knows how to heal itself but it won't be able to do it if you keep it in a constant state of anabolism and feeding. You need to lower your insulin, lower your blood sugar and allow your cells to take out the trash for you. If I got diabetes or I get diagnosed with some sort of other metabolic diseases, then I'd immediately start fasting until I'm well and then eat a very strict ketogenic diet, rather than take prescription drugs. But then again…this is not professional medical advice.

In general, if you only need to supplement the vitamins and minerals if you're actually deficient in them. More micronutrients are not definitely better, as in the example of zinc and iron absorption.

Giving your body an overabundant supply of all the nutrients isn't a good idea either because it will have conflicting reactions and may cause some other issues. It's definitely a waste of money.

Before supplementing any specific vitamin, you'd be better off by first focusing on eating real food, getting your nutrients from that, taking blood tests to see your deficiencies, and then taking those supplements you need with food.

While fasting, you're much better off by getting your electrolytes and not worrying about the other micronutrients. You won't become deficient, you'll promote the mobilization of your already existent mineral stores, you'll elicit a beneficial hormetic response, and you'll maintain your sensitivity to those nutrients while you're actually consuming food.

Chapter XVII
Metabolic Autophagy in Practice

"Everyone can perform magic, everyone can reach their goals if they are able to think, if they are able to wait, if they are able to fast."

Herman Hesse, Siddhartha

We've covered a whole lot but there are still some specifics that need to be gone through. This chapter is about the different types of intermittent fasting and how to do them. I'm also going to go into more depth of the Targeted Intermittent Fasting protocol I talked about earlier in Anabolic Fasting.

There's nothing innovative about intermittent fasting or time-restricted feeding – it's been practiced for thousands of years by different religions and people. That's why you should be careful which gurus you listen to on social media, especially those who don't really understand the underlying mechanisms of what's happening with the body.

It's simple – you just stop eating – but for optimal results, you'd want to pay attention. For Metabolic Autophagy there are some additional principles that I'll share with you as well. Here are a few of the other renown ways of doing IF:

- **24-Hour Fast.** - This is the most basic way. It doesn't even have to mean that you actually go through a day without eating. Simply have dinner in the evening, fast throughout the next day and eat dinner again. This one is also prescribed by the author of *Eat Stop Eat* Brad Pilon. The frequency of these fasts depends on the person but once or twice a week should be the golden standard. An active person who trains hard should do it less often than a sedentary person. The leaner you are the less you have to fast, but that doesn't mean you can't gain all of the other physiological enhancements from occasional abstinence.

- **16/8 Leangains Protocol** - This is one the most popular strategies, popularized by Martin Berkhan of Leangains. You fast for 16 hours and have a feeding window of 8. Simply skip breakfast and have it during lunch instead. In my opinion, this should be the minimum fasting length for everyone. You don't need to eat any earlier than that and the abstinence has many benefits. It doesn't even have to be that strict. Instead of following 16/8 we can do 14/10, 18/6, 20/4 or whatever fits the situation. The idea is to simply reduce the amount of time we spend in a fed state and to be fasting the majority of the day.

- **The Warrior Diet** is a fasting protocol created by Ori Hofmekler. The entire concept is based around ancient warrior nations, such as the Spartans and Romans, who would be physically active throughout the day and eat mostly in the evening. At daylight, they would only get a few bites here and there and would consume a lot of calories with dinner. This diet follows the 20/4 timeframe with one massive meal eaten at dinner.

- **One Meal a Day (OMAD)** – It's the simplest way of eating – just eat once a day. Usually, you fast for about 22 hours and eat your food within a 1-2 hour timeframe. This works best for weight loss because you'll be quite full and thus can effortlessly stay at a caloric deficit. If you do it with proper keto-adaptation, then you'll also preserve more muscle. However, this may not be ideal for muscle growth because of the limited protein synthesis and anabolism. Instead, the Targeted Intermittent Fasting Protocol I'm going to share with you shortly is the enhanced version of OMAD.

- **36-Hour Fasts** - In this case, you abstain from eating anything for 1 entire day + 12 hours. It's not actually difficult at all. You simply have dinner the previous night, don't eat anything in the morning, lunch nor evening, go to bed in a fasted state, wake up the next day, fast a few hours more and start eating again. Going to bed hungry sounds scary but that's what most of the world's population is doing daily. It makes you think more deeply about one's own fortune.

- **48+ Hour Fasts** – These ones will make you enter into deeper autophagy and burn a ton of body fat. In my opinion, the fastest and healthiest fat loss strategy would be to cycle between 3-5 day fasts with OMAD refeeds. You want to lose the fat as fast as possible and extended fasts are the quickest way of doing it. The next chapter will also talk about how to fast for days and days.

- **Alternate Day Fasting** - There are also approaches like *The 5:2 Diet* and *Alternate Day Fasting*, which include fasting but allow the consumption of about 500 calories on days of abstention. The Fasting Mimicking Diet falls into the same camp. Those small amount of calories are only for increasing compliance. I wouldn't recommend this, because caloric restriction won't allow all of the physiological benefits of fasting to kick in. You want to shock the body and go straight to zero for the greatest effects. Complete abstinence is a much more effective strategy for both your physiology and psychology. Eating something would neglect the entire idea behind fasting, which is to abstain and reset.

- **Dry Fasting** – Not drinking liquids is also said to have autophagic benefits although there aren't many studies putting this theory to the test. One day of dry fasting is thought to equal 3 days of water fasting. The idea is that if you deprive yourself of water, your body will start to produce its own by converting the triglycerides from the adipose tissue into metabolic water. Hydrogen gets released as a by-product of beta-oxidation. Dry fasting has been practiced in certain religious and healing practices. In general, you don't want to become dehydrated for too long. However, daily time-restricted dry fasting of 12-16 hours can be another thing to do if you need deeper autophagy. At your own risk, of course. The next chapter will talk about dry fasting a bit more.

After having done intermittent fasting since high school, I wish I started sooner, and I don't really plan to ever return to the high-frequency style of eating. It doesn't really matter what kind of IF you do as long as you do it in some shape or form. As long as you stick to one of these from the list, you'll be getting about 80% of the benefits.

However, the optimal balance between performance and longevity, in my opinion, can be accomplished with Targeted Intermittent Fasting. It's what the Metabolic Autophagy Protocol actually centers around.

Targeted Intermittent Fasting

Like with many things in life, there's the right time and place to do anything, especially in nutrition. There's also the optimal amount for fasting and building muscle. These two – anabolism and catabolism – cannot co-exist in the exact moment. When you fast, you're breaking down. When you eat, you're building up.

To overcome the challenging situation of wanting to spend more time in a fasted state while still being able to train hard I like to use targeted intermittent fasting. Here's what it looks like:

- Fast for the majority of the day as long as you can before working out.
- Consume only water and zero calorie teas or coffee all the way up until 18-20 hours of fasting.
- You can even add a window of dry fasting for 12-16 hours into it before introducing liquids and calories.
- When starting to workout at 18-20 hours, consume a protein shake with 20-30 grams protein. You can even stay within 15-20 grams. It's preferable to drink it during the actual workout and use quality protein powders that don't have artificial sweeteners or other additives.
- For even greater performance you can use a bit of D-ribose, exogenous ketones, or some beta-alanine.
- In the post-workout scenario eat the rest of your calories within 2-3 hours or in a single meal. Make sure you still get enough protein after training.

This enables you to minimize the time spent in a fed state while still gaining the benefits of a deep fasted state. Is it rational? Would it be better if you just ate real food a few hours earlier and then trained? Maybe…maybe not…The idea is to simply still eat once a day but to supplement some of the protein you'd benefit from prior to working out.

The limiting factor of intermittent fasting is that you won't be able to train as hard as you potentially could. Of course, you'll have more energy to push yourself and get stronger if you ate something before that. However, that protein supplementation minimizes the length of how long before you have to do it.

You don't need to be consuming any calories all the way up until the point you're about to do resistance training. At that moment, you'd want to have some protein and amino acids to protect you against excessive catabolism. The protein shake contains amino acids and other building blocks that will help with performance. Thanks to having fasted for about 18-20 hours before that, your body will utilize that protein quite rapidly and it's not going to cause any issues. Before lifting weights you'll be in deep ketosis and you don't need any exogenous nutrients to fuel your daily low-intensity activities.

Targeted Intermittent Fasting should be done only with intense resistance training that depletes muscle glycogen and damages muscle fibers. It's not necessary for cardio, yoga, or something else like that because your body isn't under such demanding conditions. Also, it's not advisable to be doing TIF continuously every day throughout the year. You want to use it during periods of more physical activity and anabolism but not when prioritizing autophagy.

The same principles apply to the targeted ketogenic diet – you're going to pull the trigger only when it's beneficial and yields a positive response. That's why you don't always want to be zero carb and sometimes it's better to time your calories more specifically.

It doesn't even matter how long you fast for – you can still do this type of targeted fasting and working out fasted. You simply fast all the way up until the workout, then you drink a protein shake during the workout, continue to fast until your post-workout meal, and then finish eating for the rest of the day. It's generally best to combine TIF with OMAD because a lower eating frequency is still more beneficial. On 16/8 it's not necessary because you haven't been fasting for that long.

When doing this type of targeted fasting, you should be careful with what kind of protein you use. Definitely avoid the generic supplements with artificial sweeteners, colorings, and other questionable flavorings because they promote insulin resistance and gut problems. Instead, you'd want to find a quality brand with natural ingredients and zero fillers. Also, for the best anabolic effect, you'd benefit most from either whey or rice protein. Whey is more anabolic whereas rice is slightly less. Other plant-based proteins like hemp or pea are also okay but they may lack the complete amino acid profile and they may be too high in fiber. The 20-30 grams of protein in the shake is taken to reach peak muscle protein synthesis activation. On easier workouts, you can stay within 15-20 grams and gain the same effect.

Taking BCAAs in conjunction with the protein shake isn't necessary although not detrimental either. A quality whey protein would already have enough aminos, which makes the additional BCAAs an overkill. Also, BCAAs are quite expensive as well and would work only if you're not getting adequate amounts of protein from your diet. If you're nourishing yourself with good nutritious foods during refeeds then you have nothing to worry about.

Other ingredients you can add to the TIF shake include creatine, beta-alanine, cordyceps, and D-ribose. If you're planning on having a real killer workout, then maybe taking some target keto carbs from dextrose or a banana can also be good. At that point, you simply have to know what the purpose of your training is. I myself simply stick to protein and creatine – it's all you really need. In total, it would include about 100-150 calories, which is very low compared to the amount of work you'll be still able to produce.

After the TIF workout, you should wait about 1-2 hours before having the post-workout meal. This will lower your cortisol and promotes proper digestion. To trigger additional MPS, you'd want to be eating complete proteins high in leucine, such as eggs. Honestly, eggs are the best post-workout muscle building food because the cholesterol is going to promote healing and the leucine will stimulate MPS. If you'd like to really build muscle, then taking 2-3 grams of leucine powder can also work. Next to that, meat, fish, and some veggies sounds like a good meal. If you're eating carbs like potatoes or rice, then you'd want to avoid these fattier cuts of meat and eggs. Instead, combine them with lean proteins like white fish and chicken breast.

Figure 86 How to Do Targeted Intermittent Fasting

You want to avoid autophagy boosting and anti-inflammatory compounds after resistance training. The reason is that they can block some of the adaptations you're body's going through. You shouldn't take a cold shower or an ice bath after lifting weights either – they can blunt the hormetic response. Instead of taking a bunch of turmeric and medicinal herbs, you should relish more with the anabolic foods.

Honestly, the targeted intermittent fasting should be the default way of eating whatever diet you follow. Here's why:

- You want to be in a fasted state relatively often. At least with low blood sugar and ketosis
- You only really need to eat if you're doing something physical. Most of the day doesn't require eating if you already ate within the last 24-48 hours.
- You would want to have some protein before doing resistance training for performance and recovery. Taking the protein during the workout will be digested quite rapidly and it won't even feel like you've fasted prior to that. It'll also prime muscle growth faster.
- You don't need to be eating any more than once a day. Even I who's working out almost every day can maintain physical performance and even build muscle. Most people shouldn't eat more than twice a day but OMAD is still more optimal.
- Eating at a slight caloric deficit or at least around maintenance promotes longevity. By doing TIF you're circumventing the limiting factor on muscle performance and are still able to build muscle despite eating extra calories.

The only time you should follow a higher meal frequency would be during pregnancy, malnourishment, some medical condition, severe metabolic damage, elderhood, or when you're doing a ton of physical exercise. At other times it's better to become keto-adapted and fast.

How Long Should I Fast

The leaner and more physically active you are, the less you need to fast. First of all, there's not much benefit to gain from that frequent abstinence. You'll hit a point of diminishing returns quite quickly after which there are no great advantages. Secondly, if you want to work-out and keep making progress in your exercise performance, then you simply won't be able to do so as much. Fasting is powerful but good quality nutrition is still the foundation for getting stronger and building healthy muscle.

What I recommend is to have therapeutic water fasts that last for 3-5 days 2-4 times per year. Preferably every quarter. You should have a 48 hour fast at least once within 1-3 months and a 24 hour one every 2 weeks.

You can choose your window of fasting depending on the situation you're in. If you don't have access to good food or simply feel like skipping a meal, then do the 24 hour fast. At other times you can do less and stick to 18/6. While traveling it's so easy to do this. You don't want to be consuming all of that processed food sold in airports anyway and IF helps to avoid that effortlessly. Whenever I'm flying overseas, I'm always doing either a 48 or 72 hour fast. Helps with jet lag.

Chances are if you're used to eating 4-6 meals a day you may find transitioning over to intermittent fasting difficult. Having less frequent meals forces your body to adapt and may take some time until your ketogenic pathways get reinvigorated. You can "ease into it," by starting to eat less often at first and then shortening your feeding window even further.

For me, changing my eating schedule happened quickly. First, I started off with the 16/8 approach and pushed my breakfast until 10 AM. Dinner was at 6 PM and because of that, I tended to go to bed hungry. This didn't work for long and I kept pushing my first meal later into the day. It was a lot more sustainable and satisfying, as I got to eat more food and sleep satiated. You have to find out what works for you.

The best advice I can give you is to simply start. Jump in head first and expect nothing but the best. You can ease into it and start with skipping breakfast at first, but in my opinion, it's a lot better to simply shock yourself completely. This will wake your metabolism up and tells the liver that it's time to start producing those ketone bodies. That's what I did. I read about IF and after a couple of days started doing it. I haven't looked back ever since. The same happened with my first 24- and 48-hour fast. I always wanted to do it but had never gotten around to do so. One day I was having dinner and thought to myself: *"what the hell, might as well get it over with."*

The key is to practice IF daily, in some shape or form. Even when you schedule a 100-hour fast but can only make it until 48 hours, you should still consider it a great success. You don't want your feeding window to oversize the time you spend fasted. 12-12 isn't ideal either. It's better to always be on the negative side of things if you get my point. This way your body will be predominantly on the side of autophagy and ketosis instead of unwanted growth.

Metabolic Autophagy Cycle

Both autophagy and mTOR can be good as well as bad in some situations. That's why the context of when and where you express these pathways is critical for determining the final outcome on longevity.

If you have inadequate autophagic clearance, then you'll prevent the dysfunctional components from being recycled. Likewise, not having enough mTOR, especially after resistance training, may lead to a decrease in muscle and strength. Too much of each, however, may actually cause some health problems because of over-expression. The body functions best around homeostasis.

This equilibrium can be created on a daily basis, week-by-week, within a month, or throughout the year. The overall balance between anabolism and catabolism is what matters. Your goals, whether to build muscle or promote longevity, also have to be taken into account in choosing what kind of foods you'll end up eating.

Here's an example Metabolic Autophagy Routine for 4 weeks:

DAY	FASTING	TRAINING	METABOLIC GOAL	FOOD EATEN
WEEK ONE - ANABOLIC				
Monday	16/8 2MAD	Resistance Training	Anabolic Muscle Growth	Ketogenic ModTOR Foods
Tuesday	20/4	Walking and Yoga + Sauna	Recovery and Repair	Low Carb nTOR with a Few Servings of ModTOR Foods
Wednesday	Targeted IF	Resistance Training	Anabolic Muscle Growth	Low Fat ModTOR With Some Carbs

Thursday	Targeted IF	Short Low-Intensity Cardio and Walking	Cardio and Recovery	Low Carb Moderate Fat ModTOR Foods
Friday	22/2 OMAD	Long Walking and Yoga + Cold-Heat Exposure	Recovery and Repair	Low Carb nTOR Foods With Some ModTOR
Saturday	Targeted IF	Resistance Training	Anabolic Muscle Growth	Ketogenic ModTOR Foods
Sunday	22/2 OMAD	Long Walking and Yoga	Recovery and Repair	Low Carb nTOR Foods With some ModTOR
WEEK TWO – AUTOPHAGIC				
Monday	Targeted IF	Resistance Training	Muscle Maintenance	Ketogenic ModTOR Foods
Tuesday	20/4	Walking and Yoga + Sauna	Recovery and Repair	Low Carb nTOR with ATG Foods
Wednesday	Targeted IF	Resistance Training	Anabolic Muscle Growth	Ketogenic ModTOR Foods
Thursday	20/4	Short HIIT and Walking	Cardio and Recovery	Low Carb nTOR With Some ModTOR
Friday	22/2 OMAD	Long Walking and Cardio + Cold-Heat Exposure	Deeper Autophagy	Plant-Based Low Carb nTOR With ATG Foods

Saturday	Targeted IF	Resistance Training	Anabolic Muscle Growth	Ketogenic ModTOR Foods
Sunday	23/1 24-Hour Fast	Long Walking and Sauna	Deeper Autophagy	Plant-Based Low Carb nTOR With ATG Foods
WEEK THREE – ANABOLIC				
Monday	18/6 2MAD	Resistance Training	Anabolic Muscle Growth	Low Fat ModTOR with Some Carbs
Tuesday	20/4	HIIT + Walking + Sauna	Cardio and Recovery	Low Carb nTOR with Some ModTOR
Wednesday	Targeted IF	Resistance Training	Anabolic Muscle Growth	Low Fat ModTOR with Some Carbs
Thursday	20/4	Walking and Yoga + Cold-Heat Exposure	Recovery and Repair	Low Carb nTOR with Some ModTOR
Friday	Targeted IF	Resistance Training	Muscle Growth	Ketogenic ModTOR
Saturday	20/4	Walking and Yoga + Sauna	Recovery and Repair	Ketogenic ModTOR
Sunday	22/2 OMAD	Walking + Short Tabata and Sauna	Cardio and Recovery	Plant-Based nTOR with ATG Foods
WEEK FOUR – HOMEOSTASIS				
Monday	Targeted IF	Resistance Training	Anabolic Muscle Growth	Ketogenic ModTOR + nTOR Foods

Tuesday	20/4	Low-Intensity Cardio + Sauna	Cardio and Recovery	Low Carb nTOR with Some ModTOR
Wednesday	Targeted IF	Resistance Training	Muscle Maintenance	Ketogenic ModTOR
Thursday	22/2 OMAD	Walking and Yoga + Cold-Heat Exposure	Recovery and Repair	Plant-Based Low Carb nTOR With ATG Foods
Friday	18/6 2MAD	Resistance Training	Muscle Maintenance	Ketogenic ModTOR
Saturday	22/2 OMAD	Low-Intensity Cardio and Sauna	Cardio and Recovery	Ketogenic ModTOR
Sunday	Targeted IF	Resistance Training	Anabolic Muscle Growth	Ketogenic ModTOR

I have also created a separate meal plan + workout routine called the Metabolic Autophagy Diet Program that includes a 4-week step by step walkthrough of what foods to eat in what amounts at what time and how to train as well. It has two 4-week meal plans for both low calorie and high calorie intakes and it includes exact ingredients, amounts, and macros of the foods to eat. If you're interested, then head over to https://siimland.com/metabolic-autophagy-diet-program/ and check it out.

The periods of anabolism and catabolism reflect in both the weekly cycles as well as during the 24-hour period. On days where you're not lifting heavy and are resting, then it's better to go into deeper autophagy and ketosis by limiting your carbs and mTOR stimulating foods. On these autophagic days, you can also eat at a slight caloric deficit to promote longevity and make yourself more sensitive to anabolic growth the next day.

METABOLIC AUTOPHAGY FLOW CHART

HOW LONG HAVE YOU BEEN FASTING?

Less Than 14-16 Hours — Continue Fasting

16-24 Hours

24+ Hours

DID YOU WORKOUT?

Resistance Training — ModTOR Foods — Optional Carbs

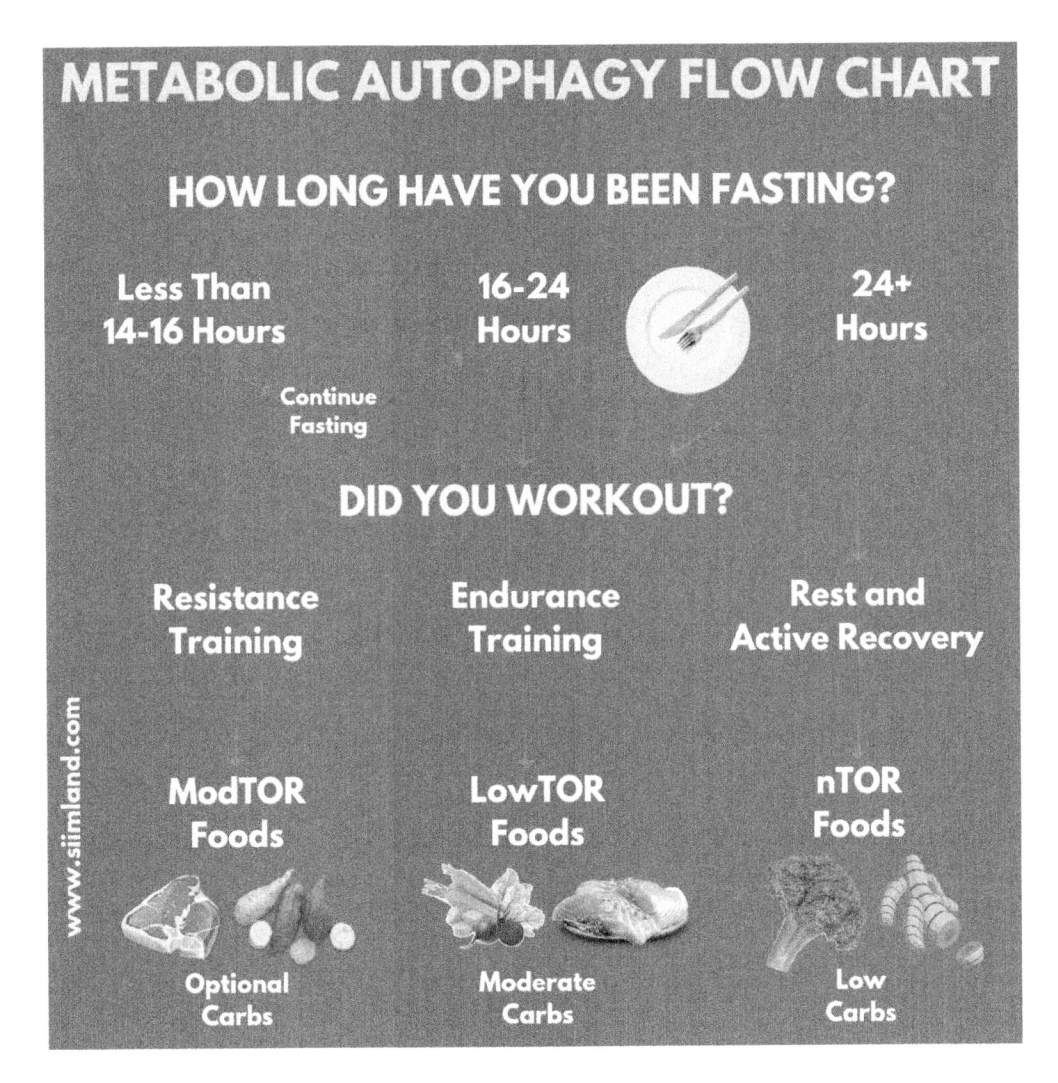

Endurance Training — LowTOR Foods — Moderate Carbs

Rest and Active Recovery — nTOR Foods — Low Carbs

www.siimland.com

Figure 87 Metabolic Autophagy Flow Chart

If you're about to have a heavy resistance training workout, then using the targeted intermittent fasting protocol is perfect for giving yourself some energy and amino acids. This can lift up your strength and muscle homeostasis even if you're coming off from a long 22-hour fast. That small amount of protein you get will make you feel like you haven't fasted for that long while still stimulating anabolism post-workout.

On days you're resting and not moving that much, it's a good idea to consume some of the ATG-boosting foods, such as turmeric, ginger, herbs, polyphenols, etc. This can deepen autophagy and promotes clearance of dead cells. If, however, you're working out, then it's not advised to fill yourself up with a ton of antioxidants or these calorie mimetics because they may blunt the anabolic response from training. That's why on workout days you should eat primarily ModTOR foods and then have the ATG ones on rest days.

Principles of Metabolic Autophagy

With the knowledge we've learned so far about different metabolic pathways and physiological effects of food, I think you're starting to form quite a firm picture of what's optimal to eat. The purpose of this book isn't to give you a very strict and confining protocol that you have to stick to no matter what or otherwise you'll become obese.

There is no one-size-fits-all solution to nutrition and health because the context of the situation is going to determine what's the final outcome. That's why I've tried to give you different guidelines and tricks to remember. These strategies are just one of the many tools in your toolbox that you can then use according to your circumstances.

Nevertheless, we can also safely say that there are still a few principles and tenets we know will exponentially increase our results. What your final outcome looks like depends upon the choices you've made.

Here are the principles of the Metabolic Autophagy Diet:

- **Time-Restricted Feeding as Long as You Can Every Day** – This is probably the most cost-effective thing you can do to improve your health and longevity. I mean, by simply not eating and fasting instead is one of the easiest ways to promote longevity and health. There isn't a real physiological reason to be eating any more than twice a day. Hell, most people will do perfectly fine with a single meal, unless they're under some special requirements. Whatever the case is, the minimum for daily time-restricted feeding is the 16/8 hour window, even when trying to build muscle. Instead of eating for distraction, you should leverage the fasted state as long as you can and then eat to support your physical conditioning.
- **Lift Heavy Things and Do Resistance Training** – The goal of your exercise should be to promote muscle growth and maintenance. As it turns out, having more lean tissue is one of the best things for healthy aging and longevity. That's why you want to predominantly resistance training instead of cardio. At minimum 2-3 and up to 4-6 times per week. If you're working out more intensely more frequently, then you may have to adjust your fasting window to make sure you're

not getting weaker or losing muscle. Some days should still be kept for cardiovascular training and full-on recovery but they aren't the main focus.

- **Get a Sweat on Daily** – It's incredibly important to keep your lymph system flowing and more active. Modern life is already quite sedentary and that can cause stagnation within the body. A lot of digestion issues and toxicities occur because of not clearing out the lymph fluids. Exercising and moving around is one of the best lymph node stimulators but any form of sweating whether by going to a sauna, doing yoga or running is great.

- **Maximize Nutrient Density** – This means eating high-quality foods that have an abundance of micronutrients, minerals, and other co-factors. Your purpose isn't to eat as many calories as you can get away with but to get more nutrition out of fewer calories. Mild caloric restriction and eating around maintenance is beneficial for longevity. Eat nose to tail, get adequate electrolytes, cover your essential nutrients, supplement your deficiencies, incorporate some superfoods into your diet, and cycle between different food groups. You shouldn't deprive yourself of nutrients either. Doing intermittent fasting doesn't mean you're starving yourself. Quite the opposite. You'll be getting more than enough nutrition. Just in a time-restricted manner.

- **Eat Whole Foods (A Lot of Plants)** – In terms of nutrient density, you'd have to focus on eating a lot of vegetables that have many vitamins and other beneficial compounds but not a lot of calories. Animal foods have their place but you shouldn't overconsume them. Most of what you eat in terms of volume should still be plant-based. Meat, eggs, fish, and fats simply have more calories. The antioxidants and polyphenols from cruciferous, veggies, berries, and other plants are pro-longevity because of being nTOR as well.

- **Control Blood Sugar and Insulin** – This is one of the best ways to ensure stable energy levels, avoid health problems, and maintain a more effective state of nutrient partitioning. It's just not a good idea to have high levels of blood sugar or insulin all the time and the research supports that. Instead, your goal should be to keep them relatively low the vast majority of time and only raise them where the body is more sensitive.

- **Don't Combine High Carb High Fat Foods** - Avoid processed inflammatory foods that are low protein, high carb, and high fat because it's a recipe for insulin resistance, diabetes, and over-eating. This change will drastically enable you to avoid most metabolic disorders. The short dopamine rush may feel good but it's not optimal in the long run. You can apply the 80/20 rule but do it at your own responsibility.

- **Limit Evolutionary Trade-Offs** – Avoid the „natural diet" fallacy both in the context of eating too much protein and animal fat as well as the plant-based approach. It's not a wise idea to go into the extremes and think that you're somehow immune to all disease. Who knows how your body individually will react to different foods. Maybe you're not as insulin sensitive as you think you are to justify that carb-up. Likewise, don't roll the dice with eating things that will potentially yield negative results but come with zero benefits. I'm talking about lectins, fruit, dairy, grains, vegetable oils, too much saturated fat, and carcinogenic meat...

- **Stimulate mTOR and Anabolism Only When It's Useful** – You don't want to be spiking insulin or mTOR just for nothing. To avoid any trade-offs in longevity, you want to eat ModTOR foods only after resistance training to support muscle homeostasis. At other times it's better to stick to nTOR and autophagy-like compounds. This is relevant mostly when you're eating more than twice a day. In the case of 2 meals a day, you'd want to make the first meal very low in anabolism and smaller in calories. The second one should be post-workout wherein you're more sensitive to mTOR and insulin. If you're not working out, then you'd be better off by limiting your protein intake and focusing on autophagy. You also don't want to be eating a lot of meat, eggs, and fish every day. Most of your food should still be plant-based because of their nTOR qualities and polyphenols. Eat meat only after heavier resistance training workouts and not in excess. This way you'll stimulate mTOR and anabolism only when it's useful and without consequences on longevity.

- **Cycle Between Anabolism and Catabolism** – Don't stay in either state for too long. If you're anabolic too long, you may accelerate aging. If you're catabolic too long, you may lose your muscle. Both aren't optimal for longevity nor performance. That's why you'd want to cycle between periods of being at a small surplus with staying around your maintenance and even dropping into a deficit.

The human body evolved under constant energy stress and it's what we thrive under. Never be stagnant or dysfunctional.

- **Expose Yourself to Hormetic Stressors** – Nutrition and exercise aren't the only components of longevity. You also want to trigger hormetic adaptation outside of the gym. To live a longer and healthier life you have to become more resilient against stress and adaptable to the ever-changing conditions of the natural environment. Of course, modern life allows us to maintain homeostasis in everything we do whether that be our core temperature, daily routines, food consumption, or physical challenges but they're elusive. To not be swept away by some unexpected circumstances, you want to follow a lifestyle that involves voluntary hormesis. Take cold showers, swim in icy lakes, turn off the central heating, burn some fat at the sauna, practice stress management, fast for 5 days a few times per year, and do something tough.

The purpose of these principles is to live a more challenging life that would ultimately make it more fulfilling and enjoyable. It might sound paradoxical that not eating for days upon end can make you a happier person but trust me it can.

Understanding the mechanisms of the human body and acknowledging to yourself the benefits of fasting, working out, or taking ice baths will make them seem that much more appealing. They require a shift from the hedonic mindset into a more stoic one. It might seem grim at first but you'll actually become more grateful for everything you have and actually yearn for more minimalism. Like Siddharta said: „*I can think. I can wait. I can fast.*"

Once you make Metabolic Autophagy a part of your lifestyle, you'll never want to go back to a normal way of being. You may adjust it according to the situation and your preference but the core tenets will remain the same.

Chapter XVIII
What Breaks a Fast

"Every fool can fast, but only the wise man knows how to break a fast."

George Bernard Shaw

You can find many questions on Google about what breaks a fast while intermittent fasting: *"Can I drink coffee while fasting? Will lemon water kick me out of a fasted state? Do 100 calories break a fast? What about eating Snickers' bars and cursing?"* Those are all great questions and you may get many different answers from different people. There are several ways people do intermittent fasting and for many reasons.

What counts as breaking a fast depends on why you're fasting for and what you're expecting to gain from it. In this chapter, I'm going to go through all the scenarios and questions in regards to *does this break a fast?*

The biggest beneficial effects of fasting come from 3 things: autophagy, ketosis, and hormesis. IF promotes all of them to a certain degree, depending on how long you've been in a fasted state. The key trigger is energy deficit and glycogen depletion.

Eating just 50 calories or even as little as 2-3 grams of leucine can already raise mTOR and put you into an anabolic state.

Fat doesn't raise insulin significantly and it keeps mTOR suppressed in small amounts. Endogenous ketone bodies from your own body fat will stimulate autophagy, which can promote brain macroautophagy as well. However, high amounts of ketones and fatty acids in the blood can still make you raise insulin. If there's too much energy circulating the body, then that's a signal to stop autophagy and trigger mTOR. Exogenous ketones can also be insulinogenic[631].

In a study done on rats, they found that exogenous ketones promote insulin secretion when blood glucose was greater than 5.0 mMol/L or 90 mg/dl[632]. If you're a healthy person who is very lean and low on body fat, then your blood glucose will probably drop below that. If you already have excess energy stored in your body fat then it's a signal that there's plenty of energy already around and any form of calories, whether that be from Bulletproof coffee or exogenous ketones will most likely inhibit autophagy.

It means that you shouldn't have small snacks like an almond. Okay, probably a single almond won't break autophagy but you'd still be better off by skipping it altogether.

Things That Keep You In a Fasted State

There are some ways to trigger autophagy that you can consume while fasting.

- **Green tea** has polyphenols and other ingredients like epigallocatechin-3-Gallate (EGCG) that stimulate autophagy. Black tea, herbal teas, and others like chamomile are also okay.

- **Coconut oil or MCT oil** can also stimulate autophagy in very small amounts by raising ketones[633][634]. However, as we found out, too much fat and too much energy in the system will raise insulin. Any more than 1 tsp will probably have a counter-balancing effect.

- **Medicinal Mushrooms** - Reishi mushroom extracts are shown to increase autophagy and inhibit breast cancer growth as do the others like chaga and cordyceps. You can take maybe 1 tsp of reishi or chaga but not any more.

- **Coffee stimulates lipid metabolism** through the autophagy-lysosomal pathway in mice[635][636]. However, excess caffeine consumption may lead to higher blood glucose and insulin levels because of over-stimulating cortisol. That's why drinking too much coffee may actually interfere with the fast. You shouldn't drink any more than 2-4 cups a day.

- **Apple Cider Vinegar has trace amounts of micronutrients and bacterial residue** but it most likely isn't enough to kick you out of a fasted state. It may actually boost autophagy by promoting the cleaning process. You can use the distilled ACV during the fasting window and opt for the one with the mother when you start eating.

To be safe, you could just drink water and salt. However, things like green tea and black coffee are great for enhancing the effectiveness of fasting by promoting autophagy. That's why it's okay to be consuming them in moderation.

Artificial Sweeteners

There's a lot of conflicting research around artificial sweeteners. Some are shown to be horrible for your health while others can cause an insulin response. Can you add some artificial sweeteners to your drinks while fasting?

- **Aspartame doesn't have any effect on the insulin response whether alone or combined with food**[637][638]. When protein produces a significant insulin response, then aspartame doesn't seem to have any effect[639]. However, aspartame is thought to be linked with many cancers and tumors. I'd skip the diet coke.

- **Neither aspartame or saccharin seem to raise insulin**. However, in one study they took people in a fasted state and made them swish 8 different taste solutions in their mouth for 45 seconds and then spit them out. Sucrose and saccharin were the only ones that activated an insulin response[640]. However, in another study, they didn't get the same results[641]. Maybe the subject's mind created their own placebo response by raising their blood sugar by will. You should remember that the next time you look at cake or think about sweets because it may raise insulin.

- **Direct transfusions of Acesulfame K increased insulin in rats**[642]. It's definitely not the same as drinking diet soda but it still matters.

- **Sucralose aka Splenda activates certain taste receptors that in some studies may stimulate insulin.** However, one study found that infusing sucralose straight into the gut didn't stimulate the hormones that raise insulin[643].

- **Stevia can lower postprandial insulin levels compared to aspartame and sucrose**[644]. It's the only natural sweetener that's fine to consume in moderation.

A recent review on low-calorie sweeteners concluded that they don't seem to have any effects on insulin in vivo[645]. Zero calorie artificial sweeteners aren't most likely going to raise insulin but they may. The bottom line is to know how they affect YOU specifically by looking at your blood sugar levels.

Even if artificial sweeteners may not spike your insulin, it doesn't mean they can't inhibit autophagy. They're also shown to have a negative effect on gut microbiome so it's better to avoid them entirely. The evolutionarily sustainable strategy would be to not be bothered with these things and avoid the potential costs.

The *cephalic phase response* describes the process of gastric secretion by your stomach before eating food. It's caused by sights, smells, tastes, and even thoughts of consuming something good. The hungrier you are the greater the stimulation and most of it is learned behavior. Thinking about biting into a lemon makes you salivate not because your brain knows what it feels like but because you've done it before. This serves as a pre-emptive mechanism for having enough insulin around when you do eat. In an environment of caloric scarcity, it's useful but not so much in a society where you can see empty calories all around you. That's why you have to avoid tricking your mind with placebo sensations of eating and attain an indifferent mindset towards sweetness. Don't use artificial sweeteners during fasting.

Chewing Gum

If you've ever fasted, then you can probably notice the utility of chewing gum. It helps to stave off hunger, keeps your mind engaged, and tastes good. But I hate to break it to you…

Sugar-free and calorie-free gum actually have calories. *#Gasp! Shocking I know*…Basically, because the amount of calories in there is so low, the production company doesn't have to list it as such. The label says something along the lines of: „*Not a significant source of calories*".

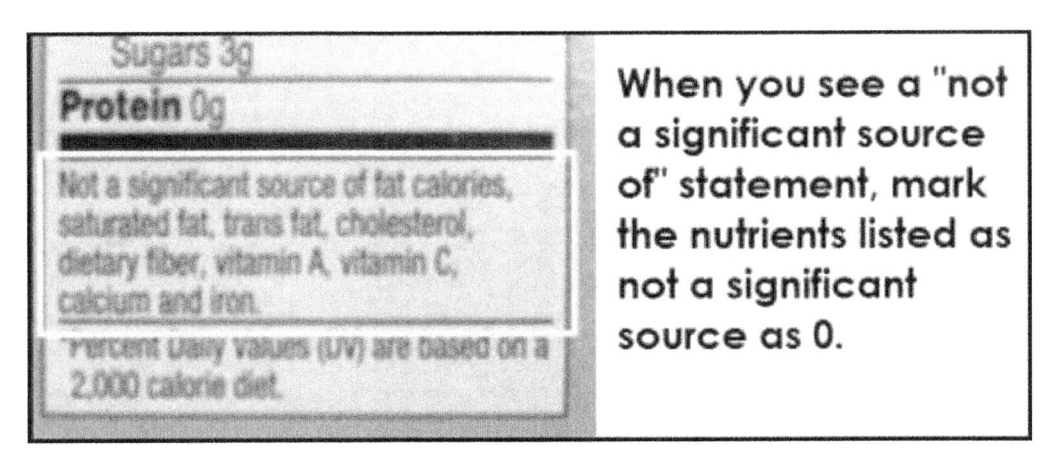

Figure 88 Chewing gum actually has some calories and sugars

Horray! Does this mean I can eat all the gum my heart desires and still continue fasting? Well, that caloric quality refers to a single portion and one piece of gum probably has 2-5 calories. Mostly coming from sugar alcohols and carbohydrates.

You would definitely want to avoid gums that have corn syrup, fructose, high fructose corn syrup, and aspartame because we know these are quite bad for your health and gut flora. Of course, if you're only consuming like 1-2 pieces of gum with these ingredients, then you won't have a negative side effect on your health, granted you're eating a healthy diet. However, if you're taking in like 4-5 and even more gum while fasting, then you'll not only break a fast but you may also end up with some small microbiome issues down the line.

What about nicotine gum while fasting? Will that break a fast?

Nicotine in cigars and other forms of tobacco can have some nootropic effects that increase your alertness and cognitive performance. However, smoking is still bad for you and promotes atherosclerosis.

I don't consume nicotine gum because I don't need it but I'd suggest that nicotine will boost your mental performance in very small doses like 1-2 mg-s. One piece of nicotine gum usually has 2-4 mg-s of nicotine, which is the equivalent of nicotine in 1-2 cigars.

Fasting itself also has nootropics effect and I'm sure everyone who's fasted for at least a day can attest to the mental clarity and sharpness. When in a fasted state, you raise cortisol, adrenaline, and ketones, which will increase your energy and acuity.

The concluding answer is that 1-2 pieces of gum won't break a fast in most cases. Brushing your teeth once is also okay. You just have to make sure you don't swallow the saliva, not follow it up with 2 cans of diet soda, and not stimulate insulin by starting to crave for more food.

Lemon Water

Putting a few slices of lemon, lime, orange, cucumber, or mint leaves into your water has zero effect on your blood sugar, autophagy, or fasting.

Lemons and other citrus, however, contain calories, namely fructose, which stimulates the liver in a way to break the fasted state.

- If you were to eat some lemons, then you're going to absorb the fructose, digest it, and thus inhibit the fasting. Whether or not it's enough to break the fast completely depends on how much lemon you swallowed and how does it affect your blood sugar. I dare to say that 1-2 slices of lemon have a negligible effect on the fasted state.

- If the lemon slices are simply sitting in the water, then they're maintaining the fructose inside the cellular matrix without releasing them into the liquid (unless you squeeze them empty, of course). In that case, you're not absorbing the fructose either, which won't affect the fasting.

The general guideline is that just drinking lemon flavored water while fasting won't break the fast as long as it's not lemon juice or some sort of a sugar-filled lemon sports drink.

Glauber's Salts

If you're doing fasting for health purposes and to promote cellular cleansing, then you can also consume Glauber's salt. It's commonly named as sodium sulfate decahydrate.

In medicine, Glauber's salt is used as a mild laxative that triggers bowel movements. If you add 5-20 grams of Glauber's salt to water, you can remove constipation, reduce bloating and clean the digestive tract. Any more than that may cause diarrhea and can lead to dehydration so don't overconsume it.

Bulletproof Coffee

Well, the rationale is that because Bulletproof Coffee consists of only fat it's not going to raise blood sugar or insulin and thus keeps you in a fasted state.

It's true that adding butter to your coffee will keep you in ketosis and maintains somewhat of a fasted state but I'm afraid it's still going to inhibit autophagy. This is not necessarily a bad thing as you'll get energy and stay in ketosis but you'll be missing out on some of the detox health benefits. Here you have to think about why you're doing intermittent fasting for.

- If you're doing fasting for weight loss purposes and adding butter or MCT oil to your coffee helps you to make it through the fast then go for it. However, do remember that you'll still need to consume fewer calories and putting an entire stick of butter into your cup will give you at least a few hundred calories.

- If you're fasting to thoroughly clean your body from toxic proteins and inflammation, then I'd advise you to not consume anything at all and do strict water fasting with mineral water with these salts and teas.

But does Bulletproof Coffee stop autophagy completely? That would depend on the amount of calories consumed and how your body responds to it. If you add like 500 calories of fat then that's definitely going to kick you out of a fasted state and stop autophagy. If you consume maybe like 100 calories then you may get away with it. It also matters what's the overall energetic status of your body.

- **If you're sitting on a couch and drinking bulletproof coffee in the morning after having slept for 8 hours, then your body is in a state where it doesn't need excess energy.** You've been sedentary and there are no real energy demands on your muscles that you can't cover with endogenous production of liver ketones. In that case, drinking that coffee will have a much bigger effect on autophagy because your homeostasis for nutrient signaling is much lower. You will get a bigger nutrient signaling effect from smaller amounts of calories because your body's energy demands are much lower.

- **If you were to take that same bulletproof coffee and drink it maybe at noon time when you've already moved around, taken a long walk, maybe had a workout or done some chores, then it's going to have less of an effect because your body will be under higher energy demands**. In that case, your ceiling for nutrient signaling is higher because the calories you'd consume would be allocated into use much more effectively and they'll be burnt off faster. You may still interfere with autophagy a little bit but not to the extent as you would when you drink that same coffee with the same amount of calories in the morning when your body doesn't need that much energy.

It's not that black and white...definitely not black and white coffee. This can be taken even further.

If you do intermittent fasting for longer and then you have a much bigger meal later in the day like 1500 calories, then you'll definitely interfere with autophagy much less than if you were to eat 2 meals of 700 calories each.

By the end of the day, your body will be that much more depleted and the food you do eat won't have that much energetic load to inhibit autophagy completely. It's almost like your body is still shocked from the catabolic stressor of fasting and because of that it's not going to turn off autophagy completely either whereas eating 2 meals is enough to say that *„okay we've got enough nutrients and we don't need to keep recycling our own cells through autophagy."* The same effect can occur with having some protein during your workout when doing targeted intermittent fasting after being in a fasted state for over 18 hours.

In the case of eating once a day, I would say that you'd still maintain higher levels of autophagy during the meal and you'll maintain it after the meal as well because the body will use those calories for the mere essentials of energy homeostasis. Of course, if you eat too many excess calories you'll stop autophagy but if you do it in a very small time frame like with OMAD you'll go back into it much faster as well. Now, I don't have any evidence or studies to prove these claims but this is my hypothesis and it makes sense from the perspective of nutrient signaling.

BCAAs

There are some calories in BCAAs although they're used for muscle protein synthesis. One gram of BCAAs usually has about 4-5 calories and a single serving of a standard amino drink will have about 20 calories. Therefore, BCAAs will kick you out of a fasted state, but whether or not it's a bad thing depends on many things.

BCAAs play the biggest role in supplying amino acids and energy with muscles specifically. At any given time, the liver is releasing a constant supply of amino acids to skeletal muscle for maintaining blood sugar levels and supporting cellular protein homeostasis. This mechanism is supposed to keep your cells nourished with enough building blocks needed for survival.

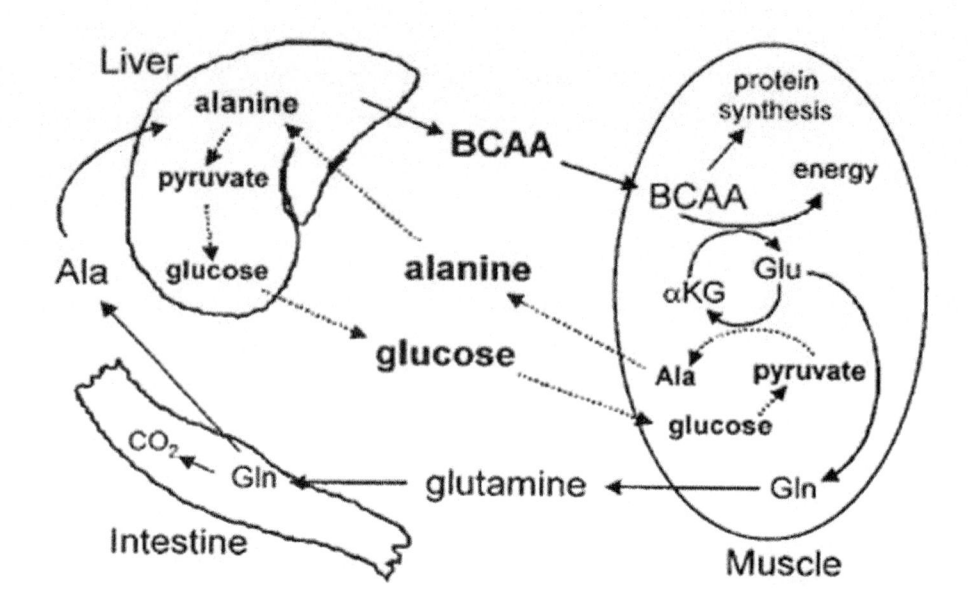

Figure 89 The transport and utilization of BCAAs

The most common said benefit of BCAAs is that they help you to preserve muscle while at a caloric deficit or when fasting.

One study found that a group of wrestlers who ingested 52 grams of BCAAs a day retained more muscle and lost more fat than the control group who didn't supplement[646]. However, the subjects were eating only about ~80 grams of protein at a mean body weight of 150 pounds, which is probably not ideal for someone engaged in such heavy physical activity. If they were to be consuming 0.8-1.0 g/lb of protein, the effects of BCAA supplementation would've probably been non-existent as all of the participants would've been on an even keel in terms of getting enough amino acids. What's more, studies have shown how increased fatty acid concentrations have a muscle sparing effect[647]. This is most likely due to the body using more fat and ketones for energy production instead of amino acids and protein.

If you're coming from a diet that doesn't restrict carbohydrates and maintains a glucose-dependent metabolism, then, of course, you're going to lose muscle tissue while fasting. That's why you want to get into ketosis as fast as possible if you're doing intermittent fasting. Don't even try to fast for extended periods of time unless you've gotten into ketosis because you're going to lose a whole lot of muscle. It means that if you're in a deep fasted state with elevated level of ketones, you're more protective of lean muscle tissue than if you were to be consuming amino acids and negating the positive effects of fasting.

Leucine and BCAAs acids do spike insulin but they increase plasma levels of insulin levels temporarily and they have almost no effect on glucose or urea nitrogen[648].

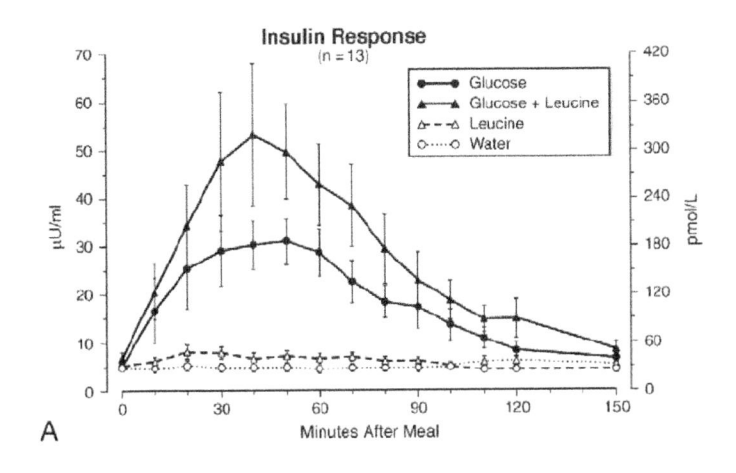

Figure 90 Leucine doesn't spike insulin when taken with water

There are many occasions outside of nutrition that can raise insulin and glucose, such as sitting in traffic, getting anxious, feeling angry, or even sitting on the couch. These examples may elevate cortisol and in so doing raise insulin and glucose as well but do they kick you out of a fasted state? Such insulin responses like that of BCAA supplementation or screaming at someone are short-term and not the same as eating a loaf of bread or potatoes.

There are also different degrees of fasting ketosis and keto-adaptation. Every person's metabolism is unique and has different levels of insulin sensitivity. It all depends on muscle mass, your level of physical activity, body fat percentages, genes, time of the day, how many ketones you have in the blood, and how many BCAAs you're actually consuming.

What's more, some amino acids can actually be converted into ketones to further promote ketosis, although not in significant amounts[649]. With that being said, a good supply of nutrients can still be very useful in some rare occasions. Working out fasted is definitely more catabolic than working out fed.

- If you've eaten a few hours before a workout, then you have nothing to worry about in regards to amino acids and muscle loss because your blood will be filled with BCAAs from food.

- If you're working out fasted and not having eaten anything for over 12-16 hours, then BCAAs can have an additional protective effect.

The key is to time your BCAA intake in a way that doesn't kick you out of a fasted state.

- Taking BCAAs before a workout or when being sedentary will stop the fast and make you catabolic.

- However, if you're working out, then you're not necessarily in ketosis anyway. Heavy exercise, especially resistance training, releases muscle glycogen into the blood, thus raising insulin and lowering concentrations of ketones. That's why a hard workout can cause a transient loss of ketosis.

Consuming a small amount of BCAAs while working out fasted, like in the example of targeted intermittent fasting, will probably have a negligible effect on the fast as you'll dip straight back into fasting ketosis right after the workout.

Therefore, the only time you'd ever want to take BCAAs is during a heavy resistance training workout in a fasted state. Whether or not you should do it at that time depends on several additional factors.

Mineral Water

And of course, you can drink regular water, sparkling water, and mineral water. The reason you'd want to drink mineral water is that when you're fasting, you're flushing out a lot of water and this may lead to electrolyte imbalances and mineral deficiencies.

To prevent that, you can simply add a pinch of sea salt or pink Himalayan sea salt to your water. Also, consuming ionized rock salt will give you some iodine as well, which promotes thyroid functioning and prevents hypothyroidism.

But that's basically it...

All in all, you'd want to know the purpose of your fast. If you're fasting just for burning fat, then you simply want to maintain a caloric deficit. I wouldn't bother with trying to trick your body into thinking its fasting by drinking bulletproof coffee or taking exogenous ketones.

If you're already overweight or have excess body fat, then your body is already carrying around enough energy. You don't need to elevate your ketones after you adapt to fasting ketosis. BHB salts or other supplements aren't a fat loss tool – they're simply ways of increasing your blood ketones but as we found out too much energy in the system can have a counter-balancing effect.

If you're planning to fast for 16-20 hours, then you'd want to boost autophagy with these compounds I mentioned earlier. You'd want to skip all calories and drink only some coffee, teas, and at the most take 1 tsp of MCT oil with compounds like ginger, turmeric, and ginseng. This will help you to maintain a caloric deficit while still cleaning your cells.

If you're fasting for over 24 hours then I recommend you to stick to only coffee, apple cider vinegar, and tea because you don't need the extra energy from fat or the herbs as your own endogenous ketone body production is through the roof.

How to Break a Fast

Breaking the fast with a lot of carbohydrates may cause an abrupt weight gain. The reason for that is sodium retention. While fasting you excrete a lot of water and refeeding on carbs causes antidiuresis of potassium and sodium. You'll get bloated but you may also have an energy crash of insulin.

Therefore, you want to be eating something that's easier to digest and doesn't put too much strain on the gut. It also depends on how long you've been fasting for. If you're coming off a 5 day fast, then you need to be more patient than you would when just doing daily intermittent fasting with 16 hours fasted.

Let's start off with the average situation where you've been fasting for about 16-18 hours and you're nearing the finish line. During your fast, you can drink non-caloric beverages like water, tea or coffee.

Now that you're about to break the fast, you want to consume something that stimulates the digestive tract without releasing insulin. Apple cider vinegar is perfect for this – it balances healthy pH levels, kills off bad bacteria in your gut, stabilizes blood sugar and improves overall health. You can drink a glass of apple cider vinegar during your fast. But it's a great drink for breaking it as well. Here's what I do before eating any meals.

- 2 tbsp of raw apple cider vinegar
- 1 half of lemon squeezed into hot water
- 1 pinch of cinnamon for better blood sugar stabilization
- 1 pinch of sea salt

Alternatively, you can do this without apple cider vinegar as well and just use hot lemon water. In any case, you want the citric acid from lemons to promote the creation of digestive enzymes in the gut before you eat.

After that, you can also drink some bone broth. Bone broth is amazing and super good for you because it has a ton of electrolytes but it's also packed with collagen.

- Collagen protein is what most of your body is made of – your joints, nails, hair, skin – it helps to keep you more youthful and elastic.
- Drinking bone broth after your fast will also help you to absorb the other electrolytes and minerals a lot better. Your gut has been cleaning house for a while and is now ready to utilize the nutrients from the bone broth as well as the foods you'll be consuming afterward.

If you've been fasting for over 20 hours, then it's a good idea to drink some bone broth or some soup before you eat anything solid. If you've fasted for just 16 hours then it's not that important but still a good idea. *BONUS TIP: Add cinnamon to bone broth – it's magnificent.*

An alternative for bone broth is also fish broth. You cook up all the leftover bits of the fish, like the head, the fins and bones so you could get all those extra omega-3 fatty acids. Don't just throw away these foods when there are still so many unused nutrients on them. When cooking fish, avoid boiling temperatures and overcooking because it may damage the fats. It would be best to steam the fish in some medium-heat water for many hours and make a soup out of it.

If you don't have access to bone broth but you still want to give yourself some immediate energy without crashing, then you can also consume some MCT oil. MCTs get converted into energy faster because the fatty acids in it bypass the liver and go straight to the bloodstream. This is beneficial because you'll also prolong the effects of fat burning and stay in ketosis for longer even after having consumed calories.

After you've drunk your lemon water and maybe bone broth, then you should wait for about 15-30 minutes to let your gut absorb the nutrients. You might feel your intestines waking up, which is a good thing.

What to Eat When Breaking a Fast

To prevent a sugar crash, I recommend your first meal should be something small and low glycemic no matter what diet you're on. This will keep you in a semi-fasted state because of the non-existent rise in blood sugar. Some examples would be:

- 2-3 eggs, half an avocado, some nuts and spinach
- 1 can of sardines with some salad and olive oil

I wouldn't recommend you break your fast with red meat because it's more difficult to digest than eggs or fish. Meat products should be eaten as your second meal.

The first meal should be still relatively small. In total, it should be around 500 calories. Unless you're coming straight from a workout. In that case, you'd want to spike muscle protein synthesis with more protein and calories.

On your second meal, you should eat based upon your desired anabolic-catabolic condition. If you're trying to build muscle or recover from exercise, then it's a good idea to pick something from the ModTOR group. When in maintenance mode or wanting deeper autophagy, stick to more nTOR foods and eat primarily plant-based.

Breaking Your Fast With Fruit

The best time to eat fruit ever is when your liver glycogen is empty i.e. when fasting or while exercising. If your liver is already full and you're eating fruit, then that fructose will be stored as fat and you may even get fatty liver disease. Fruit for dessert isn't a good idea.

But there aren't many benefits to fruit in general. If your liver glycogen gets depleted, you'll be burning ketones and you don't need to replenish your stores with fructose. You'll be actually better off with continuing to stay in this fat burning state, rather than breaking it with sugar that you don't need.

If you do choose to consume fruit, then still make it low glycemic and opt for fruit with more fiber, like apples, berries or pears. Some citrus like grapefruit and oranges can be good as well. The high carb fruits like bananas and mangos should be considered more like performance fuel because they're more insulinogenic.

Whatever type of fasting you're coming from, you shouldn't eat any more than twice a day. Consuming your calories within a smaller time-frame is what this entire concept of metabolic autophagy is all about. When doing the targeted intermittent fasting, you can have meals that span 1-2 hours but it's generally better to stay in a semi-fasted state most of the time.

Chapter XIX
How to Fast for Days and Days

*"You never know what a good meal tastes like
until you haven't had one in a long time"*

Siim Land

When someone was to ask you: *"When are you going to have your next 5-day fast?"* then you probably don't have an immediate answer to give – *"Ummm...Yea...Let's do it right now!"*

One day I was about to face 4 days of traveling and I thought why not practice some hardcore fasting during that trip? I figured: *"Alright, I haven't had a longer fast in a few months so let's take advantage of this opportunity."* Sold!

The reason I decided to have a 5-day fast was that it kinda fit into my schedule quite nicely. I had to be on the road and I wouldn't have had access to quality food sources. For me, that's a perfect situation as I'll get in my extended fast and I'll avoid all the potentially harmful calories that I don't need.

In fact, I prefer fasting and staying hungry rather than eating some random unhealthy meal on the go without much nutrition. Your body from a physiological and metabolic perspective would also be better off with avoiding *all calories all together* as to drive yourself into deeper ketosis and autophagy rather than nibbling on some dried out lettuce wrap or croissante from a fast food restaurant.

The more keto-adapted you are, the easier it is for your body to burn its own stored fat for fuel, and the easier it is for you to fast. With contemporary eating habits, like 4-6 small meals a day, very carbohydrate rich menus, highly palatable foods, and snacks, the average person's body is almost incapable of burning fat and ketones for fuel.

Caloric restriction on a non-ketogenic diet with plenty of carbohydrates also prevents you from establishing ketosis and reaping the benefits of a fat burning metabolism. You'll burn a little bit of fat but because your body is so used to glucose, it'll also compensate for that deficit by converting some of your muscle tissue and vital organs into glucose through gluconeogenesis. Gluconeogenesis occurs only if your body isn't in ketosis and it can't cover its energy demands with fat and ketones. This happens because the metabolism hasn't adapted to using your fat stores.

Therefore, the key to a healthy and successful fast is to establish ketosis as soon as possible and to prevent your body from entering into a semi-starved state. It also means that if you're trying to lose weight, then it would be better to avoid all calories and do strict fasting until you're lean rather than follow a calorically restricted diet that keeps you malnourished for several months.

With fasting, you can get into ketosis with a few days and you'll trigger the other benefits that can increase your longevity whereas with restricting your food intake will downregulate your metabolic rate and thyroid. This makes you moody and lowers your energy levels to the point of complete exhaustion.

Day One - Just the Routine

I started my 5 day fast like I start any other day of my week – the standard morning routine of meditation, cold thermogenesis, and red light therapy.

During the day I'm going to be drinking very sparingly and only when I feel thirsty. The reason being is that keeping your electrolytes in check is one of the most important things you have to do while fasting.

You NEED TO Get Enough Electrolytes! Fasting with nothing but pure water will make you flush out all of the minerals in your body, which can lead to muscle cramps, lethargy, exhaustion, heart palpitations, and elevated cortisol.

- **Daily sodium minimum is at about 1500-2300 mg-s, which is 1-2 tsp of salt.** If you're physically active or are sweating a lot, then aim for up to 3000-4000 mg-s, which is 3-4 tsp of salt. Use quality sea salt or pink rock salt. Mix it with water for better absorption.

- **RDA for potassium is 4700 but once you get enough sodium, you don't have to worry about getting any more than 2500 mg-s of potassium.** Sodium is potassium-sparing and your cells will lose potassium only if you lose sodium. You can use potassium salts like NuSalt to get more potassium with less sodium.

- **Other electrolytes like magnesium, calcium, phosphorus, and chloride aren't that important during 3-5 day fasts** because most of them are actually stored inside the body. Magnesium is much better absorbed through the skin than from supplements. That's why taking an Epson salt bath with magnesium flakes can be a very good idea. However, 300-450 mg of magnesium can do wonders. Taking a multivitamin is necessary only if you fast for longer than 5-7 days. Your ability to absorb vitamins is much lower without food anyway so it's easier to not take any supplements.

Additional sources of electrolytes can be baking soda and apple cider vinegar.

- **Baking soda** is 100% sodium bicarbonate and 1 tsp of baking soda has 1259 mg of sodium. It's great for balancing the body's pH levels, fixing digestive issues and healing the kidneys. Baking soda is a natural antibacterial and antifungal agent that promotes alkalinity in the body. You shouldn't take around meal times because it'll interfere with stomach acid production.

- **Apple Cider Vinegar** – ACV is zero calories but it has trace amounts of potassium and other minerals so it's great to consume while fasting. During the fasting window, you should use the distilled vinegar without the mother.

That's why I consume 1 cup of water with ½ tsp of baking soda, 1-2 tsp of ACV once or twice a day. This will also prevent kidney stones and ulcers while fasting, which may happen because of inadequate production of digestive enzymes.

Besides drinking salted water all day, you can also consume some black coffee, sparkling water, mineral water, non-caloric teas, and that's practically it.

The danger of drinking coffee or drinking too much water, in general, is that you may have to go to the bathroom more often. If you're peeing very frequently, then you'll also excrete out more electrolytes. A loss of minerals will make you feel more tired and actually predisposes you to dehydration. That's why dry fasting can be much more effective than water fasting because you'll hold onto your salts and you'll enter into ketosis faster.

Bone broth or collagen protein powders will definitely kick you out of a fasted state because they contain a lot of amino acids that in the presence of no calories can be insulinogenic, thus kicking you out of a fasted state.

If you want to establish a ketogenic state faster, then you would want to deplete your liver glycogen stores to promote the production of ketone bodies. That's why it's a good idea to make your last meal before fasting low carb. By the next day, you'll be in a semi-fasted state already with ketones running through your veins. Start the day with a long walk and you're already there.

The most difficult part of any extended fast is the first 24 hours. Because your body hasn't fully switched over into a fasted state, you may experience more hunger and tiredness. This can make it difficult to fall asleep in the evening. However, if you make it through the first night you're off to the races and it gets a whole lot easier. Some tips for surviving the first night would be to drink a small cup of water with 1 tsp of ACV before going to bed. This will reduce hunger and lowers your blood sugar in the morning.

PRO TIP: If you want to ramp up your metabolism and raise body temperature, then you can have 1/2 tsp of Cayenne pepper in your water. This will heat up your body and can stave off hunger. It won't break a fast and the capsaicin in Cayenne pepper actually stimulates autophagy.

Day Two - Staying Active

In the morning of day two, you'll probably feel slightly hungry but not to the point of excruciating pain.

There is no difference in hunger levels between fasting for 20 hours or 5 days – your hunger levels stay the same and you won't get increasingly more hungry the longer you fast.

Once your body begins to show the first signs of ketosis, it starts to use its own stored fat for fuel, which gives plenty of energy to the brain and muscles. Your mental acuity and alertness will actually increase because of elevated cortisol and adrenaline as well.

While fasting for longer periods of time, it's important to stay active and keep moving. You don't want your body to start wasting away your muscle tissue just because of inactivity. It's a good idea to either do a few sets of push-ups or use resistance bands to keep the stimulus active.

When I'm fasting, I don't sit on the couch or something. On that particular 5-day fast, I averaged about 8-10 km of walking a day, which is twice the amount people walk even when they're eating enough calories.

Walking is also a great way to reduce hunger. My hypothesis is that because you'll be converting more body fat into energy, you'll feel as if you've consumed some actual calories because on a metabolic level that's exactly what happens. Moving around will also get your mind out of its own rut by exposing yourself to fresh air and environmental stimuli.

The problem is that although you may be this courageous fasting warrior who's willing to go hungry for several days in a row, the people around you don't want to hear anything about it. If the average person can't even skip their breakfast or morning coffee, then we shouldn't even mention not eating for 3-5 days.

It's true that social pressures and environmental temptations can make a novice faster break very easily. They go to the mall -> they smell fresh doughnuts -> they get the cravings -> they instantly give in to the temptation and break their fast. Afterward, they'll get a sugar crash that makes them feel exhausted and they'll start loathing themselves...

How do you deal with other people cooking food around you and eating right in front of your eyes? This is purely a matter of willpower and awareness.

- First, you have to realize that the reason you get the cravings to eat something is purely due to you being exposed to that particular sensory stimulation (the aroma of doughnuts, the sizzling sound of steak, the pleasurable looks of your peers enjoying their food).

- Secondly, once you recognize that you've been tricked into craving food, then you have the opportunity to manifest your willpower. The key is to start associating more pleasure with staying indomitable and to associate pain with giving in. This way you'll feel better about yourself if you do manage to hold your discipline.

Saying no to temptations and cravings is a skill as well as a muscle that gets stronger the more you test it. Initially, if your fat-adaptation is in its infancy, you may find it more difficult but eventually you should reach expertise that resembles putting pieces of meat in front of hungry wolves but still not giving in. If you're able to stare at other people eating birthday cake right in front of your eyes and still continue to fast, then you're truly an expert in self-control.

It's much easier to fall asleep at day two than it is during your first night because you've got more used to fasting.

Day Three - Things Are Starting to Get Good

After day two, it's much easier to fast as well because you should be in deep ketosis. My daily liquid consumption usually looks something like this:

- Rinse my mouth with salt water after waking up and drink a cup of water with a pinch of sea salt.

- During the day, 1 cup of water with ½ tsp of baking soda, 1-2 tsp of ACV, drank over the course of 2 hours.

- When I feel thirsty, I sip on some regular water.

- At noon, I drink 1-2 cups of coffee.

- In the afternoon, I drink a cup of green tea with a pinch of sea salt.

- If I'm feeling hungry in the evening, I'll make a cup of decaf coffee.

- In the evening, I drink a cup of water with 1 tsp of ACV and ½ tsp of sea salt.

In total, this will add up to 5-8 cups of liquids a day and about 2000-3000 mg-s of sodium, depending on the day.

It's not actually a good idea to be drinking a lot of water while fasting because it'll make you urinate more thus excreting more electrolytes. You'll feel generally better with sipping on small amounts of liquids and then refilling yourself occasionally rather than force-feeding more water than you need.

The key is to pay attention to your hydration levels by monitoring the color of your urine and the context of your saliva.

- If your urine is crystal clear and you're going to the bathroom every hour, then you're edging on the side of overhydration and should dial down your liquid consumption.

- If your urine is dark yellow, brown, purple, or pure dark, then you're very dehydrated and have to find something to drink.

- If your mouth is all dry and lips cracked, then you're starting to get dehydrated and should get a few sips of water to drink.

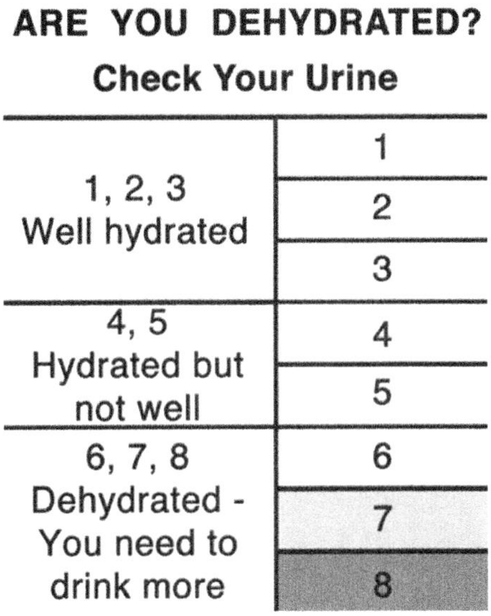

ARE YOU DEHYDRATED?
Check Your Urine

	1
1, 2, 3 Well hydrated	2
	3
4, 5 Hydrated but not well	4
	5
6, 7, 8 Dehydrated - You need to drink more	6
	7
	8

Figure 91 Hydration Chart

Short periods of mild dehydration can be good for the body by triggering a hormetic response. Dry fasting for several days is a class of its own and should never be done without prior experience but it can boost the effectiveness of the fast quite a bit if you follow it up with proper re-hydration.

Dry Fasting

Dehydration can inhibit mTOR and stimulate more autophagy because the body is forced to obtain the water it needs from other cells[650]. During the entire time, you'll still keep urinating and excreting liquids as if you were drinking water.

There's a difference between a soft and a hard dry fast.

- Soft dry fast allows you to come into contact with water on your skin such as showering or mouth hygiene
- Hard dry fast limits all contact with hydrogen whether through the air, skin care products, or fumes.

This is an important differentiation because while dry fasting, your body starts searching for more water and the pores of your skin become more susceptible to absorbing water.

That's why hard dry fasting is more effective than soft dry fasting but this can't be achieved entirely if you live in a rural environment. The best place to do dry fasting is somewhere in nature with clean air and to not allow your skin to absorb water while showering, taking a swim or brushing your teeth.

Dry fasting is an advanced version of intermittent fasting and it shouldn't be tried without any prior experience with extended water fasting. Prolonged periods of dehydration may become dangerous to your health and cause long-term damage. So, how to do it safely?

- To prevent dangerous dehydration, you'd want to drink adequate amounts of water, sodium, and vegetables the day before to fill up your mineral stores.
- During the day, don't consume any liquids from water, tea, coffee, or anything else. It'd be better to avoid strenuous physical activities as well to prevent muscle cramps. You don't want to expose yourself to too much heat either like with saunas, baths, or steam rooms.
- If you're dry fasting only during the day then you can „break the dehydration period" with a glass of salted water in the evening. Then I'd continue with more lemon juice water and apple cider vinegar. After that, you can eat a low glycemic meal that maintains stable blood sugar levels and insulin.
- If you're dry fasting for longer than 24 hours, then you would simply go to bed without eating or drinking anything. In that case, you'd have to be cautious and take full responsibility over your decisions. Again – do your research and prepare yourself in advance.

At first it may sound crazy like: *"Really?!? First no food and now no water?"* It's true that anything in excess can be harmful. However, the human body has developed many of these adaptive mechanisms like hormesis and AMPK for dealing with periods of deprivation and stress.

If you're not that extreme as to do dry fasting, then you should still try out regular intermittent fasting. In fact, you can gain about 80% of the results with just daily IF. Back to my 5-day fast story.

Day Four - Things Getting Beautiful

By day four you should be nearing the 100-hour mark of fasting. This is a significant milestone both in terms of psychological achievement as well as the physiological state.

It can be useful to measure your blood glucose and ketones to see where you lie. Ketosis itself begins around 0.5 mmol-s and within 4 days of fasting, you can expect to fall somewhere between 2.0-5.0 mmol-s. If you hit numbers above 6.0-7.0 and you're worried about ketoacidosis, then instead of bailing halfway through go for a long walk or exercise.

The reason you register so many ketones has to do with not burning them for energy and thus keeping them in the bloodstream for longer. That's why you also have to look at your blood glucose. Generally, less glucose entails more ketones, unless you're borderlining ketoacidosis. If your ketones and glucose both stay high for way too long, then consider breaking the fast or consult a physician.

By day four fasting also becomes the routine. You've gotten quite used to not eating that it has also escaped your focus of attention. Although you may start thinking about what you'll break your fast with, the sensory temptations from your environment become less effective and you may begin to actually enjoy fasting.

Being in a fasted state is a unique and profound experience. You're not feeling particularly hungry or tired but still somewhat limited in energy. It's a blissful state where you can enjoy clarity of mind and heart – free from cravings, free from attachment to food or the desire to consume something mindlessly.

Fasting not only resets your taste buds from the imprisonment of processed food but also liberates your mind from being dependent on certain types of food that you tend to crave the most. Even the healthy kind.

Not eating for several days actually changes your relationship with food for the better. You can begin to appreciate the absence of food as you know that it makes you appreciate the presence of food afterward even more. It's like a way of reminding yourself how grateful you can be for having access to your pantry and fridge at all times.

Fasting also has a spiritual component – I feel like being in this semi-transcendental state where I'm not being influenced by any attachment. To put it bluntly, it feels like I'm about to transcend the primordial soup of existence and float around in a different dimension of mind and spirit.

By the end of day four, I had actually returned from my travels and was at home. I could've easily stopped right then and there as I had already accomplished quite a lot by fasting for over 100 hours.

When I arrived home it was already 7 PM and I didn't want to go through the hassle of properly breaking a fast. It's important to follow the right procedure when stopping fasting as your gut needs some time to get used to food again. Instead of randomly eating whatever and staying up all night eating, I wanted to savor the moment by doing everything in the correct manner. Given I'd already fasted for 4 days, what difference would it make to continue on for another day?

Day Five - Preparing for the One Meal to Rule Them All

That's why I decided to change my initial 4-day fast into a 5-day one. I was already in a deep fasted state and past the most difficult part. It was much easier for me to continue until day 5 and really reap the benefits of stem cells.

Fasting is great for purifying your body and mind but it can't be done indefinitely. It's not a get out of jail free card either with which you can excuse yourself into binging.

If you're already eating a very clean diet with a ton of autophagic compounds, medicinal herbs, and staying in a slight caloric deficit, then you're already getting most of the health benefits of longer fasts. The difference is that with strict fasting you'll ramp up autophagy and other longevity pathways even more.

The morning routine was the same – ACV with baking soda and water, some coffee at noon, and smalls sips of water throughout the day. Funny enough, I actually had to go for number 2 in the bathroom. It wasn't a massive dump but it was surprising that I still had some excrements after 5 days. But let's not analyze this much longer.

We talked about how to break a fast in the previous chapter but generally, you want to start off with easily digestible liquids like hot lemon water, bone broth, and stews. After waiting about 30 minutes, you move on with some fermented foods and steamed vegetables. The first protein you'd consume should be easy on digestion like eggs or fish. Too much meat can put unwanted stress on the gut.

Day Six - The Aftermath

On day 2 of eating again, you should eat slightly above your maintenance as to nourish the body and restore its energy. A small surplus of 200 calories is more than enough to rev up your metabolism and thyroid again.

What you do on day 3 of eating again depends on your body composition and goals.

- If you're overweight and have more than 20 pounds to lose, then you can immediately go for another extended fast for 3-5 days. Your body fat is food and the second fast will be that much easier.

- If you're already lean and around 10% body fat, then you can swap over to the one meal a day or warrior diet with a daily intermittent fasting window of 16-22 hours fasted. Physically very active people can adopt targeted IF. Your next extended fast of 3-5 days for longevity-boosting reasons should be in a few months.

Whatever goals you have, you don't need to be eating any more than 2 meals a day. The 16-8 fasting schedule is a perfect sweet spot for daily use. I even think it should be the minimum time window everyone should fast for every day.

Fasting is also a great tool for overcoming any form of addictions, whether that be sugar, processed food, ice cream, alcohol, coffee, or just eating in general.

After day 3, I simply lost all attachments to everything – it was indeed like I was transcending the primordial soup of existence because I was in such a blissful state. I had low levels of energy and I carried out this transient feeling of an empty stomach but other than that I could've easily dropped dead at that spot and not give a damn. It was like everything was okay – it was okay that I hadn't eaten in days, it was okay that I didn't get to workout, it was okay that I didn't get to make more content online, it was okay that I felt cold and tired.

I think fasting shouldn't be done just for healing your body but for also healing your mind. It's important to „fast" from everything you're attached to whether that be social media, sex, alcohol, tobacco, central heating, entertainment, and even water, just so you could become independent of those things. Just so you could remain unaffected by their absence and appreciate their presence. That's why I'm definitely going to practice daily intermittent fasting as well as implement these longer 3-5 day fasts a couple times per year.

So, my question to you is: *When will you have your next long fast? What's the longest you've actually gone without food? Have you actually abstained from eating any longer than a day?*

Whatever the case may be, fasting can be used as a very effective tool for improving body composition and health. It is one of the easiest miracle drugs that will heal your body and mind.

Before you start practicing these longer fasts, you have to come to terms with your medical condition and consult your physician first. All the responsibility is on you and not your doctor, your parents, not me, or your cat - everything is on your shoulders.

Chapter XX
When Not to Fast

"Nothing earthly will make me give up my work in despair."

David Livingstone

When you first hear about intermittent fasting, then the first questions that may arise have to do with how to actually do it or what the long-term effects may be. By now you may agree that fasting is one of the most powerful health interventions and disease prevention methods there is.

However, there are some situations when it's not a good idea to fast. This doesn't necessarily mean that you stop time-restricted feeding or fasting altogether. You can simply make some minor adjustments and in so doing maintain both the health benefits of fasting as well as the metabolic flexibility.

Although fasting heals the body, it's still a catabolic stressor that can become too stimulating. It's a hormetic stressor that has a dose-specific effect on your metabolism and endocrine system.

Fasting raises certain stress hormones in the body such as cortisol, adrenaline, and norepinephrine that will heighten your awareness and mental clarity. Chronically elevated levels of these stress hormones may promote fat gain, low thyroid, exhaustion, and adrenal fatigue.

When doing intermittent fasting or any type of demanding activity, it's essential to remember how big of a stressor it is and how well adapted you are. Fasting itself won't necessarily make you stressed out or weaken your immune system. It's always context dependent and the other lifestyle factors need to be kept in mind.

If your cortisol levels get too high because of combining fasting with other stressors, then it's natural that you may be overdoing it. In that case, you'd have to assess what things are causing me most harm and whether or not I want to keep them.

Here are a few common stressors in your life that will make fasting much worse than it is:

- Being stressed out because of work makes you more prone to overeating and too much cortisol.

- Not sleeping enough promotes weight gain and insulin resistance, which will negate some of the benefits of fasting.

- Feeling anxiety and mental turmoil may cause the same stress as does physical pain and discomfort.

- Combining too much exercise with a lot of fasting may lower your thyroid functioning and lead to many plateaus.

- Poor food quality and not enough nutrients will lead to some deficiencies that make you more tired and can cause health problems.

My recommendation for all of those stressors is to deal with the root cause, which usually involves imbalances in lifestyle and habits. Keep the fasting but eliminate the noise.

Let's now move on with more specific situations about when not to fast.

Should You Fast with Low Thyroid

Low thyroid functioning is most commonly caused by too much stress on the body. If there are too many stress hormones circulating the blood, you'll lower the thyroid hormones as well as a means of self-defence.

Hypothyroidism is caused by too low amounts of thyroid hormones in the blood and it usually happens in people who've had Hashimoto's disease, thyroiditis or have had their thyroid removed.

Thyroid cells absorb iodine found in some foods and combine it with the amino acid tyrosine, which is used to create thyroid hormones: thyroxine (T4) and triiodothyronine (T3). They are then released into the bloodstream to affect your body temperature, your daily caloric needs, your heart rate and what's your metabolic rate.

T3 and T4 are regulated by the hypothalamus and the pituitary gland inside your brain.

- When thyroid hormones decrease, the pituitary gland releases thyroid releasing hormone (TRH) to signal the thyroid gland to produce more T3 and T4

- With high levels of T3 and T4 in the blood, the pituitary gland releases less TSH so the thyroid gland could slow down

Nutritional Regulators of Thyroid Hormone

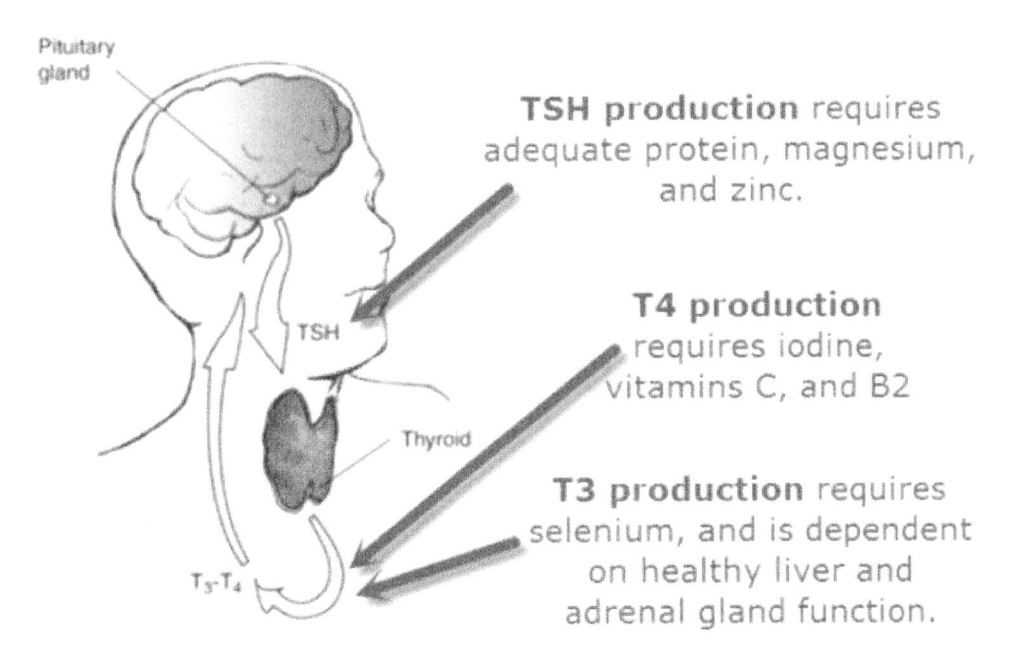

TSH production requires adequate protein, magnesium, and zinc.

T4 production requires iodine, vitamins C, and B2

T3 production requires selenium, and is dependent on healthy liver and adrenal gland function.

Figure 92 For thyroid health consume more magnesium, zinc, iodine, and selenium

Most of the thyroid hormones comprise of T4 – nearly 80%. T3 is the active form of T4 and it's thought to be more potent in regulating energy metabolism. There are also T0, T1, and T2, which act as hormone precursors and byproducts.

To prevent low thyroid from fasting you want to make sure you get enough calories during your eating window. A slow metabolism isn't the result of time restricted feeding *per se* but more by not giving the body adequate nutrients.

If you already suffer from low thyroid and experience the symptoms of hypothyroidism, then it would be a good idea to shorten your fasting window a little bit to allow your body recover. Simply have your food a few hours earlier and you'll speed up your metabolic processes.

Here's what you can do to overcome low thyroid without medication:

- **Get More Iodine** – The thyroid needs iodine to produce thyroid hormones. Even small doses of 250 micrograms of iodine a day can help the thyroid. You can get iodine only from diet and rich sources for that are sea vegetables, algae, wild caught fish, eggs, seaweed, shellfish, iodized salt, and some dairy.

- **Limit Goitrogenic Foods** – Goitrogens are compounds that can affect the thyroid gland if consumed in large amounts. Foods high in goitrogens are things like cruciferous vegetables, broccoli, cabbage, kale, and some fruit. However, the benefits of these foods far outweigh the downside. They'd become a problem only if you eat too many raw vegetables. If you cook or steam them lightly then you'll lower the number of goitrogens in them.

- **Tyrosine-rich foods that support the thyroid are pumpkin seeds, beef, poultry, almonds, avocados, eggs, and fish.** They also have B12 and selenium. You should aim for organic meats because they're not injected with antibiotics and other harmful hormones.

- **A Healthy Insulin Spike** – If you eat some carbohydrates like a sweet potato or rice, then you can boost your metabolic rate. Especially, if you're eating a low carb diet. Prolonged periods of dieting will slow down your metabolism and doing some carb cycling can be very beneficial.

- **Eat More Calories** – Dieting and restricting calories for too long will slow down your metabolism and thyroid functioning. Having a diet-break for a few days where you eat slightly above your maintenance can help you to boost your metabolic rate. However, you don't want to be eating inflammatory foods like processed carbs, pastries, pizzas, or ice cream. Instead, you'd want to get more of the thyroid supporting foods we've talked about here.

Now for the things you don't want to be doing if you suffer an under-reactive thyroid.

- **Avoid All Gluten** – Gluten is a common allergen that activates an autoimmune response on the thyroid. People with Hashimoto's or hypothyroidism tend to be sensitive to gluten as well. Gluten is found everywhere – not just in bread, pasta, and cookies – it's in almost all packaged foods, sauces, meats, skin care products, and the particles can even float around in the air. So, I'd not spend too much time in front of bakeries or pastry shops.
- **Don't Drink Tap Water** – Most tap water contains fluorine and chlorine that inhibit iodine absorption and dampen your pineal gland's functioning.
- **Limit other potential allergens** as well such as dairy, nuts, shellfish, eggs, fish, soy, or meat if you are sensitive to these foods.
- **Don't Fast for Too Long** – It'll lower your thyroid functioning further and can decrease metabolic rate. Definitely, you shouldn't fast for longer than 24 hours if you have low thyroid. Until you've healed your thyroid and metabolism, somewhere between 16-18 hours is fine.
- **Reduce Your Stress** – It's going to make you feel more drained and exhausted. Don't start exercising harder or get more stressed out about work but try to relax more and allow your adrenals to recover. You should also avoid stimulants like coffee and sugar because they'll activate the adrenal glands.
- **Avoid Environmental Toxins** such as pesticides, mold, BPA found in plastics, isoflavones found in soy, potassium perchlorate, cigarette smoke, and keep your home clean from these toxins. Heat saunas and sweating can help you to eliminate these toxins from your body.

Having a decreased metabolism and lower thyroid hormones itself aren't innately bad or harmful for your health. It's just that the quality of your life will be lessened if you're feeling tired and lethargic all the time. This can lead to obesity and other metabolic diseases.

An ounce of prevention is worth a pound of cure and it's so much easier to avoid the bad habits all together rather than try to fix the symptoms with medication and continue to suffer. Nothing won't work if your stress levels are too high. Therefore, proper stress management that involves meditation, relaxation, sleeping, and easy exercise are key here as well.

Should You Do Intermittent Fasting When Trying to Bulk

What if you're trying to gain weight and build muscle with fasting? Should you fast then as well?

In mainstream fitness advice, it's recommended to eat very frequently – up to 4-6 small meals a day to avoid the catabolism of fasting. However, intermittent fasting can help you to actually build muscle in many ways.

- Short-term fasting has been shown to increase Leutenizing Hormone (LH), which is a precursor to testosterone. In a study done on obese men, LH increased by 67% after 56 hours[651]

- Another study found that obese men saw a 26% increase in GNHR (Gonadotropin-releasing hormone), which is another testosterone stimulant[652]. The same study found that men who were working out saw a 67% increase in GNHR, which led to a 180% boost in testosterone.

- Occasional extended fasts increase nutrient partitioning and can help with absorption rates. Cycling your food intake is probably one of the best ways of maintaining metabolic flexibility.

The best thing about doing intermittent fasting while trying to gain weight is that it makes you healthier than if you were to not do fasting. A healthier body will build healthier muscles and promotes quality weight gain, not fat gain.

For building muscle, you need to do resistance training, eat slightly more protein, get a small surplus of calories, sleep enough, and stay consistent.

The only thing to keep in mind would be to restrict your feeding window a little bit so you would have some more time to increase protein synthesis. Simply shorten the fasting window a little bit and you'll see great results.

When trying to build muscle, then fasted workouts aren't ideal because they may limit your strength potential. The key to growing muscle is getting stronger and that's why you may find the targeted intermittent fasting protocol or the 16/8 Leangains protocol perfectly suited for this. Any more frequent eating will definitely promote more anabolism but they're not necessary.

Should Pregnant Women Fast

What if you're trying to grow another living being inside of yourself? Should pregnant women fast?

The fetus requires a lot of nutrients and growth factors to ensure its growth. That's why it's even more important for the mother to make sure they're getting good quality nutrition and avoid the potential toxins or bad ingredients.

There are some studies that say that Muslim women who fast during Ramadan may get babies with lower body weight, they're shorter, thinner, or more prone to premature labor. However, these differences are quite small.

Whatever the case may be, the mother should focus on being as healthy as they can to promote the health of her child in the future. Quality food, stress management, enough movement, and sunlight are all essential.

Fasting can be a good addition but only in a dose-specific manner and definitely not all the time. I would still recommend having a daily fast of at least 14-16 hours to promote the general health of both organisms. Then once a week you may have a 20 hour fast but definitely not any longer than that during pregnancy.

Women may react to time-restricted feeding more negatively than men because of their hormonal sensitivity. It makes sense physiologically as well because they need to be the caregivers and have to actually give birth to offspring.

Both men and women can fast just fine. However, to avoid any potential side-effects, women should have slightly shorter fasts. A good rule of thumb is to lengthen the eating window by about 2 hours and break the fast a little bit sooner. The end result would be the same.

If the woman is also fasting in fear of getting fat during pregnancy, then the focus should yet again be on the other pillars of health, not their body image.

Should Children Do Intermittent Fasting

What if the child has already been born? Should children do intermittent fasting? That depends again on their age, health status, lifestyle factors, and situation.

A growing child needs adequate amounts of IGF-1 and mTOR to promote the proper development of skeletal muscle, bone strength, jawline, brain growth, and the right hormones. That's why it's even more important to make sure young kids get things like quality grass-fed meat, fish, eggs, vegetables, and even whole milk, preferably from the mother's breast. It's not a good idea to deprive children of these essential nutrients, especially during the most critical periods of growth.

Children already produce a lot of growth hormone, which helps them to build new tissue, muscles, and develops their brain. As you age this surge of growth begins to drop. Fasting is an amazing way to promote growth hormone production and increase longevity. But this may not be necessary for young children. However...

When you look at the body composition and health of most children today then you can see a growing concern with diabetes, obesity, insulin resistance, and other problems. Fasting can cure a lot of those diseases but it's still not going to compensate for an unhealthy lifestyle.

Children should learn how to follow their hunger more intuitively and recognize when they actually need fuel and when they're simply craving junk food. This teaches them to be more flexible while still dipping into a fasted state.

Deliberately enforcing time-restricted feeding onto children isn't necessary and parents should simply eliminate behaviors of snacking, binging, sedentarism, and cravings. If there's one thing you could do to benefit your child's future metabolic health, then it would be not teaching them to eat sugar and candy. It's just conditioning that can be easily prevented by replacing it with only whole foods. Your kids learn their behavior from you and those around them. So, make sure you are a good role model first.

Should Old People Do Intermittent Fasting

Moving on with the elderly. Should old people do intermittent fasting? There is no real physiological reason why older people can't do intermittent fasting. They'd gain the same health benefits.

The only problem with fasting when you're older is that you may be more predisposed to muscle loss and catabolism. This may make you more prone to metabolic disease and more aging. As you lose muscle you're more predisposed to bone fractures, insulin resistance, neurodegeneration, and cardiovascular disease.

In general, the trend already shows that all the anabolic hormones like HGH and testosterone drop alongside with a reduction in lean body mass. Part of it has to do with becoming more sedentary and not doing resistance training but it's also partly because of becoming less anabolic.

Leucine resistance, in particular, makes it more difficult for old people to maintain muscle, not to mention build it. That's why longer periods of fasting may lead to accumulated sarcopenia. However, that can be easily alleviated by becoming more active and stronger.

To circumvent sarcopenia, you'd have to keep lifting weights, eat adequate amounts of protein, and avoid sedentarism. Increasing protein intake up to 35-40% can also be a good idea as to negate some of the leucine resistance. To stimulate MPS after exercise, you can even take 3-5 grams of leucine to promote muscle homeostasis.

It's not recommended for old people above their 60s to have extended fasts for 3-5 days either. Going without eating for such a long time may have side-effects on muscle mass. The elderly go catabolic more easily because of low anabolic hormones. That's why even on the daily IF schedule they should focus more on the 16/8 time frame instead of a very tight OMAD meal.

You can safely do intermittent fasting even when you're old as long as you consume enough calories and protein to maintain your lean muscle tissue.

Should You Fast Every Day

Another thing to consider is changing up your intermittent fasting routine every once in a while.

Your body is an adaptation machine and it's constantly trying to become as efficient as possible with the stimuli it gets exposed to. Exercise, cold stress, heat, and fasting are prime examples of this. Adapting to fasting through keto-adaptation and stress tolerance is a good thing – it's going to help you survive longer periods of energy deprivation and promote your longevity. However, it may have a downward trend after a while.

If you jeopardize your metabolic rate too much, or if you cause too much stress with fasting, then you'll end up harming your health by making yourself more predisposed to metabolic issues. Fasting may not be always good for you and sometimes it'd be better to be more anabolic i.e. eat more frequently and consume higher calories. As long as you use it to promote quality muscle growth you'll be improving your health.

That's why you should change up your fasting window every once in a while by either extending your eating window a few hours, or going for a longer 3-5 day fast, and then refeeding. The idea is to cycle fasting as you do with your overall nutrition.

Additional Fasting Mistakes to Avoid

- **Don't Become Dependent of Coffee** – Caffeine is an amazing appetite suppressant that helps to continue fasting. At the same time, it can turn into a powerful drug. Don't drink any more than 1-3 cups a day and periodically cycle off from it. There's a bonus chapter at the end of this book that talks about strategic coffee consumption.
- **Don't Get Dehydrated** – Although short dry fasting is beneficial, there's the danger of not hydrating properly afterwards. Just drinking water may not be enough, which is why you should eat plenty of vegetables and occasional fruit to hydrate the cells better.
- **Don't Neglect the Electrolytes** – Insufficient sodium, potassium, and magnesium may not only cause cramping but will also raise cortisol. In fact, not getting enough salt can raise insulin and keep you in fight or flight mode. Use a bit of minerals in your water but also season the food properly.
- **Don't Bring In Additional Stressors** – Make sure your life doesn't get in the way of fasting. Get enough sleep, practice meditation, don't over-work yourself, have downtime, do something fun, spend time with family, and don't combine long fasts with a lot of caloric restriction and exercise.
- **Don't Freeze to Death** – During fasting, you may feel slightly colder than normally. This is because of limited calories and lower heat production. If your fingers and toes are getting numb or blue, then you should stop and cover yourself up. Dress warmly and don't push yourself too hard.
- **Don't Fast for Too Long Too Often** – Be mindful of how well you can handle fasting. Start off with daily time-restricted feeding. Then have a 24-hour fast, then a 48-hour one, and then a 3-day fast. Some signs of too much fasting are constant headaches, fatigue, feeling like being hit with a club, not sleeping well, shivering and feeling very cold despite wearing a lot of clothes.
- **Don't Be Afraid of Hunger** – Most people have manic fear of going hungry. It's more psychological than physical. Fasting can actually help you to reconceptualize hunger and associate it with more vigor and success.

- **Don't Make It a Big Deal** - If you think that you may potentially damage yourself, then you probably will, because of creating a stress response with your mind. Choose to see it as something that empowers not harms you and you'll start to feel amazing.
- **Don't Gorge After Fasting** – It doesn't matter how long you fast if you jeopardize it all by eating excessive amounts of calories and still gaining weight. Moderation is key, especially in fasting and eating.

In this chapter, we talked about different situations when you should not do intermittent fasting but all gets tied back to still doing it. There are no real reasons why you shouldn't do intermittent fasting and there's always a way to make it work. You just have to know what kind of a signal are you sending to your body and how to leverage it according to your goals.

The concluding message would be that you should fast and time restrict your eating despite the day or what your condition is. Just modify it. Don't become dogmatic either and you can safely have random bouts of eating and fasting as you wish as long as you just do it.

Chapter XXI
Circadian Rhythms and Autophagy

"We can easily forgive a child who is afraid of the dark;
the real tragedy of life is when men are afraid of the light."

Plato

Many life forms follow certain patterns of behavior that are linked with the day and night cycles of the planet. Virtually all animals have some sort of a sleeping routine. There's also evidence for the presence of internal clocks that regulate the body's biological processes based on daytime. However, these biological rhythms appear to be present even in the absence of external cues, such as light and temperature.

In 1729, a French scientist Jean-Jacques d'Ortous de Mairan saw how the 24-hour patterns in the movement of the leaves of *Mimosa pudica* remained constant even when the plant was kept in continuous darkness[653]. These endogenous cycles were first independently discovered in fruit flies in 1935 by two German zoologists, Hans Kalmus and Erwin Bünning[654].

The term *circadian* was coined by Franz Halberg in 1959 who said:

> *The term "circadian" was derived from circa (about) and dies (day); it may serve to imply that certain physiologic periods are close to 24 hours, if not exactly that length. Herein, "circadian" might be applied to all "24-hour" rhythms, whether or not their periods, individually or on the average, are different from 24 hours, longer or shorter, by a few minutes or hours.*[655]

In 1977, the International Committee on Nomenclature of the International Society for Chronobiology adopted the official definition for circadian rhythms, which goes like this:

> *Circadian: relating to biologic variations or rhythms with a frequency of 1 cycle in 24 ± 4-h; circa (about, approximately) and dies (day or 24 h).*

Note: term describes rhythms with an about 24-h cycle length, whether they are frequency-synchronized with (acceptable) or are desynchronized or free-running from the local environmental time scale, with periods of slightly yet consistently different from 24 hours[656].

In 2017, three scientists Jeffrey C. Hall, Michael Rosbash and Michael W. Young were awarded the Nobel Prize in Physiology or Medicine for their discoveries of molecular mechanisms controlling the circadian rhythm[657]. During their research, the men identified a gene in fruit flies that controls their circadian rhythms. They named this gene *period*, which encodes a protein called PER. PER accumulates during the night and degrades during the day, thus it oscillates over a 24-hour cycle, in synchrony with the circadian rhythm.

Figure 93 The 2017 Nobel Prize winners in Physiology or Medicine

Circadian rhythms enable living organisms to prepare for and adapt to environmental changes that happen on a regular basis. For instance, the coming winter, nighttime, seasonality of certain food sources, faunal mobility, and fluctuations in climate. By now we know that almost every organ and individual cell has its own biological clock that's regulated by the master clock in the brain[658].

In regards to health and nutrition, circadian rhythms play an immense role. They're going to dictate what kind of physiological processes to commence and which hormonal patterns to follow. That's why it's a vital component to the regulation of autophagy, growth hormone, fat loss, insulin sensitivity and much more.

Let's go through some of the basics. What are circadian rhythms then?

Circadian Rhythm Basics

Circadian rhythms are biological rhythms inside your body that are connected with the day and night cycles of the environment. Humans are diurnal creatures, which means we're active during the daytime and sleep at night. Rats and owls are nocturnal – they're active at night and sleep during the day.

With these circadian patterns come distinctive physiological processes that have evolved over the course of eons. They're evolutionary adaptations of creatures living in a certain way that promoted their survival and evolution.

That's why there's some genetic variance to every person's circadian code. Think of night owls and morning larks who sleep at different times. As hunter-gatherers, some people were more suitable for guarding the camp when others were sleeping etc. However, that difference is very small and differs maybe like a few hours. Humans are still evolved to be diurnal and when the sun's out we better be sleeping.

There is no one whose natural circadian rhythm would be to be awake after midnight and sleep until noon. Those things are the result of living in a modern world with different type of circadian disruptors and lifestyle factors. Shift work, playing video games until the morning, and partying under fluorescent light is unnatural and one of the worst things for your health.

Early research in humans speculated that most people's circadian rhythm is closer to 25 hours when isolated from external cues[659]. However, these results were misleading because the participants weren't shielded from artificial light. In 1999, a Harvard study found that the human circadian rhythm is about 24 hours and 11 minutes, which is closer to the solar day[660].

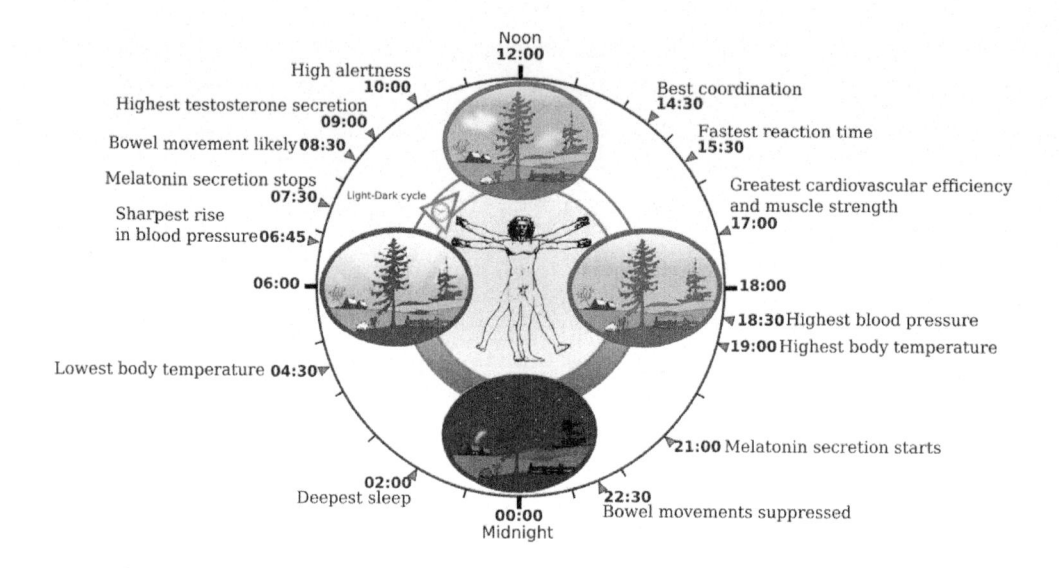

Figure 94 Circadian Rhythms in Humans

What Affects Circadian Rhythms

There are 3 main signaling factors that affect the circadian rhythms – light, movement, and food. Most of the circadian signaling is transmitted through your eyes. When light enters the retinas and gets transmitted into the brain it stimulates the suprachiasmatic nucleus (SCN) (See Figure 95). The SCN is the master circadian clock in your body that regulates all the other biological rhythms and clocks. There are many different types of clocks and it's thought that most organs like the liver, heart, and pancreas actually have their own circadian clock. That's why all these different factors like sunlight, physical exercise, and eating affect the entire circadian rhythm of your body.

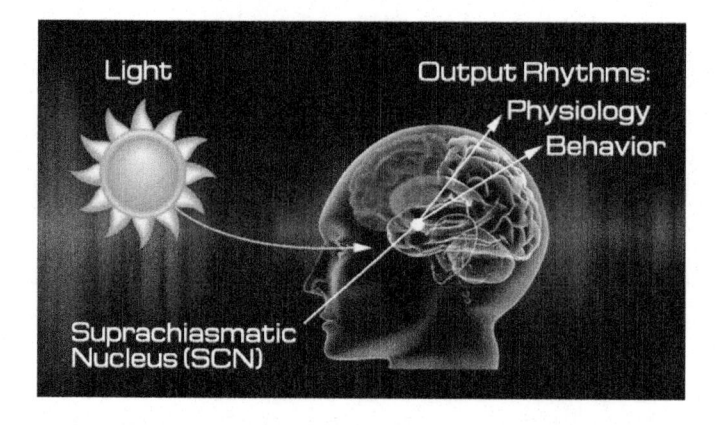

Figure 95 The Master Clock in the Brain is the Suprachiasmatic Nucleus or SCN

Light is made of many electromagnetic particles or photons that travel through space in a wavelength form. They emit energy and are represented by different colors (Figure 96). Sunlight's wavelength is called the solar spectrum and it contains ultraviolet, visible, and infrared wavelengths.

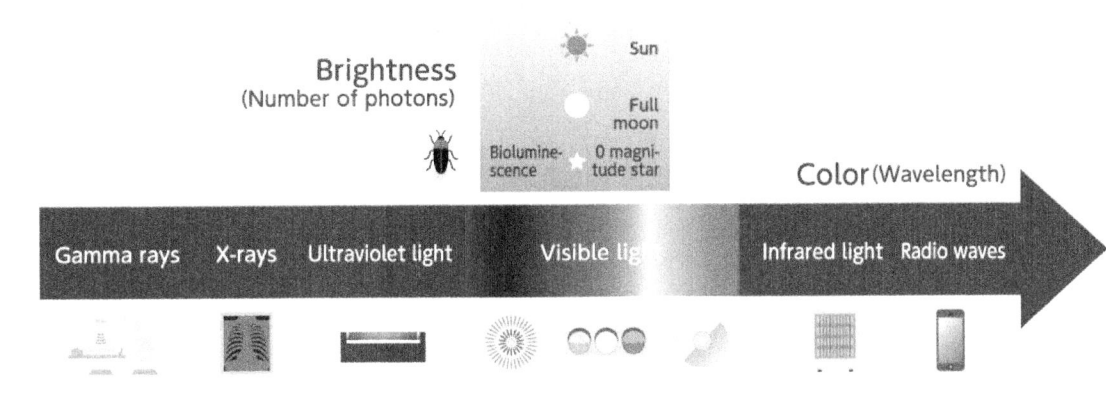

Figure 96 The Light Spectrum

The human eye can only detect visible light which is seen as either violet, indigo, blue, green, yellow, orange, or red light.

Blue light is part of the natural environment and can be seen almost everywhere. The reason why the sky is blue is actually that the blue light coming from the Sun collides with the air molecules and makes the blue light scatter everywhere.

Blue light exposure to the eyes plays a very important role in regulating your circadian rhythms and day and night cycles. It has antibacterial properties, boosts wakefulness, increases alertness, and can adjust the circadian clock. Too much blue light at the wrong time can damage your mitochondria, promote insulin resistance[661], cause insomnia, depression, and increase inflammation.

Blue light has a short wavelength of 380-500 nM, which makes it produce higher amounts of energy. Naturally, you wouldn't get exposed to much blue light aside from the early to afternoon parts of the day. However, ever since the invention of the light bulb, our environment has many additional sources of blue light. Because of technology and new gadgets, we're getting exposed to more blue light for longer periods of time which can offset the circadian rhythm and cause damage to our health.

Blue light exposure at night and circadian mismatches are linked to many types of cancer, diabetes, obesity, heart disease, and Alzheimer's[662]. You wouldn't think that it has such a huge role but after you learn about how light affects your body's biological processes you'll realize how serious this actually is.

Melatonin is the sleep hormone and a powerful antioxidant that helps to conduct many repair processes in the brain and body. If you inhibit melatonin secretion because of blue light at night, then you're going to lower growth hormone, which makes it more difficult for you to burn fat and build muscle, and you'll also prevent the brain from clearing out the toxins that get accumulated there during the day.

There are these proteins called beta-amyloids that are associated with Alzheimer's and Parkinson's and if you don't remove them during sleep with autophagy they begin to accumulate there. Keep in mind that these neurodegenerative diseases happen over the course of decades and you can begin to show the first signs of Alzheimer's 10-20 years already before you actually get the disease.

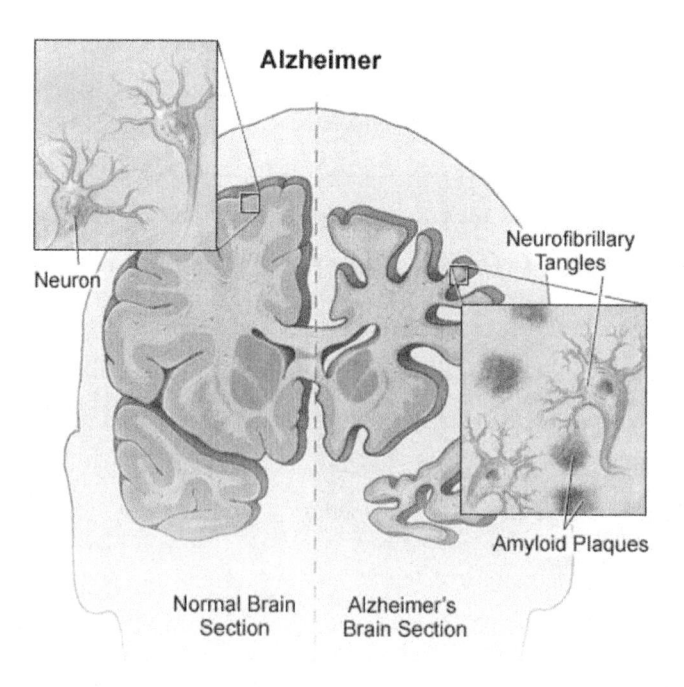

Figure 97 The Difference Between a Normal and an Alzheimer's Brain

So, if you've noticed yourself having brain fog, forgetfulness, or the inability to focus, then you should start taking your circadian rhythms and sleep quality a lot more seriously. I dare say nearly 70% of modern diseases are actually rooted in circadian rhythm mismatches not just eating too much and moving too little.

If you're working at night shift or tend to stay up late all the time, then you have to seriously reconsider what time do you go to bed and what time do you wake up because it's literally killing you.

Protect Your Circadian Rhythms

However, not all blue light is bad. The timing of when you get exposed to it matters a whole lot more, which is why you'd have to entrain yourself to follow a proper circadian rhythm.

- **Expose yourself to the natural sunlight first thing in the morning**. This will synchronize your biological clock to the surrounding environments and maintains consistency.

- **If you live in an area where you don't get much sunlight or if it's cloudy, then use blue light emitting devices** such as the Human Charger or face lamps for 10-15 minutes.

- **On days with clear sunlight try to spend more time outside by going for a long walk.** This will raise vitamin D levels and charges up the mitochondria as well. The majority of circadian signaling happens through the eyes so try to expose them to the daylight. Don't look directly at the sun but gaze the entire blue sky. Don't wear sunglasses or hats that cover your vision either because you'll miss out on the blue light.

- **When indoors wear long sleeve clothes to protect your skin from too much blue light exposure**. A lot of the circadian signaling also happens through the skin, which is why you don't want to sit under fluorescent lights before bed.

- **Wear blue blocking glasses in the evening.** I'm using different blue blockers. The most effective ones are the Truedark ones that literally cover your entire eyes and make you look like Vin Diesel from the movie Riddick but there are less scary ones like BluBlox that actually look stylish and something you could wear in public.

- **Install a software called F.lux or Iris on your computer** to automatically match the brightness of the screen with the circadian rhythms. On Android, it's called Twilight.

- **Sleep in pitch black darkness** with blackout blinds and a sleeping mask that covers your eyes.

- **Make sure there are no hidden sources of blue or green light** in your house like the alarm clock, night lamps, red dots on the TV screen, smoke detector lights and so on.

When you do some research, then you'll find that blue light at night really is quite scary stuff and it's going to shut off melatonin production.

Food Intake and Circadian Rhythms

Another critical component to circadian rhythms is food intake. It turns out that the timing of when you eat is even more important to your health than what and how much you eat.

Time-Restricted Feeding (TEF) emerged as a concept within the context of circadian rhythms. The master clock is very connected to nutrient sensing pathways that detect the presence of calories and food[663].

In most animals, feeding is confined to a certain time period, which leaves a short period of fasting that coincides with sleep. Unfortunately, modern life not only disrupts our circadian rhythms with light but also with food. An average person in the Western world tends to spend most of the day in a fed state, leaving no time for the body to heal itself. Some people can even eat right before going to bed, sleep for about 7-8 hours, and start eating immediately after waking up. This prevents them from ever entering into fasted conditions that are so vital for longevity.

Dr Satchin Panda, who is a professor at the Salk Institute and an expert in circadian rhythms, recommends people to eat their food within a minimum of 8-10 hours. During fasting, the gut and immune system have then enough time to repair themselves and conduct other autophagic processes. Dr Panda's research has found that the average person eats over a 15-hour period, starting with a drop of milk in their morning coffee and ending with a late night snack of some nuts or chips[664].

In 2015, a study tested how eating an entire day's caloric intake within 10-11 hours affect overweight individuals. Their eating window ended around 8 PM. They lost about 4% body weight in 16 weeks and retained it for up to a year. This was accompanied by a spontaneous 20% reduction in calories just because of skipping out on random snacks or alcohol late at night. The participants also reported improved sleep and higher alertness during daytime.

Time-restricted feeding has also been shown to prevent metabolic disorders in mice who are fed a high-fat diet without reducing calories[665]. The mice who were fed their food within 8 hours didn't get obese or develop disease compared to those who ate the same amount of calories with no time restrictions.

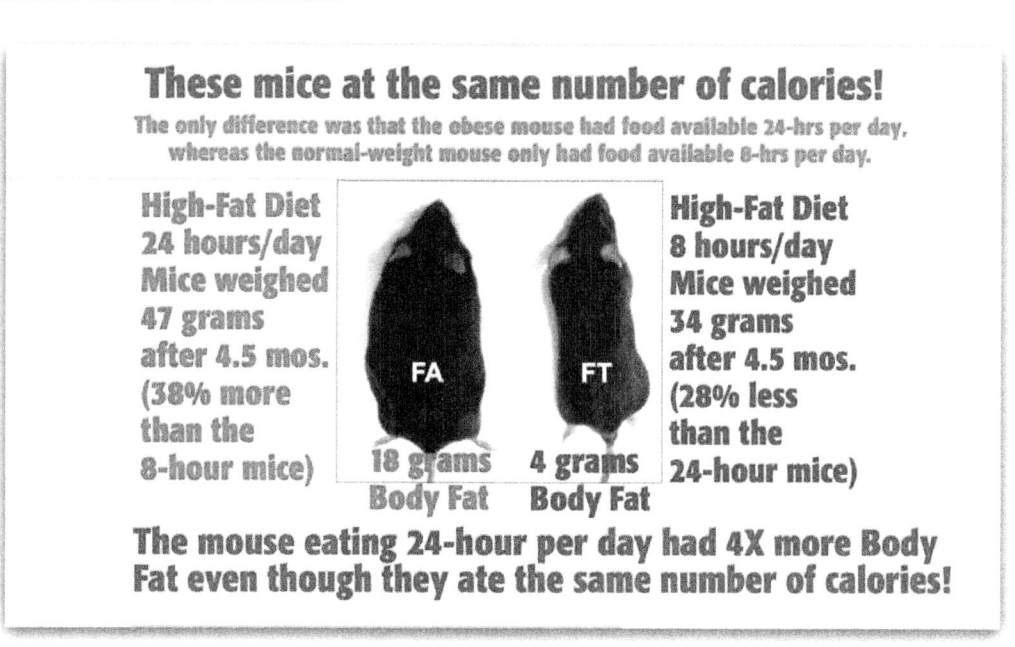

Figure 98 The power of time-restricted feeding on weight loss

By this part of the book, you don't need to be convinced about the benefits of intermittent fasting and time restricted feeding. A smaller eating frequency is in most cases more optimal for both body composition as well as the circadian rhythm. However, we still should learn about what's the optimal time to eat food.

The Best Time to Eat

Cortisol, the stress hormone, is also highest in the morning. It starts rising at about 5-7 AM and peaks around 8 AM so that we could have the energy to get out of bed. At this point, cortisol is actually beneficial because it ignites the body's fat oxidation mechanisms and initiates the circadian rhythm. However, as we've learned by now, it's not a good idea to be eating anything with elevated cortisol. Instead, the best thing to do is to postpone it by at least a few hours.

Hunger also follows circadian rhythms. A study found that despite the extended overnight fast, paradoxically, people aren't as ravenous in the morning and they tend to not want much breakfast. You'd think that the longer they've spent fasting the hungrier they'd get, but the opposite happened. No matter how long their fast had lasted, the participants still reported less desire to eat after waking up. Instead, the internal clock increased appetite in the evening, independent of food intake and other factors. Hunger ditches at 8 AM and peaks at 8 PM[666]. See Figure 99 Circadian Rhythm of Hunger.

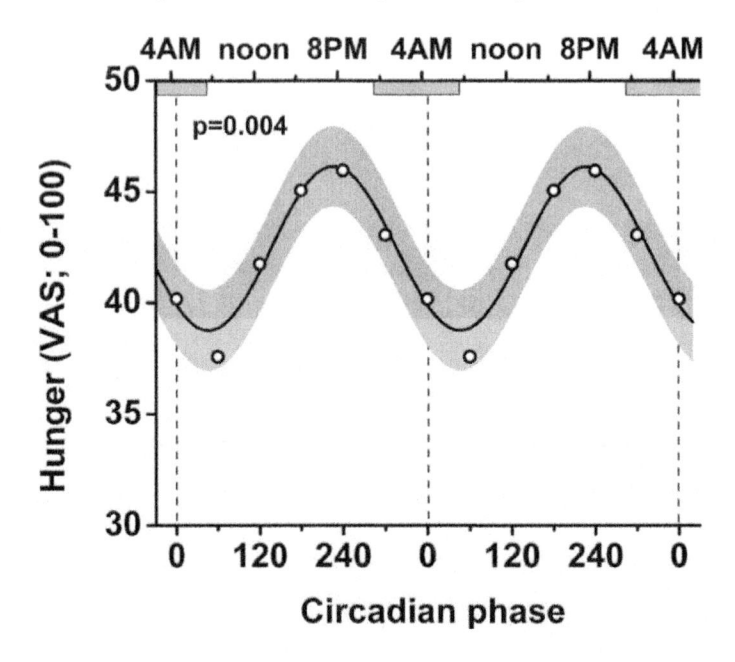

Figure 99 Circadian Rhythm of Hunger

This makes perfect sense, as after an overnight fast we're in mild ketosis and utilizing fat for fuel. Ketone bodies tend to be appetite suppressing and prevent hunger, which is why a healthy metabolism shouldn't feel ravenous right after waking up.

Based on this knowledge, even if you're not doing intermittent fasting, for optimal health, you'd want to postpone breakfast by at least a few hours. But what about other times?

During daytime, insulin production by the pancreas is much better than at night. Blood sugar control is also best during the earlier parts of the day and worse in the evening[667]. This makes perfect circadian sense, as blue light in the evening promotes insulin resistance and weight gain. Part of the reason has to do with how the circadian clocks in all of the organs and cells switch on and off synchronously at the same time of the day[668].

Melatonin levels begin to rise a few hours before habitual bedtime, given that you don't get exposed to blue light. When melatonin rises, it can bind to its receptor in the pancreas, which essentially tells the pancreas to stop producing insulin[669]. Basically, the idea is that it's not necessary to be releasing insulin anymore and it's time to sleep. If someone's having a big meal during that time, there might not be enough insulin to clear the bloodstream from glucose.

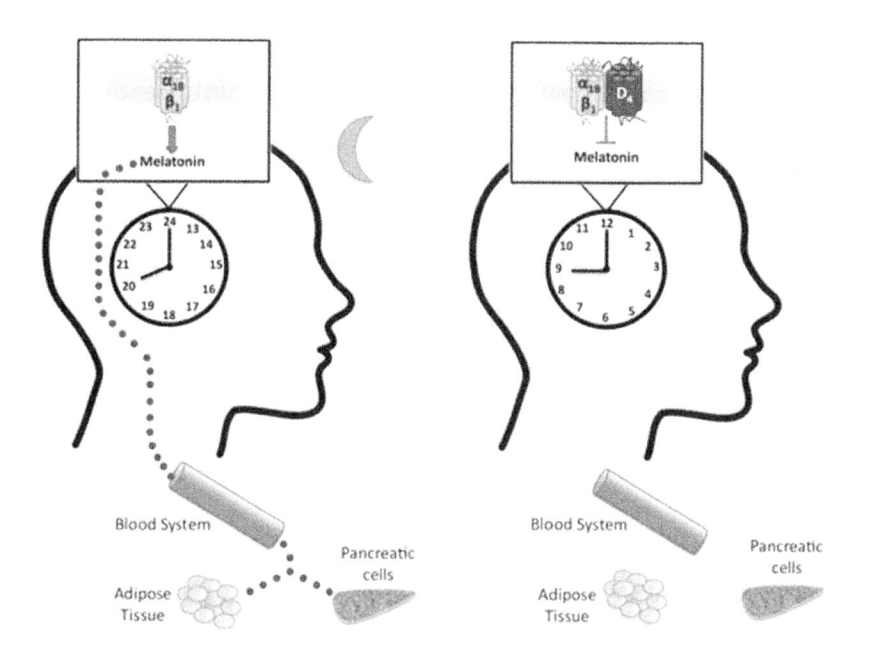

Figure 100 Melatonin at night inhibits the production of insulin. High insulin at night probably inhibits melatonin as well.

However, all these studies miss the critical aspect of extended intermittent fasting that lasts for up to 20 hours a day. In those cases, the body will be extremely insulin sensitive regardless of what time of the day it is. Even when there is a minute 5-10% reduction in blood sugar control, the effects would be insignificant due to the already high state of insulin sensitivity.

Compared to eating sporadically and with a 15+ hour eating window, time restricted feeding in the evening or later in the day is still much better for your general health and longevity than having your food without timing it.

Furthermore, resistance training also makes the body more efficient with the food you consume. The best time for heavy physical exercise like lifting weights according to the circadian rhythm is in the afternoon around 4-6 PM. That's when your coordination and strength tend to peak. Having a workout and then eating afterwards should have no significant negative side-effect on your long-term health or body composition.

Theoretically, if you spend the entire day fasting, then you're suppressing your food circadian rhythm and only activating it in the evening when you're starting to eat. However, the circadian rhythms are persistent, meaning they persist even in constant darkness with a period of about 24 hours. The clocks are always running. Whether or not you're in sync with them depends on the cues you receive from your environment.

There are different things that can offset the food circadian rhythm without having to break the fast. Drinking water and things like coffee also stimulate certain metabolic processes in the liver which can then set off the liver's circadian rhythm. You don't need to be eating a lot of calories to trigger the circadian processes or to affect it. The same is with light – even just a tiny bit of blue light at night time can already suppress melatonin production and inhibit your sleep quality. Autophagy as well - even just a small amount of calories will break autophagy and start the feeding cycle.

When it comes to OMAD and eating most of your food later in the day, then I would say it's not going to have any negative effect on the circadian rhythm directly because it's not the most influential factor. Light and movement are much more powerful signals of circadian rhythms and eating food later in the day will be used based on the physical conditions of the body i.e. how insulin sensitive you are at that point, what's your glycogen status, and how long you've been fasting.

Fasting can only have a negative impact on circadian rhythms and your health if it's going to disrupt your sleep. If you eat too much food too close to bedtime, then your gut is going to have to spend extra energy on digestion. This can make you sleep worse, preventing your brain from going into deeper stages of sleep.

During the night, the body would naturally start to cool itself down as to preserve energy and go into repair mode. However, eating and digesting food raises your core temperature, which takes away energy from the repair processes. It can also cause bloating, constipation, weight gain, and other digestive issues because in order to break down food and digest it, you need to produce a lot of stomach acid and enzymes.

If you eat a ton of food and then lay on your back for the coming night, then you may get acid reflux, you may stop producing hydrochloric acid, which may stop all digestive processes as well. While you're sleeping, the food will then start to sit there and you'll only start digesting it when you wake up the next morning. This will make you feel like you're in a food coma because (1) you didn't get enough deep sleep, (2) you're still digesting the food from the night before, and (3) certain foods may have begun to ferment in the small intestine, especially fructose and carbohydrates, because they got stuck there for the entire night, which then can cause leaky gut and brain fog.

So, it's not a good idea to be eating immediately before going to bed. That's why people eating later at night may not gain all the benefits of glucose control and insulin sensitivity. It depends on what you ate and what macronutrients there are but you should expect to digest the food for at least 4 hours. The optimal time frame to stop eating before sleep would be about 2-4 hours.

What you also want to do is spend a little bit of time moving around after eating. Walking is one of the best ways of lowering your post-prandial blood sugar, promoting digestion, increasing nutrient absorption, and helping with gastric emptying. I recommend going for a slow and steady walk for 10-15 minutes after dinner as to speed up digestive processes. This shouldn't be a type-A speed-walking stroll or anything hyped up. If you push it too much you'll trigger the sympathetic nervous system, which will actually shut down digestion again.

How to Not Disrupt the Circadian Rhythms with Fasting

Coming back to the circadian rhythms. To make sure you're not causing any mismatch with your chronobiology, you'd want to ensure you have the other circadian signaling factors on point.

- **Expose yourself to natural daylight in the morning for at least 5-10 minutes.** That's going to offset the right processes via the suprachiasmatic nucleus in the brain. You'll sleep better at night, you'll have a better mood, and you'll start the metabolism as well to a certain extent.

- **After waking up get moving** – Do some mobility exercises, go for a walk, shake yourself, rebounding, and do breathing exercises. Movement is also a powerful circadian cue.

- **During the day, restrict your caffeine intake.** The half-life of caffeine is 4-6 hours, which means that if you drink coffee at noon, 50% of it will still be in your system at 5 PM. That's why the last time you can safely ingest caffeine is 1-2 PM. After that, you may decrease your sleep quality.

    - **Also, I would actually recommend everyone to postpone their coffee intake all the way until noon** because you shouldn't need caffeine to wake yourself up in the morning. When you wake up, you should feel energized and alert because of cortisol. At least that's supposed to happen if you have proper circadian rhythms.

    - **Taking coffee will only mask the symptoms of fatigue which cover much more deeper issues related to sleep and circadian rhythms**. As I said, coffee is also a circadian signaller that offsets the liver's circadian clock. So, I'd say it's a good idea to postpone your first cup of coffee a few hours after waking up as to allow your digestion to rest and induce deeper cellular autophagy. Coffee stimulates autophagy to a certain extent but only in small amounts and not all the time. That's why the mid-noon small coffee break can be a good way to promote autophagy as well as give some energy until the evening.

- **Other circadian rhythm factors related to eating have to do with the type of food you eat as well.** Some nutrients like tryptophan and serotonin can actually make you sleep better and promote relaxation. Both carbohydrates and protein specifically have these nutrients and they're great for dinner. Getting some protein before going to bed can also be good for providing enough amino acids for your muscles and neurotransmitters.

Generally, most people find it easier and more convenient to skip breakfast and eat later in the day. I'd say it's better for the circadian rhythms as well and not eat anything at least until noon.

In conclusion, light is more important of a circadian signaller than food but you definitely don't want to eat a lot of food right before going to bed. The optimal time frame to stop eating for the night is 2-4 hours. Having food right in the morning isn't ideal either and you'd like to postpone your fast by at least a few hours after waking up.

It doesn't matter what kind of a fasting window you follow as long as you avoid eating immediately after opening your eyes, having dinner too late, and snacking all the way up to going to bed.

Chapter XXII

Sleep Optimization

"There is a time for many words, and there is also a time for sleep."

Homer, The Odyssey

This book is primarily about fasting because it provides many health and longevity-boosting benefits without having to do anything. It's simple – you just stop eating for a period of time, then you lift, and then you eat. However, as we found out in the last chapter, the circadian rhythms play an enormous role in how your body is going to respond to these actions.

Circadian rhythms and energy homeostasis are central to Metabolic Autophagy because they govern most of the proclaimed benefits of fasting, such as increased growth hormone, cellular autophagy, mitochondrial functioning, and insulin sensitivity. That's why sleep is crucial for healthy living and longevity.

All muscle growth happens during rest. Training is catabolic, which causes tissue tearing. As a result, the hypertrophic response will be augmented during anabolism – when we eat and sleep.

Sleep is probably the biggest thing that gets neglected when it comes to enhancing our performance. Lack of sleep can increase our risk of heart disease, diabetes, neurodegenerative disease, and obesity. It will also cause insulin resistance, mood swings, low testosterone, and fatigue, both physical and mental.

Human growth hormone actually gets released during the first hours of our sleep, which is incredibly important for building tissue and maintaining leanness. That 2000% boost everyone's talking about starts at 11 PM and lasts until about 2 AM, during which the body conducts its physical repair. Missing out on this makes you gain weight and lose strength. With a circadian mismatch, you may not be even getting said benefit of growth hormone secretion or autophagy.

CIRCADIAN RHYTHMS

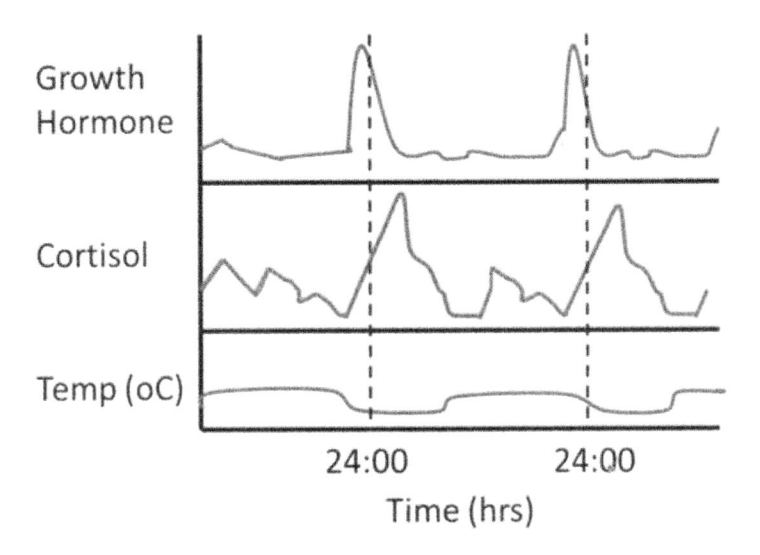

Figure 101 Circadian Rhythms of Human Growth Hormone

Sleep deprivation releases cortisol, which is the catabolic stress hormone. This will promote tissue breakdown and the accumulation of fat. That's why stressed out people tend to struggle with weight loss – chronically elevated levels of cortisol from not sleeping enough, over-caffeinating, long working hours, and emotional turmoil. It's one of the biggest things that accelerates aging and wrinkled skin. If you want to live longer and become high-performance, then you not only have to learn how to manage your stress but also get enough shut-eye.

Why Do We Sleep

Sleep is quite a paradoxical phenomenon from the perspective of evolution. Although it has many restorative benefits on the body, it's still a gambit. In nature, while your sleeping, you're putting yourself in risk of being eaten, getting killed, or missing out on some feeding opportunities. The fact that virtually all animals have some sort of a sleep-wakefulness cycle shows how important it must've been for organisms to develop this process. During sleep, the brain doesn't actually turn off but undergoes many neurological processes.

There are 2 main types of sleep with characteristic brain wave patterns and activities.

- **Non-rapid-eye-movement (NREM)** sleep is anything that is not REM or rapid-eye-movement sleep. It consists of 3 separate stages.
    - The first is NREM1 between drowsy wakefulness and sleep, in which your muscles are still quite active, you're rolling around in bed and you may occasionally open your eyes.
    - In NREM2 your muscle activity keeps decreasing and you start to slowly fade away into sleep.
    - Stage 3 NREM3 (previously 3 and 4) is the deep or slow-wave sleep (SWS) characterized by delta brain waves of 0.5-4 Hz. In here, you are cut off from the conscious world around you and irresponsive to most sounds or other stimuli.
- **Rapid-eye-movement (REM)** sleep is a phase characterized by random movement of the eyes, low muscle tone and the possibility of lucid dreaming. It allows us to learn complex tasks and motor skills. This is where the magic happens.

One cycle of REM usually takes about 70-90 minutes depending on the length of overall NREM sleep and comprises 20-25% of total night's volume. Light NREM1-2 takes about 50-60% and deep sleep NREM3 comprises up to 10-25%.

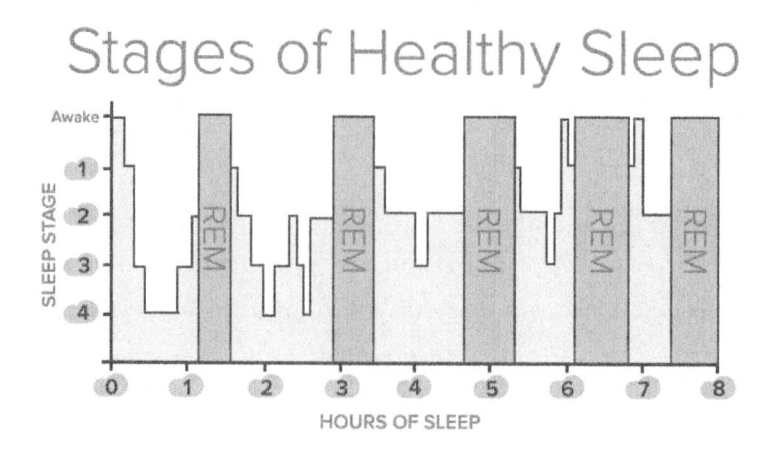

Figure 102 Sleep Stages

Physically, the most restorative stages are deep sleep and REM, which is why they should be prioritized much more. However, consistently over 25% of REM sleep may cause hyper brain activity, mood disorders and other neurological issues.

It's a very good idea to use some sort of a sleep tracker so you'd have actual data about your progress. The best non-bluetooth device out there is the OURA ring that can be kept in airplane mode. It also measures heart rate variability (HRV), body temperature, how fast you fall asleep, your physical activity, and how many times you wake up during the night. You can use the code SIIMLAND to get a $50 discount at www.ouraring.com.

How much Sleep Do We Need?

The optimal length of sleep varies between people and it's determined by genetics, lifestyle factors, circadian rhythms, and how much physical repair is needed. It's recommended that children should get about 10-12 hours a day and adults 7-9. Too much sleep, however, can also have negative side-effects, such as insomnia, restlessness, epilepsy, obesity, and other medical conditions.

But it's not the quantity of sleep that matters but the quality of it. Recovery takes place only in the deepest stages which we can enter after about 90 minutes. Sleeping more won't increase our performance. Doing it smarter will. Here's how:

- **Adjust to the Circadian Rhythm** – The most productive hours of sleep are in the early parts of the night. As you recall, growth hormone starts rising at 11 PM and reaches its peak at 2 AM. That's when melatonin should increase as well. Cortisol would also naturally start rising early in the morning between 5-8 AM. For optimal circadian rhythms, you'd want to stick to that as much as possible. This means:

    - get up early at about 6-8 AM
    - go to bed between 9 and 11 PM

- **Establish a Daily Sleep Routine** - Follow a series of habitual activities prior going to bed letting your unconscious self know that it's time to go to sleep. This will include certain activities both physical and mental. It can be anything, such as reading, brushing your teeth, stretching or whatever. Just make it a habit so that your mind knows it's time to rest. If you can't make it to bed at the optimal hours, then you should at least focus on sticking to a consistent bedtime and waking up schedule because it's critical for maintaining consistency with your circadian rhythm.

- **Sleep in Pitch Black Darkness** – Blue light in the evening is brutal for disrupting the circadian rhythms and interfering with melatonin production. That's why a pair of quality blue blocking glasses is essential but you also want to use blackout curtains and wear a sleeping mask.

- **Change Your Lightbulbs** - The UV light from ordinary lightbulbs emanates blue light. Changing to amber lights will fix that by reducing the spectrum to more red which at night resembles the sunset.

- **Use a program on your computer called F.lux** which changes the bright color of the screen to orange and relaxing.

- **Use an Acupuncture Mattress** - Get a small bedding that has little spikes on top of it. This is relatively cheap yet very effective. You can lay down before going to bed for 15 minutes or sleep on it throughout the night. I have used both options. At first, it feels like a lot of thorns are trying to penetrate your skin. After a while, the body relaxes and it becomes incredibly soothing.

- **Sleep in a Cool Environment** - The perfect temperature is 20 degrees Celsius (65 Fahrenheit). Turn down the heating and cover yourself with only the bed sheets. Extra fluffy blankets are actually counterproductive.

- **Turn Off All Electronics in Your Room** - Not only are they a possible source of blue light sneaking in but they also radiate Electro Magnetic Frequencies (EMF) which not only decrease testosterone production but also have a negative impact on our overall health including sleep. Turn off your router for the night and keep your phone on airplane mode most of the time.

- **Create White Noise in Your Room** - Whether that would be from an audio player switched to airplane mode or something less technical. I would suggest using a simple fan. Not only will the ventilation keep the air moving and cooling the temperature but the noise will contribute to the production of alpha waves while we are sleeping. White noise promotes a meditative state which will allow us to enter the deepest stages of recovery more easily.

- **Dehydrate a Few Hours Prior to Bedtime** – If you have to constantly wake up to go to the bathroom at night then it's going to interfere with getting into deep sleep. After dinner don't consume any form of liquid as it will inevitably have to come out.

- **Binaural Beats** - While awake our brain is producing mainly beta waves which is an alert state of consciousness that promotes stress and anxiety. To enter deeper stages, we have to drop lower into alpha waves. During the day it can happen while we're daydreaming or meditating. Binaural beats can help us to go from beta to alpha and then progress further into theta and delta, which resembles the natural progression of a healthy cycle.

- **Essential Oils** that emanate different aromas can be used around your bed that will improve the quality of our sleep. For instance, rose oil inhibits sympathetic nervous system activity and decreases adrenaline. Additionally, lavender enhances deep sleep, lowers our stress, blood pressure, heart rate, skin temperature, and cortisol levels. Their soothing smells will prime our body and mind for relaxation and augmentation.

You should also avoid using too much coffee. Caffeine has a half-life of 6-8 hours. To not be too alert before bedtime don't consume caffeine after 3-4 PM. It's best to drink it in the morning or at least not later than the afternoon. Green tea has some caffeine in it but in very small doses.

A good night's sleep starts the moment we wake up the day before. If we're not in tune with the circadian rhythms and not follow the activities described above, then we are not getting the most out of our downtime. Our ritual prior going to bed determines how well we're going to be resting. Sleep hygiene is very important and we need to prepare for it in advance.

Our modern world is constantly stimulating us and keeping us in the circuit. Social media, TV and working cause more stress and maintain our wakefulness. If we don't give our body to adjust to the change then we will go to bed all wired up thus impairing our recovery. In the evening we should be winding down and relaxing. If we're constantly wired up, we won't be able to shut down completely. When it's night time you have to behave accordingly. You shouldn't be running from predators or chasing prey but instead simply take it easy and soothe your mind.

The sleep ritual should start at least 2 hours before bedtime. Start winding down preemptively and incorporate the habitual habits that let your subconscious mind know it's that time of the day.

Sleep Nutrition

What we eat is also extremely important. Of course, filling up your gut right before bed will not only keep your digestive track active but will also keep you awake because of the discomfort.

What research has shown is that some protein and fat before bed decreases the amount of times people wake up a night in comparison to a high carb diet[670]. It also improves body composition and enhances muscle recovery by providing more assistance to the repair mechanisms.

To make yourself sleep better, have a nice ketogenic dinner with some good protein and fat, such as butter or coconut oil. However, don't eat too much protein, as it can spike your insulin. This will increase blood glucose and keeps you up at night. Keep it moderate.

Carbohydrates can also promote sleep by promoting serotonin production. On carb refeeds, the spike in insulin and following crash will actually make you sleep better. Best options for that are sweet potatoes and rice.

How to Track Your Readiness

There are many ways of knowing whether or not you've actually recovered from the stress of daily life and exercise.

- **Measuring your heart rate variability is a great way to know the state of your nervous system** – are you more sympathetic or parasympathetic dominant. There are many chest traps and devices out there that you can use to measure this. I myself use the OURA ring.

- **Looking at your physical strength and balance will also indicate the state of your nervous system.** If you're weaker than you were before then you haven't recovered and it would be better to have an easier recovery style workout before hitting it hard again. If you struggle to maintain balance or suffer from brain fog, then you're also under-recovered. It takes about 48-72 hours for your muscles to recover and grow but your nervous system can take up to 5-7 days, so you have to be very careful with how intense exercise you do.

- **I can tell you this simple 1-minute exercise that can give you some idea about the state of your nervous system.** You time yourself for 20 seconds and during that time you tap your finger on the table as fast as you can. You get a score which should tell you how recovered you are. Keep in mind that you have to do this over a longer period to establish a baseline of where you're currently at and you have to do it at the same time of day as well because your readiness will fluctuate between the morning and evening.

- **Tracking your mood and overall sense of well-being in a simple journal are the easiest ways of doing this.** You score yourself on a scale of 1-10. 10 being *I can run through a wall with no problems and I'm super-motivated.* And 1 would be that you're hospitalized in a bed. Lack of motivation can also mean you're still tired from your previous workout. Of course, there's a difference between just being lazy and actually having adrenal fatigue but you have to test and experiment, keep track of your numbers and then develop this intuitive knowledge about your body.

You don't want to live a completely stress-free life because it'll make you less prone to dealing with unexpected stressors but you don't want to be stressed out long-term either. Think of it as elusive cycling between stress exposure, recuperation, recovery, and adaptation. This way you'll have enough time to deal with it and get better.

What hormesis shows is that it's not the intensity of the stimuli that matters.

- You won't get fit by going to the gym once a month
- You won't lose weight by eating a salad for lunch but still have burgers for dinner

You have to stay consistent with it as to build up your tolerance to these stressors. And this applies to everything – to meditating, to working on your business, to reading and everything else.

Consistency will lead to long-term adaptation and it's the key to positive hormesis.

It is the principle of balance that we need to keep in mind and respect. Growth happens during rest. The reason why an empowered being is able to perform at such a high level is that they follow the correct pattern of going through a stressful stimulus and proper recollection of resources afterwards. It allows us to push ourselves even further the next time.

Those who say that sleep is for the weak and wicked don't know what they're talking about. It's the most important thing for our recovery as well as longevity.

Bonus Chapter
How to Drink Coffee Like a Strategic MotherF#%ka!

"I would rather suffer from coffee than be senseless"

Napoleon Bonaparte

One of the most commonly used performance-enhancing drugs in the world – coffee. It's used by many vocations and has been fueling the progress of Western society. Unfortunately, this beverage has some side-effects that can doom those who misuse it.

Besides the great taste, coffee has a ton of benefits. Long-term consumption of caffeine in the form of coffee is associated with cognitive enhancements[671], reduced risk for type-2 diabetes[672], Alzheimer's[673], and Parkinson's[674].

Caffeine travels to the brain and blocks a neurotransmitter called adenosine (See Figure 103). As a result, norepinephrine and dopamine actually increase, which hastens the firing rate of neurons[675].

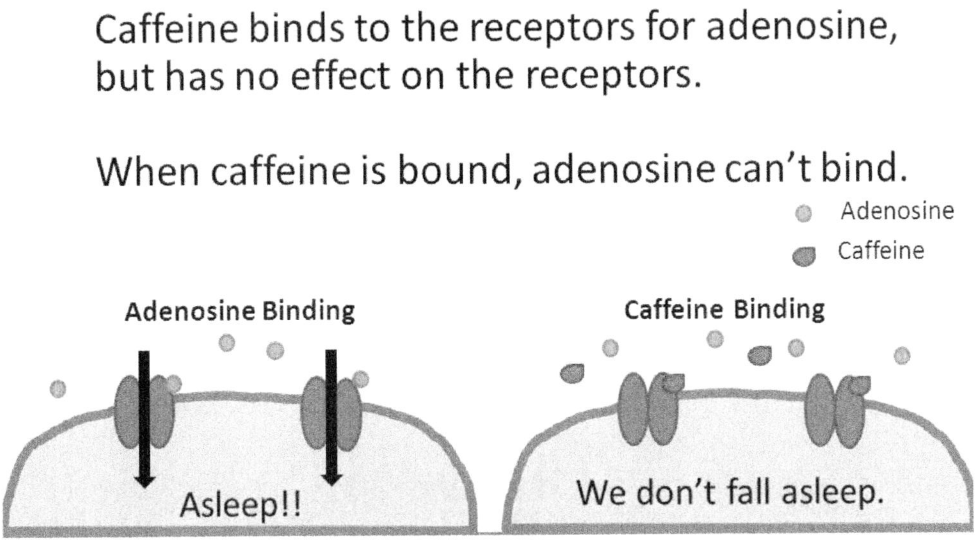

Caffeine binds to the receptors for adenosine, but has no effect on the receptors.

When caffeine is bound, adenosine can't bind.

- Adenosine
- Caffeine

Adenosine Binding — Asleep!!

Caffeine Binding — We don't fall asleep.

Figure 103 How Caffeine Blocks Adenosine

Coffee beans have a lot of antioxidants, called *quinines*, that fight disease and clean the body. After the roasting process, they become even more potent. They also contain naturally a lot of magnesium.

What about the costs? Are there any negative side-effects?

You've probably seen people who have become addicted to coffee. It's a dreadful sight – their hands are jittering and they have anxiety. The reason is that they have simply taken advantage of caffeine the wrong way.

For the wide majority of people, it's safe. However, additional side-effects can be insomnia, upset stomach, increased heart rate, and blood pressure.

In my opinion, **caffeine should be used only in certain situations when you actually need a boost.** It's just that – a performance-enhancing stimulant that gives us the right amount of energy for whatever the task might be.

There's a much healthier way to drink coffee effectively, which I'm about to share with you, that circumvents most of those issues.

The Best Time to Drink Coffee

For the ordinary person, drinking coffee immediately after waking up is the only thing that gets them going. They open their eyes, roll out of the bed and have to crawl to get their dose of java ASAP. But those are first signs of dependence and overdosing. It's not the ideal time to be consuming caffeine either. Coffee acts as a stimulant for the body that triggers some physiological processes.

Between the hours of 8-9 AM, our cortisol levels are at their peak[676]. It's the *fight or flight* hormone, that rises in the morning so we'd have increased alertness and focus. We're already supposed to be fully alert and energized after waking up. So, if we simultaneously drink coffee, we're wasting the potential benefits of caffeine and offsetting the circadian rhythm.

The best time to drink coffee is between 9:30 AM and 11:30 AM. Cortisol peaks in the early morning, but also fluctuates during the day. Other times it rises are 12 PM – 1 PM and 5:30 PM and 6:30 PM, so avoid a cup of joe at those hours as well. See Figure 104.

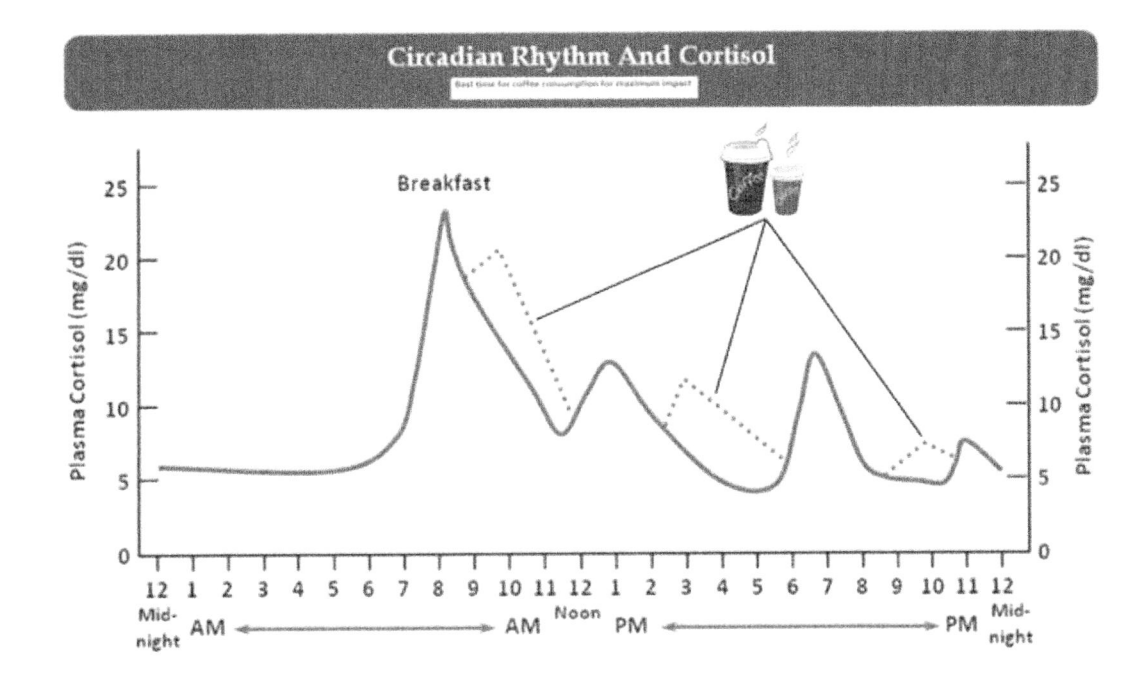

Figure 104 Drink Coffee at the Dotted Lines

When we're doing intermittent fasting, timing our coffee is even more relevant. If we were to cash in on one of our back-up cards, we would be left unarmed when hunger strikes.

Instead of drinking coffee immediately after we get hungry, we should first drink some water, then wait for about 30 minutes and only then decide whether or not it's worth it to have a nice cup of joe. In general, wait for a few hours after waking up before getting yours.

Drinking coffee at the wrong time can also keep you up and prevents you from falling asleep completely. You won't be able to get a good night's sleep and because of that wake up groggy and tired, you immediately grab another cup and the perpetual cycle continues.

The half-life of caffeine is about 5.7 hours[677], which means that if you drink coffee at 12 PM, then 50% of it will still be in your system at 6 PM. According to the circadian rhythm, the best time to go to bed is at about 9-11 PM. You should be sound asleep before midnight, because that's when the most growth hormone gets released. Ingesting caffeine in the evening will definitely keep you up at night. That's why you should stop drinking coffee after 2-4 PM in the afternoon.

How Much Caffeine is in my System?

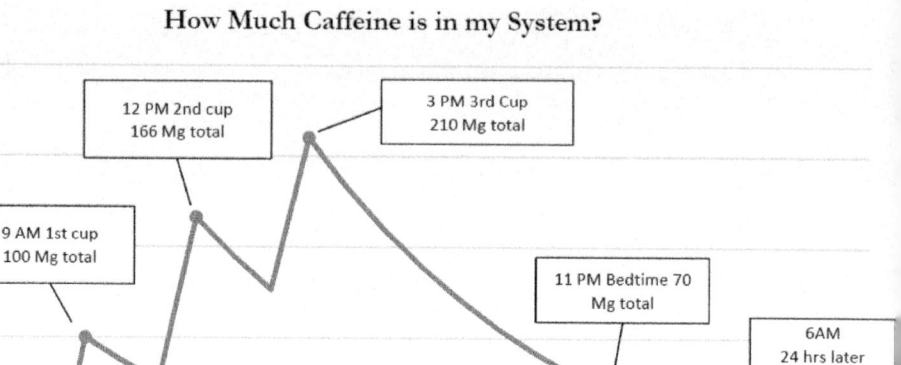

- 9 AM 1st cup — 100 Mg total
- 12 PM 2nd cup — 166 Mg total
- 3 PM 3rd Cup — 210 Mg total
- 11 PM Bedtime — 70 Mg total
- 6AM 24 hrs later — 27 Mg total

1 Cup = 8 oz & 100 Mg Caffeine | Metabolic Half Life = 6 Hours

Figure 105 Note how long caffeine actually stays in the body

But some people don't report these issues. They can drink coffee even just a few hours before going to bed and still fall asleep just fine. What gives?

Our metabolism differs between individuals and we have our own unique type, which makes us metabolize nutrients at different speeds.

- **The fast oxidizer is someone who digests food very quickly and converts it into energy rapidly.** They need to focus on eating heavier meals with more fat and protein that would keep them satiated. By the same token, they will also absorb caffeine that much faster and it will go through their system almost at an instant.

- **If you're a slow oxidizer, then you need more time to convert food into energy.** Because of that, you require more carbohydrates, rather than protein and fat. Getting the benefits of coffee will also be less rapid.

An average cup of coffee contains 100-150 mg of caffeine, but you won't get the full benefits from just one cup.

Consuming caffeine in small but frequent amounts is more advantageous. The optimal dose for cognitive functioning may be 20-200 mg per hour[678].

Small hourly doses can support extended wakefulness, by acting against the homeostatic sleep pressure, which builds up slowly throughout the day[679] and benefits the prefrontal cortex, which is responsible for higher executive functions[680].

Doses of 600 mg are often comparable to the effects of modafinil, which is a top-notch nootropic and cognitive enhancer. It's a smart drug but there are no reported advantages over large amounts of caffeine.

To avoid any unwanted side-effects, use filtered coffee. Darker roasts have less caffeine in them, due to the roasting process.

What Tasks Benefit the Most from Coffee

Drinking coffee won't make you a bad person, quite the opposite. There are also a lot of mood-enhancing benefits that will make you more enjoyable to be around. Napoleon Bonaparte said: *"The only good thing about St Helena is the coffee."*

The famous French philosopher of the Enlightenment Voltaire was said to be consuming about 40-50 cups of coffee a day. But in that era, those cups were also very small. Given what he accomplished with his writings, it's safe to say that this *"black gold"* will definitely help us to become a high performing individual.

However, caffeine works best for only some activities. It may increase our attention span, the speed at which we work, prevent us from getting side-tracked, and may even benefit recall, but it's less likely to improve more complex cognitive functions. Like with modafinil, you only get better at what you're already good at. You can't expand upon your existing cognitive limitations. The actual benefit you get is just more energy and alertness. In fact, it may actually harm tasks of higher executive functioning, such as creativity or problem solving, because large doses of caffeine may cause shivers and too much excitement.

Use caffeine to rush through the repetitive activities that require a lot of micromanagement and aren't too difficult. This way you'll waste less time doing the small stuff and can free up more space for focusing on what's more important. With or without coffee depends on your own decision.

Once you take your first zip of the day, you can immediately feel your energy levels rising. This happens because your body will release more adrenaline and dopamine. What ensues is lipolysis, which is the conversion of stored body fat into energy. However, the increased use of free fatty acids is reported to happen only in low carb/high fat diets[681]. Caffeine may be less useful on a high carb one[682].

At the same time, coffee will still increase your metabolic rate and has other physical performance-enhancing effects. Caffeine has a positive impact on muscular contraction and fatigue, which makes it a great tool for training.

When it comes to performance, then drinking a larger dose of caffeine 15-30 minutes earlier will yield some great results. Zipping on some beverage intra-workout is also viable. Even more, post-workout caffeine can also help to refuel muscles and increase fat burning[683].

How to Drink Coffee Without Getting Addicted

As great as the benefits of caffeine are, we shouldn't overdose it by any means. Consuming it daily will increase our body's tolerance to it, which eventually leads to the receptors in our brain to becoming resistant to coffee. After some time, it stops working and we need a lot more to get the same effect.

Theodore Roosevelt drank a gallon of coffee a day. His son said that the president's mug was *"more on the side of a bathtub."* Even though the amounts consumed by Teddy and Voltaire might sound encouraging, don't try to drink as much as them.

Herein lies the point where people get addicted to coffee. They simply have developed resistance towards caffeine and don't even feel like they've consumed it. To keep themselves awake, they reach out for another cup, crash and burn, and get another one, while getting stuck in the vicious cycle again.

To prevent that from happening, **you have to habitually cycle off caffeine.** For at least 1 week of the month, you should allow your body's receptors to reset and become sensitive again.

Another option would be to drink coffee only on days where you most need it, say during a hard workout or while doing repetitive tasks. This doesn't mean you can't drink coffee every day. You can. Simply swap out the caffeinated version with decaf. The taste is the same and you can get almost all of the benefits. If not the increased energy, then at least you'll still use it as an antioxidant and a mood enhancer.

What to Combine Coffee With

The effects of caffeine will also depend on what else is in your system at that time. Your metabolic type will already influence your rate of absorption but other nutrients will do so as well.

There are some benefits to consuming caffeine with glucose, which may improve cognition not seen with either alone[684]. Additionally, grapefruit juice can keep caffeine levels in the bloodstream for longer[685]. If you're a slow oxidizer, then you may find adding these ingredients useful. Because we'll be fasting most of the time, this option isn't viable, unless you mask your drink with some MCT or coconut oil.

If you're a fast oxidizer, then adding sugar will only hasten your downfall. You may get an immediate boost, but that short high will be followed by a steep low. To not crash and burn, you can add fat to the mix.

Dave Asprey's Bulletproof Coffee is probably the latest coffee drinking trend. Is it hype or does it actually work? Probably some of both. Adding butter to your cup of joe will definitely have some positive effects. It decreases the rate of absorption, gives you long-lasting energy, keeps you satiated for hours and tastes incredible. Whatever the case might be, you should try it.

During fasting, you'd want to avoid any extra calories in your coffee because it may reduce some of the autophagy benefits. Coffee itself boosts autophagy in some amounts but not if you over-do it or add too many other ingredients. The only thing allowed would be medicinal mushroom powders like chaga, reishi, or lion's mane but not over 1 tsp. If you have some adrenal issues or experience too much cortisol in the morning, then it's not recommended to drink coffee because it may simply overshoot the stress hormones. To not become even more stressed out and catabolic you should have some tea instead.

It's also advisable to drink coffee only in the fasted window because it usually coincides with daytime. Most people spend the early parts of the day fasting and eat primarily lunch or dinner. Having caffeine in the system at that time may cause some circadian mismatches and sleep issues.

Use it as you may but never become dependent on coffee. It should be seen as a performance-enhancing stimulant that can increase your physical and mental performance. There are too many caffeine addicts in the world already so we don't need another one.

Conclusion

*„It is not death a man should fear,
but he should fear never beginning to live"*

Marcus Aurelius

We've reached the end of this book. There are definitely many things that could fit between these pages because a lot of nuances could be covered further. Unfortunately, that would require a more specific understanding of the context in a particular situation.

Nevertheless, I think the main ideas of Metabolic Autophagy should be quite clear in terms of what it means, how to do it, and when to practice intermittent fasting.

There is no way to know what's the optimal length to fast, how often to do it, when is the best time to eat, how much autophagy is too much, and when it's better to be more anabolic. Even if we did have a lot more information about these processes, what works for you may not work for me etc. The context of the situation and the conditions of the individual are still the most important variable when determining what's good to eat, in what amounts, and when. What this book gave you was the main principles of these cycles and knowing in what direction you're heading towards with your metabolism.

There is no such thing as a complete balance either because that would entail stagnation. A perfect equilibrium inside the body and in other complex systems involves being severely out of balance at one time and then getting back into homeostasis. In order to grow and expand, you have to first contract and vice versa.

The mythological creature Ouroboros manifests itself not only in cellular autophagy but in every other area of your life as well. In order to evolve as a person, you have to bite your own tail – to devour yourself – and leave behind the previous version of yourself. From the energy of this process, you shall be reborn without even knowing it. These pathways are constantly in action inside your body. It's the silent path of living and growing, which you can only seek to optimize with your daily practice.

The goal of the book wasn't to give you a blueprint to living over 100 or to become immortal. It might even be that you'll die prematurely because of some unknown reason. You can never know what happens in the future. However, what you can do is prepare yourself in advance by honing your body and mind.

Our own individual desires and ideas about the world lack the greater perspective of millions of years of evolution. When we do zoom outward and try to comprehend all the inner workings of our body as well as the larger scheme of things, we're forced to realize how little we still are.

I got asked once: „Do I believe in an afterlife?"

Contemplating about an afterlife and supernatural entities forces you to think about death. That's the biggest reasons people think about these possible scenarios in the first place – there must be something more – that I can't just die one day and that's it – that I'm going to disappear and vanish away into nothingness. Well, if that's true then you don't really have nothing to worry about it because if there is indeed nothing after death then you won't be there to experience it either – you'll just be dead.

What I do believe in is epigenetics – that what you do right now to yourself will have an effect on the future generations and your offspring. Even if you die, your genes will continue to live. That's how evolution has been working so far and that's how it will carry on for at least the foreseeable future.

How your great great grandparents treated their epigenome has affected the genetic makeup of your grandparents which influenced your parents' genes all the way up to your own blueprint.

I think that this is also the closest way of proving the existence of some Psychosomatic Karma or causality. Your personality and your perspective of the world is very much determined by your genes and heritage. If you have some predisposition to some disease, then you'll behave differently and if your family has certain values or beliefs such as religious practices, being a hard worker, or creativity, then you're more likely to adopt those same traits yourself. It's the genetic transmission of certain characteristics that go through their own process of natural selection.

What you as a person have to ask yourself is are these the genes the ones I want to keep carrying further? Maybe I need to adopt a new set of beliefs and values instead of following the old dogmas and doctrines of what my parents taught me, of what their parents taught them, of what their grandparents learned from their grandparents and so on...

You have to also ask, what kind of a genome do I want to leave for future generations because a lot of these issues can be alleviated to a certain extent. If you follow good habits and live a healthy lifestyle, then you're improving your epigenetic makeup and in so doing promoting the health and vitality of your children and their children and so on.

It really is something to think about extremely carefully. Bad lifestyle habits like smoking, alcohol, junk food, not sleeping enough, being stressed out all the time, environmental toxins, and afflictive thought patterns are all changing your physiology on an epigenetic level and it's going to re-manifest itself in the future. Your children are more predisposed to exhibit those same traits whether they be related to diet, beliefs, cellular functioning, and emotional processing.

Instead of selfishly thinking about how to make sure you'll have a painless afterlife with no suffering, you should worry about how to not cause additional suffering for those people who are left behind. Don't leave a heritage of disease and sickness when it could've been avoided by eating better, exercising a little bit more, sleeping better, sticking to the circadian rhythm, and paying more attention to the little things.

When doing Metabolic Autophagy, you're never stagnant and motionless. Instead, you're always moving between the anabolic and catabolic cycles of life on a daily basis. This knowledge is invaluable in terms of giving you actual information about where you're steering your body. You don't want to leave yourself surprised or ignorant of how to control your own metabolic processes. In so doing you'll also be able to combat entropy more effectively and promote your longevity.

On this note, I'm going to end the book. May your fasts be autophagic and refeeds anabolic. Whichever side of the coin gets flipped more often will have to depend upon your own individual choices and goals.

Don't know where to start with intermittent fasting and optimizing autophagy for your goals? Afraid of making mistakes or jeopardizing your health in the long run?

As a final cliffhanger, you can check out the Metabolic Autophagy Diet Program. It includes a 4-week step by step walkthrough of what foods to eat in what amounts at what time and how to train as well.

There's a two 4-week meal plans for both low calorie and high calorie intakes and it includes exact ingredients, amounts, and macros of the foods to eat. If you're interested, then head over to https://siimland.com/metabolic-autophagy-diet-program/ and check it out.

If you want to learn more about the methods of such Metabolic Autophagy, then make sure you check out my blog at http://www.siimland.com/blog/ and reach out to me on social media. I'm Siim Land on all the platforms.

Extras

Glycemic Index

It measures how quickly foods breakdown into sugar in your bloodstream. High glycemic foods turn into blood sugar very quickly.

The GI tells you how fast foods spike your blood sugar. But the GI won't tell you how much carbohydrate per serving you're getting. That's where the Glycemic Load is a great help. It measures the amount of carbohydrate in each service of food. Foods with a glycemic load under 10 are good choices—these foods should be your first choice for carbs. Foods that fall between 10 and 20 on the glycemic load scale have a moderate effect on your blood sugar. Foods with a glycemic load above 20 will cause blood sugar and insulin spikes

Food	Glycemic Index (Glucose=100)	Glycemic load per serving
Bakery and Bread		
Bagel	72	25
Baguette	95	15
Barley bread	34	7
Hamburger bun	61	9
White wheat flour bread	71	10
Whole wheat bread	71	9
Whole grain bread	51	7
Pita bread	68	10
Corn tortilla	52	12

Beverages		
Coca-Cola	63	16
Fanta	68	23
Apple juice, unsweetened	44	33
Gatorade	78	12
Orange juice, unsweetened	55	12
Tomato juice, canned	38	4
Breakfast cereal		
All-bran	55	12
Coco puffs	77	20
Cornflakes	93	23
Cream of wheat	66	17
Grapenuts	75	16
Muesli	66	16
Oatmeal	55	13
Grains		
Sweet corn on the cub	60	20
Couscous	65	9
Quinoa	53	13
White rice	73	43
Brown rice	68	16
Bulgur	48	12

Cookies and crackers		
Wafers	77	14
Rice cakes	82	17
Dairy		
Ice cream	57	6
Milk, full fat	41	5
Milk, skim	32	4
Reduced fat yogurt	33	11
Fruit		
Apple	39	6
Banana, ripe	62	16
Dates, dried	42	18
Grapefruit	25	6
Grapes	59	11
Orange	40	4
Pear	38	4
Raisins	64	28
Beans and nuts		
Baked beans	40	6
Blackeye peas	33	10
Black beans	30	7
Chickpeas	10	3

Chickpeas, canned in brine	38	9
Navy beans	31	9
Kidney beans	29	7
Lentils	29	5
Soybeans	15	1
Cashews, salted	27	3
Peanuts	7	0
PASTA and NOODLES		
Fettucini	32	15
Macaroni	47	23
Macaroni and Cheese	64	32
Spaghetti	46	22
SNACK FOODS		
Corn chips	42	11
Fruit Roll-Ups	99	24
M & M's	33	6
Microwave popcorn	55	6
Potato chips	51	12
Pretzels	83	16
Snickers Bar	51	18
VEGETABLES		
Green peas	51	4

Carrots	35	2
Parsnips	52	4
Baked russet potato	111	33
Boiled white potato	82	21
Instant mashed potato	87	33
Sweet potato	70	21
Yam	54	17
Tomato	38	2
Broccoli	0	0
Cabbage	0	0
Celery	0	0
Spinach	0	0
Mushrooms	0	0
MISCELLANEOUS		
Hummus	6	20
Chicken nuggets	46	
Pizza	80	0
Honey	61	22

Insulin Index

The insulin index (II) is different than the glycemic index (GI). The GI shows the relationship between how glucose (sugar, carbs) raise insulin. The II shows how "other" foods raise insulin.

Food	Insulin Index
Butter	2%
Olives/Olive oil	3%
Coconut oil	3%
Flax oil	3%
Heavy cream	4%
Pecans	5%
Macadamia nuts	5%
Avocado	6%
Coconut meat	7%
Cream cheese	8%
Sour cream	8%
Bacon	9%
Walnut	9%
Pine nut	9%
Pepperoni	10%
Pork	11%
Peanut butter	11%
Codfish	12%

Duck	12%
Peanuts	13%
Pumpkin	14%
Almonds	14%
Cheddar cheese	15%
Sunflower seeds	15%
Chia seeds	15%
Egg yolk	15%
Blue cheese	16%
Pistachios	19%
Coleslaw	20%
Swiss cheese	21%
Whole egg	21%
Turkey	23%
All-bran	24%
Chicken	24%
Low fat cream cheese	25%
Pasta	29%
Fish	33%
Whole milk	40%
Low fat Swiss	43%
Berries	47%

Beef	51%
Popcorn	54%
Egg whites	55%
Scallops	59%
Potato chips	61%
Brown rice	62%
Apple	75%
Low-fat yogurt	76%
Fat-free pretzel	81%
Banana	84%
Crackers	87%
Whole wheat bread	96%
White bread	100%
Baked beans	100%
Sweetened yogurt	115%
Potatoes	121%
Jelly beans	160%

About the Author

Siim Land is a bestselling author, content creator, podcaster, high-performance coach and a self-empowered being.

He creates content about Body Mind Empowerment which is human life enhancement through optimizing your physiological potential and becoming the greatest version of yourself.

The phrase *"Stay Empowered"* refers to showing up to the greatest version of yourself and earning your laurels every day.

Stay Empowered

Siim

Contact me at my blog: http://siimland.com/contact

Facebook Page: https://www.facebook.com/thesiimland/

YouTube Channel: http://www.youtube.com/c/SiimLand/

Instagram Page: https://www.instagram.com/siimland/

Twitter Page: https://twitter.com/iamsiimland

Podcast Page: http://siimland.com/podcast

Performance Store: http://siimland.com/store/

Books and Products: http://siimland.com/books-and-products/

Audiobooks: http://siimland.com/audiobooks/

Blog Page: http://siimland.com/blog/

More Books From the Author

You can check out all the printed and audio versions of these books at http://www.siimland.com/books-and-products/ and http://www.siimland.com/audiobooks/

https://siimland.com/metabolic-autophagy-diet-program/

https://siimland.com/keto-if-fasting/

https://siimland.com/keto-fit-program/

Keto Fasting

479

Keto Bodybuilding

Keto Cycle the Cyclical Ketogenic Diet Book

Simple Keto the Easiest Ketogenic Diet Book

Target Keto the Targeted Ketogenic Diet Book

Intermittent Fasting and Feasting

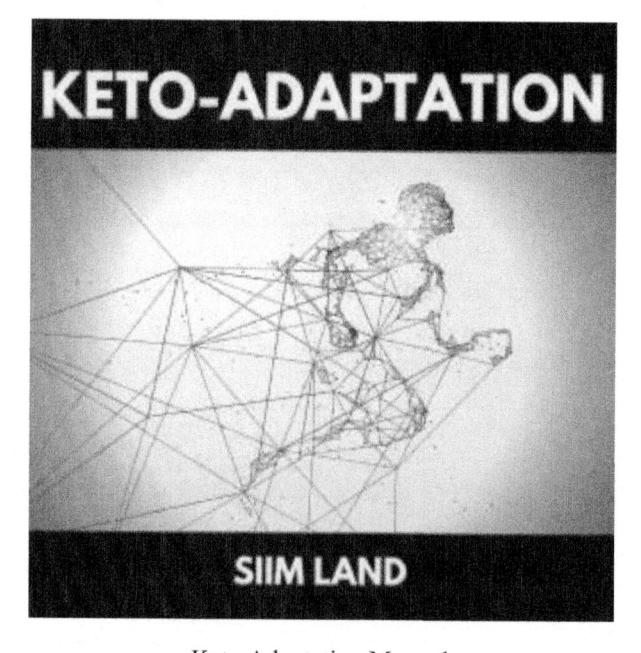

Keto Adaptation Manual

References

Here are all the references and studies mentioned in this book.

---

[1] Bernard C. (1878) 'Leçons sur les phénomènes de la vie communs aux animaux et aux vegetaux,' Paris, Bailliere JB, editor.

[2] Starling, EH. (1923) 'The Wisdom of the Body', British Medical Journal, October 20;2(3277):685-90.

[3] Cannon WB: Wisdom of the Body, New York, W. W. Norton & Company; Rev. and Enl. Ed edition (April 17, 1963).

[4] Martinez, D.E. (1998) 'Mortality Patterns Suggest Lack of Senescence in Hydra', Experimental Gerontology, Vol 33(3), p 217-225.

[5] Beck, S. and Bharadwaj, R. (1972) 'Reversed Development and Cellular Aging in an Insect', Science, Vol 178(4066), p 1210-1211.

[6] López-Ótin C. et al. (2013) 'The Hallmarks of Aging', Vol 153(6), p 1194-1217.

[7] Harman, D. (1956) 'Aging: A Theory Based on Free Radical and Radiation Chemistry', Journal of Gerontology, Vol 11(3), p 298-300.

[8] Harman, D. (1972) 'The biologic clock: the mitochondria?', Journal of the American Geriatrics Society, April 20(4), p 145-147.

[9] Schriner, SE. et al (2005) 'Extension of murine life span by overexpression of catalase targeted to mitochondria', Science, Vol 308(5730), p 1909-1911.

[10] Miguel, J. et al (1980) 'Mitochondrial role in cell aging', Experimental Gerontology, Vol 15(6), p 575-591.

[11] Trifunovic, A. et al (2005) 'Somatic mtDNA mutations cause aging phenotypes without affecting reactive oxygen species production', Proceedings of the National Academy of Sciences of the United States of America, Vol 102(50), p 17993-17998.

[12] Wei, YH. et al (2001) 'Mitochondrial theory of aging matures--roles of mtDNA mutation and oxidative stress in human aging', Zhonghua Yi Xue Za Zhi (Taipei), Vol 64(5), p 259-270.

[13] Fontana, L. et al (2013) 'Dietary Restriction, Growth Factors and Aging: from yeast to humans', Science, Vol 328(5976), p 321-326.

[14] Pérez, V. et al (2009) 'Is the Oxidative Stress Theory of Aging Dead?', Biochimica et Biophysica Acta, Vol 1790(10), p 1005-1014.

[15] Van Raamsdonk, J.M. and Hekimi, S. (2009) 'Deletion of the Mitochondrial Superoxide Dismutase sod-2 Extends Lifespan in Caenorhabditis elegans', PLOS Genetics, Vol 5(2).

[16] Bjelakovic, G. et al (2007) 'Mortality in Randomized Trials of Antioxidant Supplements for Primary and Secondary Prevention', JAMA, Vol 297(8), p 842-857.

[17] Boffetta, P. et al (2010) 'Fruit and Vegetable Intake and Overall Cancer Risk in the European Prospective Investigation Into Cancer and Nutrition (EPIC)', JNCI: Journal of the National Cancer Institute, Vol 102(8), p 529-537.

[18] Halliwell, B. (2012) 'Free radicals and antioxidants: updating a personal view', Nutrition Reviews, Vol 70(5), p 257-265.

[19] Tapia, P. (2006) 'Sublethal mitochondrial stress with an attendant stoichiometric augmentation of reactive oxygen species may precipitate many of the beneficial alterations in cellular physiology produced by caloric restriction, intermittent fasting, exercise and dietary phytonutrients: "Mitohormesis" for health and vitality', Medical Hypotheses, Vol 66(4), p 832-843.

[20] Geolotto, G. et al (2004) 'Insulin generates free radicals by an NAD(P)H, phosphatidylinositol 3'-kinase-dependent mechanism in human skin fibroblasts ex vivo', Diabetes, Vol 53(5), p 1344-1351.

[21] Goldstein, BJ. et al (2005) 'Redox paradox: insulin action is facilitated by insulin-stimulated reactive oxygen species with multiple potential signaling targets', Diabetes, Vol 52(2), p 311-21.

[22] Tatar, M. et al (2003) 'The endocrine regulation of aging by insulin-like signals', Science, Vol 299(5611), p 1346-1351.

[23] Kenyon, C. et al (1993) 'A C. elegans mutant that lives twice as long as wild type', Vol 366(6454), p 461-464.

[24] Lin, K. et al (1997) 'daf-16: An HNF-3/forkhead family member that can function to double the life-span of Caenorhabditis elegans', Science, Vol 278(5341), p 1319-22.

[25] Lin, K. et al (2001) 'Regulation of the Caenorhabditis elegans longevity protein DAF-16 by insulin/IGF-1 and germline signaling', Nature Genetics, Vol 28(2), p 139-145.

[26] Lee, SJ. et al (2009) 'Glucose shortens the life span of C. elegans by downregulating DAF-16/FOXO activity and aquaporin gene expression', Cell Metabolism, Vol 10(5), p 379-391.

[27] Tatar, M. et al (2001) 'A mutant Drosophila insulin receptor homolog that extends life-span and impairs neuroendocrine function', Science, Vol 292(5514), p 107-110.

[28] Clancy, DJ. et al (2001) 'Extension of life-span by loss of CHICO, a Drosophila insulin receptor substrate protein', Vol 292(5514), Vol 104-106.

[29] Hwangbo, DS. (2004) 'Drosophila dFOXO controls lifespan and regulates insulin signalling in brain and fat body', Nature, Vol 429(6991), p 562-566.

[30] Holzenberger, M. et al (2003) 'IGF-1 receptor regulates lifespan and resistance to oxidative stress in mice', Nature, Vol 421(6919), p 182-187.

[31] Blüher, M. et al (2003) 'Extended longevity in mice lacking the insulin receptor in adipose tissue', Vol 299(5606), p 572-574.

[32] Longo, VD (2003) 'The Ras and Sch9 pathways regulate stress resistance and longevity', Experimental Gerontology, Vol 38(7), p 807-811.

[33] Tissenbaum, HA. and Guarente, L. (2001) 'Increased dosage of a sir-2 gene extends lifespan in Caenorhabditis elegans', Vol 410(6825), p 227-230.

[34] Murphy C.T., Hu P.J. 'Insulin/insulin-like growth factor' (December 26, 2013), WormBook, ed. The C. elegans Research Community.

[35] Altintas, O. et al (2016) 'The role of insulin/IGF-1 signaling in the longevity of model invertebrates, C. elegans and D. melanogaster', BMB Reports, Vol 49(2), p 81-92.

[36] Klass M. and Hirsh D. (1976) 'Non-ageing developmental variant of Caenorhabditis elegans', Nature, Vol 260(5551), p 523-525.

[37] Kaeberlein, M. et al (1999) 'The SIR2/3/4 complex and SIR2 alone promote longevity in Saccharomyces cerevisiae by two different mechanisms', Genes & Development, Vol 13(19), p 2570-2580.

[38] Guarente, L. (2007) 'Sirtuins in aging and disease', Cold Spring Harbor Symposia on Quantitative Biology, Vol 72, p 483-488.

[39] Kanfi, Y. et al (2012) 'The sirtuin SIRT6 regulates lifespan in male mice', Nature, Vol 483, p 218-221.

[40] Mostoslavsky, R. (2006) 'Genomic instability and aging-like phenotype in the absence of mammalian SIRT6', Cell, Vol 124(2), p 315-329.

[41] Sedelnikova, OA. et al (2004), 'Senescing human cells and ageing mice accumulate DNA lesions with unrepairable double-strand breaks', Nat Cell Biology, Vol 6(2), p 168-170.

[42] Chung, S. et al (2010) 'Regulation of SIRT1 in cellular functions: role of polyphenols', Archives of Biochemistry and Biophysics, Vol 501(1), p 79-90.

[43] Wang, RH. et al (2008) 'Impaired DNA damage response, genome instability, and tumorigenesis in SIRT1 mutant mice', Cancer Cell, Vol 14(4), p 312-323.

[44] Lee, H.I. et al (2008) 'A role for the NAD-dependent deacetylase Sirt1 in the regulation of autophagy', PNAS, Vol 105(9), p 3374-3379.

[45] Yang, NC. et al (2015) 'Up-regulation of nicotinamide phosphoribosyltransferase and increase of NAD+ levels by glucose restriction extend replicative lifespan of human fibroblast Hs68 cells', Biogerontology, Vol 16(1), p 31-42.

[46] Houtkooper, R.H. and Auwerx, J. (2012) 'Exploring the therapeutic space around NAD+', Vol 199(2), p 205.

[47] Colman, RJ. et al (2009) 'Caloric restriction delays disease onset and mortality in rhesus monkeys', Science, Vol 325(5937), p 201-204.

[48] Ingram, D.K. and Roth, G.S. (2015) 'Calorie restriction mimetics: Can you have your cake and eat it, too?', Ageing Research Reviews, Vol 20, p 46-62.

[49] Chen, D. et al (2005) 'Increase in activity during calorie restriction requires Sirt1', Science, Vol 310(5754), p 1641.

[50] Cantó, C. et al (2009) 'AMPK regulates energy expenditure by modulating NAD+ metabolism and SIRT1 activity', Nature, Vol 458(7241), p 1056-1060.

[51] Duan, W. (2013) 'Sirtuins: from metabolic regulation to brain aging', Frontiers of Aging Neuroscience.

[52] Srivastava, S. et al (2013) 'A ketogenic diet increases brown adipose tissue mitochondrial proteins and UCP1 levels in mice', IUBMB Life, Vol 65(1), p 58-66.

[53] Csiszar, A. et al (2009) 'Anti-oxidative and anti-inflammatory vasoprotective effects of caloric restriction in aging: role of circulating factors and SIRT1', Mechanisms of Ageing and Development, Vol 130(8), p 518-527.

[54] Gehart-Hines, Z. et al (2011) 'The cAMP/PKA Pathway Rapidly Activates SIRT1 to Promote Fatty Acid Oxidation Independently of Changes in NAD+', Molecular Cell, Vol 44(6), p 851-863.

[55] Raynes, R.R. (2013) 'SIRT1 Regulation of the Heat Shock Response in an HSF1-Dependent Manner and the Impact of Caloric Restriction', Scholar Commons, 4567.

[56] Ramis, MR. et al (2015) 'Caloric restriction, resveratrol and melatonin: Role of SIRT1 and implications for aging and related-diseases', Mechanisms of Ageing and Development, 146-148:28-41.

[57] Beek, C.B. et al (2013) 'Circadian Clock NAD+ Cycle Drives Mitochondrial Oxidative Metabolism in Mice', Vol 342(6158).

[58] Meléndez, A. et al (2003) 'Autophagy genes are essential for dauer development and life-span extension in C. elegans', Vol 301(5638), p 1387-1391.

[59] Morselli, E. et al (2010) 'Caloric restriction and resveratrol promote longevity through the Sirtuin-1-dependent induction of autophagy', Cell Death & Disease, Vol 1(1).

[60] Lamb, CA., Yoshimori, T. and Tooze SA. (2013) 'The autophagosome: origins unknown, biogenesis complex', Nature Reviews Molecular Cell Biology, Vol 14(12), p 759-774.

[61] Carlson, A.J. and Hoelzel, F. (1946) 'Apparent Prolongation of the Life Span of Rats by Intermittent Fasting: One Figure', The Journal of Nutrition, Vol 31(3), p 363-375.

[62] Wei, M. et al (2017) 'Fasting-mimicking diet and markers/risk factors for aging, diabetes, cancer, and cardiovascular disease', Science Translational Medicine, Vol 9(377).

[63] Fontana, L., Longo, VD., and Partridge, L. (2010) 'Dietary Restriction, Growth Factors and Aging: from yeast to humans', Science. 2010 Apr 16; 328(5976): 321–326.

[64] Kalsi, D.S. (2015) 'What is the effect of fasting on the lifespan of neurons?', Ageing Research Reviews, Vol 24(B), p 160-165.

[65] Ramsey, JJ. et al (2000) 'Dietary restriction and aging in rhesus monkeys: the University of Wisconsin study', Experimental Gerontology, Vol 35(9-10), p 1131-1149.

[66] Mattison, J.A. et al (2017) 'Caloric restriction improves health and survival of rhesus monkeys', Nature Communications, Vol 8(14063).

[67] Heilbronn, LK. et al (2005) 'Glucose tolerance and skeletal muscle gene expression in response to alternate day fasting', Obesity Research, Vol 13(3), p 574-581.

[68] Nemoto, S. et al (2005) 'SIRT1 functionally interacts with the metabolic regulator and transcriptional coactivator PGC-1{alpha}', Journal of Biological Chemistry, Vol 280(16), p 16456-60.

[69] Sack, MN. and Finkel, T. (2012) 'Mitochondrial metabolism, sirtuins, and aging', Cold Spring Harbor Perspectives in Biology, Vol 4(12).

[70] Lamb CA, Yoshimori T, Tooze SA. (2013) 'The autophagosome: origins unknown, biogenesis complex', Nature Reviews Molecular Cell Biology. 2013;14:759–74.

[71] Apfield, J. et al (2004) 'The AMP-activated protein kinase AAK-2 links energy levels and insulin-like signals to lifespan in C. elegans', Genes & Development, Vol 18(24), p 3004-3009.

[72] Blüher, M. et al (2003) 'Extended longevity in mice lacking the insulin receptor in adipose tissue', Science, Vol 299(5606), p 572-574.

[73] Jia, K. et al (2004) 'The TOR pathway interacts with the insulin signaling pathway to regulate C. elegans larval development, metabolism and life span', Development, Vol 131(16), p 3897-3906.

[74] Hsu, AL. (2003) 'Regulation of aging and age-related disease by DAF-16 and heat-shock factor', Science, Vol 300(5622), p 1142-1145.

[75] Hercus, MJ. et al (2003) 'Lifespan extension of Drosophila melanogaster through hormesis by repeated mild heat stress', Biogerontology, Vol 4(3), p 149-156.

[76] Lin, K. et al (2001) 'Regulation of the Caenorhabditis elegans longevity protein DAF-16 by insulin/IGF-1 and germline signaling', Nature Genetics, Vol 28(2), p 139-145.

[77] Wang, M.C. et al (2003) 'JNK Signaling Confers Tolerance to Oxidative Stress and Extends Lifespan in Drosophila', Developmental Cell, Vol 5(5), p 811-816.

[78] Martins, R. et al (2016) Long live FOXO: unraveling the role of FOXO proteins in aging and longevity', Aging Cell, Vol 15(2), p 196-207.

[79] Nakae, J. et al (2008) 'The FoxO transcription factors and metabolic regulation', FEBS Letters, Vol 582(1), p 54-67.

[80] Peng, SL. (2008) 'Foxo in the immune system', Oncogene, Vol 27, p 2337-2344.

[81] Dobson, A.J. et al (2017) 'Nutritional Programming of Lifespan by FOXO Inhibition on Sugar-Rich Diets', Cell Reports, Vol 18(2), p 299-306.

[82] Duan, W. (2013) 'Sirtuins: from metabolic regulation to brain aging', Frontiers of Aging Neuroscience.

[83] Palacios, OM. et al (2009) 'Diet and exercise signals regulate SIRT3 and activate AMPK and PGC-1alpha in skeletal muscle', Aging (Albany NY).

[84] Imae, M. et al (2003) 'Nutritional and hormonal factors control the gene expression of FoxOs, the mammalian homologues of DAF-16', Journal of Molecular Endocrinology, Vol 30(2), p 253-262.

[85] Puigserver, P. et al (2003) 'Insulin-regulated hepatic gluconeogenesis through FOXO1-PGC-1alpha interaction', Nature, Vol 423(6939), p 550-555.

[86] Kaestner, K.H. (2008) 'A Two-Step Pathway to Resist Fasting', Cell Metabolism, Vol 8(6), p 449-451.

[87] Ropelle, E. et al (2009) 'Acute exercise modulates the Foxo1/PGC-1α pathway', The Journal of Physiology, Vol 587(9), p 2069-2076.

[88] Sanchez, A. (2015) 'FoxO transcription factors and endurance training: a role for FoxO1 and FoxO3 in exercise-induced angiogenesis', The Journal of Physiology, Vol 593(pt2), p 363-364.

[89] Sanchez, AM. et al (2014) 'FoxO transcription factors: their roles in the maintenance of skeletal muscle homeostasis', Cellular and Molecular Life Sciences, Vol 71(9), p 1657-1671.

[90] Donovon, M. and Marr, M.T. (2016) 'dFOXO Activates Large and Small Heat Shock Protein Genes in Response to Oxidative Stress to Maintain Proteostasis in Drosophila', Jounral of Biological Chemistry, Vol 291(36), 19042–19050.

[91] Polesello, C. and Le Bourg, E. (2017) 'A mild cold stress that increases resistance to heat lowers FOXO translocation in Drosophila melanogaster', Biogerontology, Vol 18(5), p 791-801.

[92] Bakker WJ et al (2007) 'FOXO3a is activated in response to hypoxic stress and inhibits HIF1-induced apoptosis via regulation of CITED2', Molecular Cell, Vol 28(6), p 941-953.

[93] McClintock, B. The Nobel Prize in Physiology or Medicine 1983

[94] Blackburn, E. Greider, C.W., and Szostak, J.W. The Nobel Prize in Physiology or Medicine 2009

[95] Eisenberg, D. (2011) 'An evolutionary review of human telomere biology: The thrifty telomere hypothesis and notes on potential adaptive paternal effects', American Journal of Human Biology, Vol 23(2), p 149-167.

[96] Armanios, M. (2013) 'Telomeres and age-related disease: how telomere biology informs clinical paradigms', The Journal of Clinical Investigation.

[97] Cawton, R. et al (2003) 'Association between telomere length in blood and mortality in people aged 60 years or older', Research Letters, Vol 361(9355), p 393-395.

[98] Okuda, K. et al (2002) 'Telomere Length in the Newborn', Pediatric Research, Vol 52, p 377-381.

[99] Arai, Y. et al (2015) 'Inflammation, But Not Telomere Length, Predicts Successful Ageing at Extreme Old Age: A Longitudinal Study of Semi-supercentenarians', E Bio Medicine, Vol 2(10), p 1549-1558.

[100] Shampay, J. and Blackburn, EH. (1988) 'Generation of telomere-length heterogeneity in Saccharomyces cerevisiae', PNAS, Vol 85(2), p 534-538.

[101] Bodnar, A. et al (1998) 'Extension of Life-Span by Introduction of Telomerase into Normal Human Cells', Science, Vol 279(5349), p 349-352.

[102] Flores, I. (2006) 'Telomerase regulation and stem cell behaviour', Current Opinion in Cell Biology, Vol 18(3), p 254-260.

[103] Zhang, A. (2003) 'Deletion of the Telomerase Reverse Transcriptase Gene and Haploinsufficiency of Telomere Maintenance in Cri du Chat Syndrome', AJHG, Vol 72(4), p 940-948.

[104] Blasco, MA. (2005) 'Telomeres and human disease: ageing, cancer and beyond', Nature Reviews. Genetics., Vol 6(8), p 611-622.

[105] Zhang, X. et al (1999) 'Telomere shortening and apoptosis in telomerase-inhibited human tumor cells', Genes & Development, Vol 13, p 2388-2399.

[106] Richter, T. and Zglinicki, T. (2007) 'A continuous correlation between oxidative stress and telomere shortening in fibroblasts', Experimental Gerontology, Vol 42(11), p 1039-1042.

[107] Epel, E. et al (2009) 'Can meditation slow rate of cellular aging? Cognitive stress, mindfulness, and telomeres', Annals of the New York Academy of Sciences, Vol 1172, p 34-53.

[108] Jacobs, TL. et al (2011) 'Intensive meditation training, immune cell telomerase activity, and psychological mediators', Psychoneuroendocrinology, Vol 36(5), p 664-681.

[109] Scutte, N.S. and Malouff, J.M. (2014) 'A meta-analytic review of the effects of mindfulness meditation on telomerase activity', Psychoneuroendocrinology, Vol 42, p 45-48.

[110] Meyer, A. et al (2016) 'Leukocyte telomere length is related to appendicular lean mass: cross-sectional data from the Berlin Aging Study II (BASE-II)', Americal Journal of Clinical Nutrition, Vol 103(1), p 178-183.

[111] Radak, Z. et al (2008) 'Exercise, oxidative stress and hormesis', Ageing Research Reviews, Vol 7(1), p 34-42.

[112] Mattson MP (2012) Energy intake and exercise as determinants of brain health and vulnerability to injury and disease. Cell Metabolism 16: 706-722

[113] Arumugam TV, Phillips TM, Cheng A, Morrell CH, Mattson MP, Wan R (2010) Age and energy intake interact to modify cell stress pathways and stroke outcome. Ann Neurol. 67:41-52.

[114] Yang JL, Lin YT, Chuang PC, Bohr VA, Mattson MP (2014). BDNF and exercise enhance neuronal DNA repair by stimulating CREB-mediated production of apurinic/apyrimidinic endonuclease 1. Neuromolecular Med. 16:161-74

[115] Reznick, DN. et al (2004) 'Effect of extrinsic mortality on the evolution of senescence in guppies', Nature, Vol 431(7012), p 1095-1099.

[116] Rogina, B. et al (2000) 'Extended life-span conferred by cotransporter gene mutations in Drosophila', Science, Vol 290(5499), p 2137-2140.

[117] Hsin, H. and Kenyon, C. (1999) 'Signals from the reproductive system regulate the lifespan of C. elegans', Nature, Vol 399(6734), p 362-366.

[118] Brook, MS. et al (2015) 'Skeletal muscle homeostasis and plasticity in youth and ageing: impact of nutrition and exercise', Acta Physiologica, Vol 216(1), p 15-41.

[119] Burgers, AMG. et al (2011) 'Meta-Analysis and Dose-Response Metaregression: Circulating Insulin-Like Growth Factor I (IGF-I) and Mortality', JCEM, Vol 96(9), p 2912-2920.

[120] Smith TJ (2010) 'Insulin-like growth factor-I regulation of immune function: a potential therapeutic target in autoimmune diseases?', Pharmacological Reviews, Vol 62(2), p 199-236.

[121] Puche, J.E. and Castilla-Cortazar, I. (2012) 'Human conditions of insulin-like growth factor-I (IGF-I) deficiency', Journal of Translational Medicine, Vol 10, p 224.

[122] Owen OE, et al (1969) Liver and kidney metabolism during prolonged starvation. J Clin Invest; Vol 48, p 574-583.

[123] Eichhorn, G. et al (2011) 'Heterothermy in growing king penguins', Nature Communications, Vol 2, p 435.

[124] Eble, AS. et al (1983) 'Nonenzymatic glucosylation and glucose-dependent cross-linking of protein', Journal of Biological Chemistry, Vol 258(15), p 9406-9412.

[125] Vlassara, H. (2001) 'The AGE-receptor in the pathogenesis of diabetic complications', Diabetes Metabolism Research and Reviews, Vol 17(6), p 436-443.

[126] Gkogkolou, P. and Böhm, M. (2012) 'Advanced glycation end products: Key players in skin aging?', Dermatoendocrinology, Vol 4(3), p 259-270.

[127] Spindler, SR (2010) 'Biological Effects of Calorie Restriction: Implications for Modification of Human Aging', The Future of Aging, p 367-438.

[128] Keys A, et al (1950) 'The Biology of Human Starvation', University of Minnesota Press, Minneapolis.

[129] "Men Starve in Minnesota" (July 30, 1945). Life 19(5): 43-46.

[130] Oswalt, W H. (1976). *An Anthropological Analysis of Foodgetting Technology.* New York, NY: Wiley.

[131] Paton DN, and Stockman R (1888-1889) 'Observations on the metabolism of man during starvation', Proc R Acad Edinb; pp 121-131.

[132] Rcbins GN: The fasting man. Br Med J 1890 Jun 21; 1: 1444-1446 9. Benedict FG: A Study of Prolonged Fasting, Publication No. 203. Washington DC, Carnegie Institute, 1915

[133] Thomson TJ, Runcie J, Miller V: Treatment of obesity by total fasting for up to 249 days. Lancet 1966 Nov 5; 2:992-996

[134] Stewart WK, and Fleming LW (1973) 'Features cf a successful therapeutic fast of 382 days' duration', Postgrad Med J, Vol 49, p 203-209.

[135] Weir, H.J. et al (2017) 'Dietary Restriction and AMPK Increase Lifespan via Mitochondrial Network and Peroxisome Remodeling', Cell Metabolism, Vol 26(6), p 884-896.

[136] Liang, H. and Ward, WF. (2006) 'PGC-1alpha: a key regulator of energy metabolism', Advances in Physiology Education, Vol 30(4), p 145-151.

[137] Zong, H. et al (2002) 'AMP kinase is required for mitochondrial biogenesis in skeletal muscle in response to chronic energy deprivation', PNAS, Vol 99(25), p 15983-15987.

[138] Seok, S. et al (2018) 'Fasting-induced JMJD3 histone demethylase epigenetically activates mitochondrial fatty acid β-oxidation', JCI.

[139] Lokireddy, S. et al (2012) 'The ubiquitin ligase Mul1 induces mitophagy in skeletal muscle in response to muscle-wasting stimuli', Cell Metabolism, Vol 16(5), p 613-624.

[140] Houtkooper, RH. et al (2012) 'Sirtuins as regulators of metabolism and healthspan', Nat Rev Mol Cell Biol, Vol 13(4), p 225-238.

[141] Li, P. et al (2017) 'SIRT1 is required for mitochondrial biogenesis reprogramming in hypoxic human pulmonary arteriolar smooth muscle cells', Int J Mol Med, Vol 39(5), p 1127-1136.

[142] Krishnan, J. et al (2012) 'Dietary obesity-associated Hif1α activation in adipocytes restricts fatty acid oxidation and energy expenditure via suppression of the Sirt2-NAD+ system', Genes Dev, Vol 26(3), p 259-270.

[143] Bell, EL. and Guarente, L. (2011) 'The SirT3 divining rod points to oxidative stress', Mol Cell, Vol 42(5), p 561-568.

[144] Gomes, A.P. et al (2014) 'Declining NAD+ Induces a Pseudohypoxic State Disrupting Nuclear-Mitochondrial Communication during Aging', Cell, Vol 155(7), p 1624-1638.

[145] Yang, H. et al (2007) 'Nutrient-Sensitive Mitochondrial NAD+ Levels Dictate Cell Survival', Cell, Vol 130(6), p 1095-1107.

[146] Trafton, A. (2018) 'Study suggests method for boosting growth of blood vessels and muscle', MIT News.

[147] Fang, E.F. et al (2016) 'NAD+ Replenishment Improves Lifespan and Healthspan in Ataxia Telangiectasia Models via Mitophagy and DNA Repair', Cell Metabolism, Vol 24(4), p 566-581.

[148] Li, J. et al (2017) 'A conserved NAD+ binding pocket that regulates protein-protein interactions during aging', Science, Vol 355(6331), p 1312-1317.

[149] Tothova, Z. and Gilliland, DG. (2007) 'FoxO transcription factors and stem cell homeostasis: insights from the hematopoietic system', Cell Stem Cell, Vol 1(2), p 140-152.

[150] Mergenthaler, P. et al (2014) 'Sugar for the brain: the role of glucose in physiological and pathological brain function', Trends Neuroscience, Vol 36(10), p 587-597.

[151] Mattson, MP. et al (2004) 'BDNF and 5-HT: a dynamic duo in age-related neuronal plasticity and neurodegenerative disorders', Trends Neuroscience, Vol 27(10), p 589-594.

[152] Mattson, MP. et al (2003) 'Meal size and frequency affect neuronal plasticity and vulnerability to disease: cellular and molecular mechanisms', Journal of Neurochemistry, Vol 84(3), p 417-431.

[153] Manzanero, S. et al (2014) 'Intermittent fasting attenuates increases in neurogenesis after ischemia and reperfusion and improves recovery', JCBFM, Vol 34(5), p 897-905.

[154] Redman, LM. et al (2018) 'Metabolic Slowing and Reduced Oxidative Damage with Sustained Caloric Restriction Support the Rate of Living and Oxidative Damage Theories of Aging', Clinical and Translational Report, Vol 27(4), p 805-815.

[155] Aberg, ND. et al (2006) 'Aspects of growth hormone and insulin-like growth factor-I related to neuroprotection, regeneration, and functional plasticity in the adult brain', Scientific World Journal, Vol 18(6), p 53-80.

[156] Lavin DN, et al (2011) 'Fasting induces an anti-inflammatory effect on the neuroimmune system which a high-fat diet prevents', Obesity (Silver Spring), Vol 19(8), p 1586-1594.

[157] Willeumier, KC. et al (2011) 'Elevated BMI is associated with decreased blood flow in the prefrontal cortex using SPECT imaging in healthy adults', Obesity (Silver Spring), Vol 19(5), p 1095-1097.

[158] https://www.rose-hulman.edu/~brandt/Chem330/EndocrineNotes/Chapter_5_Glucose.pdf

[159] Guzman, M. and Blazquez, C. (2004) 'Ketone body synthesis in the brain: possible neuroprotective effects', Prostaglandins, Leukotrienes, and Essential Fatty Acids, Vol 70(3), p 287-292.

[160] Baba, H. et al (1995) 'Glycerol gluconeogenesis in fasting humans', Nutrition, Vol 11(2), p 149-153.

[161] Wyass, T.M. et al (2011) 'In Vivo Evidence for Lactate as a Neuronal Energy Source', Journal of Neruoscience, Vol 31(20), p 7477-7485.

[162] Ivanov, A. and Zilberter, Y. (2011) 'Critical State of Energy Metabolism in Brain Slices: The Principal Role of Oxygen Delivery and Energy Substrates in Shaping Neuronal Activity', Frontiers in Neuroenergetics, Vol 3, p 9.

[163] de la Monte, S.M. and Wands, JR. (2008) 'Alzheimer's Disease Is Type 3 Diabetes–Evidence Reviewed', Journal of Diabetes Science and Technology, Vol 2(6), p 1101-1113.

[164] Penn Medicine News (2012) 'Brain Insulin Resistance Contributes to Cognitive Decline in Alzheimer's Disease', Accessed: https://www.pennmedicine.org/news/news-releases/2012/march/brain-insulin-resistance-contr

[165] Arumugam, TV. et al (2010) 'Age and energy intake interact to modify cell stress pathways and stroke outcome', Annals of Neurology, Vol 67(1), p 41-52.

[166] Yale University (2015) 'Anti-inflammatory mechanism of dieting and fasting revealed', Accessed: https://www.sciencedaily.com/releases/2015/02/150216131146.htm

[167] Stafstrom, C.E. and Rho, JM. (2012) 'The ketogenic diet as a treatment paradigm for diverse neurological disorders', Frontiers in Pharmacology.

[168] Green, MW. et al (1995) 'Lack of effect of short-term fasting on cognitive function', Journal of Psychiatric Research, Vol 29(3), p 245-253.

[169] Fontana, L. et al (2010) 'Extending healthy life span--from yeast to humans', Science, Vol 328(5976), p 321-326.

[170] Cheng, CW. et al (2014) 'Prolonged Fasting reduces IGF-1/PKA to promote hematopoietic stem cell-based regeneration and reverse immunosuppression', Cell Stem Cell, Vol 14(6), p 810-823.

[171] Wei, M. et al (2017) 'Fasting-mimicking diet and markers/risk factors for aging, diabetes, cancer, and cardiovascular disease', Sci Transl Med, Vol 9 (377).

[172] Clemente, J.C. et al (2012) 'The Impact of the Gut Microbiota on Human Health: An Integrative View', Cell, Vol 148(6), p 1258-1270.

[173] Conlon, M.A. and Bird, A.R. (2015) 'The Impact of Diet and Lifestyle on Gut Microbiota and Human Health', Nutrients, Vol 7(1), p 17-44.

[174] Regan, JC. et al (2016) 'Sex difference in pathology of the ageing gut mediates the greater response of female lifespan to dietary restriction', eLife.

[175] Catterson, J. et al (2018) 'Short-Term, Intermittent Fasting Induces Long-Lasting Gut Health and TOR-Independent Lifespan Extension', Current Biology, Vol 28(11), p 1714-1724.

[176] Godinez-Victoria, M. et al (2014) 'Intermittent Fasting Promotes Bacterial Clearance and Intestinal IgA Production in Salmonella typhimurium-Infected Mice', Experimental Immunology, Vol 79(5), p 315-324.

[177] Lara-Padilla, E. et al (2015) 'Intermittent fasting modulates IgA levels in the small intestine under intense stress: A mouse model', Vol 285, p 22-30.

[178] Shen, R. et al (2016) 'Neuronal energy-sensing pathway promotes energy balance by modulating disease tolerance', PNAS, Vol 113(23), p E3307-E3314.

[179] Li, G. et al (2017) 'Intermittent Fasting Promotes White Adipose Browning and Decreases Obesity by Shaping the Gut Microbiota', Vol 26(4), p 672-685.

[180] Moreno-Navarette, J.M. et al (2017) 'Gut Microbiota Interacts with Markers of Adipose Tissue Browning, Insulin Action and Plasma Acetate in Morbid Obesity', Molecular Nutrition and Food Research, Vol 62(3).

[181] Kivelä, R. and Alitalo, K. (2017) 'White adipose tissue coloring by intermittent fasting', Cell Research, Vol 27, p 1300-1301.

[182] Ley, RE. et al (2006) 'Microbial ecology: human gut microbes associated with obesity', Nature, Vol 444(7122), p 1022-1023.

[183] Zhang, C. et al (2013) 'Structural modulation of gut microbiota in life-long calorie-restricted mice', Nature Communications, Vol 4, 2163.

[184] Deloose, E. et al (2012) 'The migrating motor complex: control mechanisms and its role in health and disease', Nat Rev Gastroentology & Hepatology, Vol 9(5), p 271-285.

[185] Chaix, A. and Zarrinpar, A. (2015) 'The effects of time-restricted feeding on lipid metabolism and adiposity', Adipocyte, Vol 4(4), p 319-324.

[186] Kanazawa, M. and Fukudo, S. (2006) 'Effects of fasting therapy on irritable bowel syndrome', International Journal of Behavioral Medicine, Vol 13(3), p 214-220.

[187] Mihaylova, M. et al (2018) 'Fasting Activates Fatty Acid Oxidation to Enhance Intestinal Stem Cell Function during Homeostasis and Aging', Cell Stem Cell, Vol 22(5), p 769-778.

[188] Paulose, JK. et al (2016) 'Human Gut Bacteria Are Sensitive to Melatonin and Express Endogenous Circadian Rhythmicity', PLOS One, Vol 11(1).

[189] Trinder, M. et al (2015) 'Bacteria Need Sleep Too?: Microbiome Circadian Rhythmicity, Metabolic Disease, and Beyond', University of Toronto Medical Journal, Vol 92(3).

[190] David, LA. et al (2014) 'Diet rapidly and reproducibly alters the human gut microbiome', Nature, Vol 505(7484), p 559-563.

[191] Karasov, WH. et al (2004) 'Anatomical and histological changes in the alimentary tract of migrating blackcaps (Sylvia atricapilla): a comparison among fed, fasted, food-restricted, and refed birds', Physiol Biochem Zool, Vol 77(1), p 149-160.

[192] Cignarella, F. et al (2018) 'Intermittent Fasting Confers Protection in CNS Autoimmunity by Altering the Gut Microbiota', Cell Metabolism, Vol 27(6), p 1222-1235.

[193] Zarrinpar, A. et al (2014) 'Diet and feeding pattern affect the diurnal dynamics of the gut microbiome', Cell Metabolism, Vol 20(6), p 1006-1017.

[194] Zarrinpar, A. et al (2014) 'Diet and Feeding Pattern Affect the Diurnal Dynamics of the Gut Microbiome', Cell Metabolism, Vol 20(6), p 1006-1017.

[195] Thaiss, C.A. et al (2014) 'A day in the life of the meta-organism: diurnal rhythms of the intestinal microbiome and its host', Gut Microbes, Vol 6(2), p 137-142.

[196] Patterson, R. and Sears, D. (2017) 'Metabolic Effects of Intermittent Fasting', Annual Review of Nutrition, Vol 37, p 371-393.

[197] Rhee, S.H. et al (2009) 'Principles and clinical implications of the brain–gut–enteric microbiota axis', Nature Reviews. Gastroenterology & Hepatology, Vol 6, p 306-314.

[198] Wang, Y. and Kasper, L.H. (2014) 'The role of microbiome in central nervous system disorders', Brain Behav Immun, Vol 38, p 1-12.

[199] Carabotti, M. et al (2015) 'The gut-brain axis: interactions between enteric microbiota, central and enteric nervous systems', Annals & Gastroenterology, Vol 28(2), p 203-209.

[200] Tunstall RJ, et al. (2002) 'Fasting activates the gene expression of UCP3 independent of genes necessary for lipid transport and oxidation in skeletal muscle', Biochemical and Biophysical Research Communications, Vol 294, p 301-308.

[201] Mansell, PI. et al (1990) 'Enhanced thermogenic response to epinephrine after 48-h starvation in humans', Am J Physiol, Vol 258 (1 pt 2), R87-93.

[202] Heilbronn, LK. et al (2005) 'Alternate-day fasting in nonobese subjects: effects on body weight, body composition, and energy metabolism', Am J Clin Nutr, Vol 81(1), p 69-73.

[203] Ho, K.Y. et al (1988) 'Fasting enhances growth hormone secretion and amplifies the complex rhythms of growth hormone secretion in man', J Clin Invest, Vol 81(4), p 968-975.

[204] Merimee TJ, and Fineberg SE (1974) 'Growth hormone secretion in starvation', A reassessment. J Clin Endocrinol Metab, Vol 39, p 385-386.

[205] Palmblad J, Levi L, Burger A, et al (1977) 'Effects of total energy withdrawal (fasting) on the levels of growth hormone, thyrotropin, cortisol, adrenaline, noradrenaline, T4, T3, and rT3 in healthy males'. Acta Med Scand, Vol 201, p 15-22.

[206] Roth J, Glick SM, Yalow RS, et al (1963) 'Secretion of human growth hormone: Physiologic and experimental modification', Metabolism, Vol 12, p 577-579.

Beck P, Koumans JT, Winterling CA, et al (1964), Studies of insulin and growth hormone secretion in human obesity'. J Lab Clin Med, Vol 64, p 654-667.

[207] Thissen, JP. et al (1994) 'Nutritional regulation of the insulin-like growth factors', Endocr Rev, Vol 15(1), p 80-101.

[208] Klionsky, D. (2012) 'Look people, "Atg" is an abbreviation for "autophagy-related." That's it.', Autophagy, Vol 8(9), p 1281-1282.

[209] Sedwick, C. (2012) 'Yoshinori Ohsumi: Autophagy from beginning to end', J Cell Biol, Vol 197(2), p 164-165.

[210] Coupé, B. et al (2012) 'Loss of autophagy in pro-opiomelanocortin neurons perturbs axon growth and causes metabolic dysregulation', Cell Metab, Vol 15(2), p 247-255.

[211] Klionsky, D. (2008) 'Autophagy revisited: A conversation with Christian de Duve', Autophagy, Vol 4(6), p 740-743.

[212] Mizushima, N. (2007) 'Autophagy: process and function', Vol 21, p 2861-2873.

[213] Kim, J. et al (2001) 'Membrane Recruitment of Aut7p in the Autophagy and Cytoplasm to Vacuole Targeting Pathways Requires Aut1p, Aut2p, and the Autophagy Conjugation Complex', Journal of Cell Biology, Vol 152(1), p 51.

[214] Takeshige, K. et al (1992) 'Autophagy in yeast demonstrated with proteinase-deficient mutants and conditions for its induction', JCB Home, Vol 119(2), p 301.

[215] Moriyasu, Y. and Ohsumi, Y. (1996) 'Autophagy in Tobacco Suspension-Cultured Cells in Response to Sucrose Starvation', Cell Biology and Transduction, Vol 111(4), p 1233-1241.

[216] Noda, T. and Ohsumi, Y. (1998) 'Tor, a Phosphatidylinositol Kinase Homologue, Controls Autophagy in Yeast', Journal of Biological Chemistry, Vol 273, p 3963-3966.

[217] Ravikumar, B. et al (2004) 'Inhibition of mTOR induces autophagy and reduces toxicity of polyglutamine expansions in fly and mouse models of Huntington disease', Nature Genetics, Vol 36, p 585-595.

[218] Mordier, S. et al (2000) 'Leucine Limitation Induces Autophagy and Activation of Lysosome-dependent Proteolysis in C2C12 Myotubes through a Mammalian Target of Rapamycin-independent Signaling Pathway', JBC, Vol 275, p 29900-29906.

[219] Djavaheri-Mergny, M. et al (2006) 'NF-κB Activation Represses Tumor Necrosis Factor-α-induced Autophagy', JBC, Vol 281, p 30373-30382.

[220] Scherz-Shouval, R. et al (2007) 'Reactive oxygen species are essential for autophagy and specifically regulate the activity of Atg4', The EMBO Journal, Vol 26, p 1749-1760.

[221] Bastholm, L. et al (2007) 'Control of Macroautophagy by Calcium, Calmodulin-Dependent Kinase Kinase-β, and Bcl-2', Molecular Cell, Vol 25(2), p 193-205.

[222] Meley, D. et al (2006) 'AMP-activated Protein Kinase and the Regulation of Autophagic Proteolysis', JBC, Vol 281, p 34870-34879.

[223] Zhao, J. et al (2007) 'FoxO3 coordinately activates protein degradation by the autophagic/lysosomal and proteasomal pathways in atrophying muscle cells', Cell Metabolism, Vol 6(6), p 472-483.

[224] Fan, J. et al (2015) 'Autophagy as a Potential Target for Sarcopenia', Journal of Cellular Physiology, Vol 231(7), p 1450-1459.

[225] Neel, B. et al (2013) 'Skeletal muscle autophagy: a new metabolic regulator', Trends in Endocrinology & Metabolism, Vol 24(12), p 635-643.

[226] Sanchez, A. et al (2014) 'Autophagy is essential to support skeletal muscle plasticity in response to endurance exercise', American Journal of Physiology.

[227] Jamart, C. et al (1985) 'Modulation of autophagy and ubiquitin-proteasome pathways during ultra-endurance running', J Appl Physiol, Vol 112(9), p 1529-1537.

[228] https://www.ncbi.nlm.nih.gov/pubmed/23964069

[229] Sanchez, A. et al (2012) 'The role of AMP-activated protein kinase in the coordination of skeletal', Am J Physiol Cell Physiol, Vol 303, C475–C485.

[230] Deretic, V. et al (2013) 'Autophagy in infection, inflammation and immunity', Nat Rev Immunol, Vol 13(10), p 722-737.

[231] Gomes, LC and Dikic, I. (2014) 'Autophagy in antimicrobial immunity', Mol Cell, Vol 54(2), p 224-233.

[232] Starr T et al (2012) 'Selective subversion of autophagy complexes facilitates completion of the Brucella intracellular cycle', Cell Host & Microbe, Vol 11(1), p 33-45.

[233] Laddha, SV. et al (2014) 'Mutational landscape of the essential autophagy gene BECN1 in human cancers', Molecular Cancer Research: MCR, Vol 12(4), p 485-490.

[234] Kenific, CM et al (2010) 'Autophagy and metastasis: another double-edged sword', Curr Opin Cell Biol, Vol 22(2), p 241-245.

[235] Gump, JM. and Thouburn, A. (2011) 'Autophagy and apoptosis: what is the connection?', Trends in Cellular Biology, Vol 21(7), p 387-392.

[236] Gump, JM. et al (2014) 'Autophagy variation within a cell population determines cell fate through selective degradation of Fap-1', Nat Cell Biol, Vol 16(1), p 47-54.

[237] Takai, N. et al (2012) 'Bufalin, a traditional oriental medicine, induces apoptosis in human cancer cells', Asian Pacific Journal of Cancer Prevention, Vol 13(1), p 399-402.

[238] Zhao, H. et al (2017) 'Blocking autophagy enhances the pro-apoptotic effect of bufalin on human gastric cancer cells through endoplasmic reticulum stress', Biology Open, Vol 6, p 1416-1422.

[239] Lum JJ et al (2005) 'Growth factor regulation of autophagy and cell survival in the absence of apoptosis', Cell, Vol 120(2), p 237-248.

[240] Takamura, A. et al (2011) 'Autophagy-deficient mice develop multiple liver tumors', Genes Dev, Vol 25(8), p 795-800.

[241] Green, DR. and Levine, B. (2014) 'To be or not to be? How selective autophagy and cell death govern cell fate', Cell, Vol 157(1), p 65-75.

[242] Rosedale, R 'Burn Fat, Not Sugar to Lose Weight', Accessed 2018 http://drrosedale.com/resources/pdf/Burn%20Fat,%20Not%20Sugar%20to%20lose%20weight.pdf

[243] Science Daily (2013) 'Rapamycin: Limited anti-aging effects', Helmholtz Association of German Research Centres.

[244] Ancel Keys et al (1972) 'Indices of relative weight and obesity', Vol 25(6-7), p 329-343.

[245] Barry, VW. et al (2014) 'Fitness vs. fatness on all-cause mortality: a meta-analysis', Prog Cardiovasc Dis, Vol 56(4), p 382-390.

[246] Goodpaster, BH. (2006) 'The loss of skeletal muscle strength, mass, and quality in older adults: the health, aging and body composition study', The Journals of Gerontology. Series A, Biological Sciences and Medical Sciences, Vol 61(10), p 1059-1064.

[247] Strasser, B. et al (2018) 'Role of Dietary Protein and Muscular Fitness on Longevity and Aging', Aging and Disease, Vol 9(1), p 119-132.

[248] Hasten, DL. et al (2000) 'Resistance exercise acutely increases MHC and mixed muscle protein synthesis rates in 78-84 and 23-32 yr olds', Americal Journal of Physiology. Endocrinology and Metabolism, Vol 278(4), E620-626.

[249] West, DW. et al (2011) 'Rapid aminoacidemia enhances myofibrillar protein synthesis and anabolic intramuscular signaling responses after resistance exercise', Americal Journal of Clinical Nutrition, Vol 94(3), p 795-803.

[250] American Heart Association (2018) 'American Heart Association Recommendations for Physical Activity in Adults and Kids', Accessed: https://www.heart.org/en/healthy-living/fitness/fitness-basics/aha-recs-for-physical-activity-in-adults#.WdIG-xO0NBw˘

[251] Volaklis, KA. et al (2015) 'Muscular strength as a strong predictor of mortality: A narrative review', European Journal of Internal Medicine, Vol 26(5), p 303-310.

[252] Kraschnewski, JL. et al (2016) 'Is strength training associated with mortality benefits? A 15year cohort study of US older adults', Prev Med, Vol 87, p 121-127.

[253] Buettner, D. (2003) 'The Secrets of a Long Life', National Geographic, November 2003, Accessed: https://bluezones.com/wp-content/uploads/2015/01/Nat_Geo_LongevityF.pdf

[254] Kent-Braun, JA et al (1985) 'Skeletal muscle contractile and noncontractile components in young and older women and men', Journal of Applied Physiology, Vol 88(2), p 662-668.

[255] Srikanthan, P. and Karlamangla, AS. (2011) 'Relative Muscle Mass Is Inversely Associated with Insulin Resistance and Prediabetes. Findings from The Third National Health and Nutrition Examination Survey', The Journal of Clinical Endocrinology & Metabolism, Volume 96, Issue 9, 1 September 2011, Pages 2898–2903.

[256] Owusu-Ansah, E. et al (2013) 'Muscle Mitohormesis Promotes Longevity via Systemic Repression of Insulin Signaling' Cell, Vol 155(3), p 699-712.

[257] Brand, MD. (2000) 'Uncoupling to survive? The role of mitochondrial inefficiency in ageing', Experimental Gerontology, Vol 35(6-7), p 811-820.

[258] Keipert, S. et al (2013) 'Skeletal muscle uncoupling-induced longevity in mice is linked to increased substrate metabolism and induction of the endogenous antioxidant defense system', The American Physiological Society, Vol 304(5), p E495-E506.

[259] Pedersen, BK. (2011) 'Muscles and their myokines', The Journal of Experimental Biology, Vol 214(Pt2), p 337-346.

[260] Newman, AB. et al (2006) 'Strength, but not muscle mass, is associated with mortality in the health, aging and body composition study cohort', The Journals of Gerontology. Series A, Biological Sciences and Medical Sciences, Vol 61(1), p 72-77.

[261] Goodpaster, BH. et al (1985) 'Attenuation of skeletal muscle and strength in the elderly: The Health ABC Study', Journal of Applied Physiology (1985), Vol 90(6), p 2157-2165.

[262] Leong, DP. et al (2015) 'Prognostic value of grip strength: findings from the Prospective Urban Rural Epidemiology (PURE) study', Lancet, Vol 386(9990), p 266-273.

[263] Kim, Y. et al (2017) 'Independent and joint associations of grip strength and adiposity with all-cause and cardiovascular disease mortality in 403,199 adults: the UK Biobank study', American Journal of Clinical Nutrition, Vol 106(3), p 773-782.

[264] Sayer, AA. and Kirkwood, T. (2015) 'Grip strength and mortality: a biomarker of ageing?', The Lancet, Vol 386(9990), p 226-227.

[265] Bouchard, DR. et al (2011) 'Association between muscle mass, leg strength, and fat mass with physical function in older adults: influence of age and sex', Journal of Aging and Health, Vol 23(2), p 313-328.

[266] Bautmans, I. et al (2009) 'Sarcopenia and functional decline: pathophysiology, prevention and therapy', Vol 64(4), p 303-316.

[267] Zebis, MK. et al (2011) 'Implementation of neck/shoulder exercises for pain relief among industrial workers: a randomized controlled trial', BMC Musculoskeletal Disorders, Vol 12, p 205.

[268] Pereira, LM. et al (2012) 'Comparing the Pilates method with no exercise or lumbar stabilization for pain and functionality in patients with chronic low back pain: systematic review and meta-analysis', Clinical Rehabilitation, Vol 26(1), p 10-20.

[269] Caelle, MC. and Fernandez, ML. (2010) 'Effects of resistance training on the inflammatory response', Nutrition Research and Practice, Vol 4(4), p 259-269.

[270] Clark, J. (2015) 'Diet, exercise or diet with exercise: comparing the effectiveness of treatment options for weight loss and changes in fitness for adults (16-65 years old) who are overfat, or obese; systematic review and meta-analysis', Journal of Diabetes & Metabolic Disorders, Vol 14(31).

[271] Fleg, JL. et al (2005) 'Accelerated longitudinal decline of aerobic capacity in healthy older adults', Circulation, Vol 112(5), p 674-682.

[272] Tabata, I. et al (1996) 'Effects of moderate-intensity endurance and high-intensity intermittent training on anaerobic capacity and VO2max', Medicine and Science in Sports and Exercise, Vol 28(10), p 1327-1330.

[273] Robinson, M. et al (2017) 'Enhanced Protein Translation Underlies Improved Metabolic and Physical Adaptations to Different Exercise Training Modes in Young and Old Humans', Clinical and Translation Report, Vol 25(3), p 581-592.

[274] Ropelle, ER. et al (2009) 'Acute exercise modulates the Foxo1/PGC-1α pathway', Journal of Physiology, p 2069-2076.

[275] Cantó C. et al (2010) 'Interdependence of AMPK and SIRT1 for metabolic adaptation to fasting and exercise in skeletal muscle', Cell Metabolism, Vol 11(3), p 213-219.

[276] Costford, SR. et al (2010) 'Skeletal muscle NAMPT is induced by exercise in humans', American Journal of Physiology. Endocrinology and Metabolism, Vol 298(1), p E117-26.

[277] Weichhart, T. (2012) 'Mammalian target of rapamycin: a signaling kinase for every aspect of cellular life', Methods in Molecular Biology, Vol 821, p 1-14.

[278] Wullschleger, S. et al (2006) 'TOR Signaling in Growth and Metabolism', Cell, Vol 124(3), p 471-484.

[279] Kim, DH. et al (2002) 'mTOR Interacts with Raptor to Form a Nutrient-Sensitive Complex that Signals to the Cell Growth Machinery', Cell, Vol 110(2), p 163-175.

[280] Bond, P. (2016) 'Regulation of mTORC1 by growth factors, energy status, amino acids and mechanical stimuli at a glance', JISSN, Vol 13, p 8.

[281] Adegoke, OA. et al (2012) 'mTORC1 and the regulation of skeletal muscle anabolism and mass', Applied Physiology, Nutrition, and Metabolism, Vol 37(3), p 395-406.

[282] Sarbassov, DD. et al (2004) 'Rictor, a Novel Binding Partner of mTOR, Defines a Rapamycin-Insensitive and Raptor-Independent Pathway that Regulates the Cytoskeleton', Current Biology, Vol 14(14), p 1296-1302.

[283] Yin, Y. et al (2015) 'mTORC2 promotes type I insulin-like growth factor receptor and insulin receptor activation through the tyrosine kinase activity of mTOR', Cell Research, Vol 26, p 46-65.

[284] Hassay, N. and Sonenberg, N. (2004) 'Upstream and downstream of mTOR', Genes & Dev, Vol 18, p 1926-1945.

[285] Tokunaga, C. et al (2004) 'mTOR integrates amino acid- and energy-sensing pathways', BBRC, Vol 313(2), p 443-446.

[286] Sarbassov, DD. et al (2004) 'Rictor, a novel binding partner of mTOR, defines a rapamycin-insensitive and raptor-independent pathway that regulates the cytoskeleton', Current Biology, Vol 14(14), p 1296-1302.

[287] Yoon, MS. and Choi, CS. (2016) 'The role of amino acid-induced mammalian target of rapamycin complex 1(mTORC1) signaling in insulin resistance', Exp Mol Med, Vol 48(1), e201.

[288] Shimobayashi, M. and Hall, MN. (2014) 'Making new contacts: the mTOR network in metabolism and signalling crosstalk', Nat Rev Mol Cell Biol, Vol 15(3), p 155-162.

[289] Myers, MG. et al (1992) 'IRS-1 activates phosphatidylinositol 3'-kinase by associating with src homology 2 domains of p85', Proc Natl Acad Sci USA, Vol 89(21), p 10350-4.

[290] Bodline, SC. (2006) 'mTOR signaling and the molecular adaptation to resistance exercise', Med Sci Sports Exerc, Vol 38(11), p 1950-7.

[291] Wu, M. et al (2011) 'Akt/protein kinase B in skeletal muscle physiology and pathology', J Cell Physiol, Vol 226(1), p 29-36.

[292] Sanchez, AM. et al (2014) 'FoxO transcription factors: their roles in the maintenance of skeletal muscle homeostasis', Cell Mol Life Sci, Vol 71(9), p 1657-71.

[293] Zhang, X. et al (2011) 'Akt, FoxO and regulation of apoptosis', Biochim, Biophys Acta, Vol 1813(11), p 1978-86.

[294] Rabinowitz, JD. and White, E. (2010) 'Autophagy and metabolism', Science, Vol 330(6009), p 1344-8.

[295] Rodriguez, J. et al (2014) 'Myostatin and the skeletal muscle atrophy and hypertrophy signaling pathways', Cell Mol Life Sci, Vol 71(22), p 4361-71.

[296] Trendelenburg, AU. et al (2009) 'Myostatin reduces Akt/TORC1/p70S6K signaling, inhibiting myoblast differentiation and myotube size', Am J Physiol Cell Physiol, Vol 296(6), p C1258-70.

[297] Dunlop, EA. and Tee, AR. (2013), 'The kinase triad, AMPK, mTORC1 and ULK1, maintains energy and nutrient homoeostasis', Biochem Soc Trans, Vol 41(4), p 939-43.

[298] Hardie, DG. et al (2012) 'AMPK: a nutrient and energy sensor that maintains energy homeostasis', Nat Rev Mol Cell Biol, Vol 13(4), p 251-62.

[299] Sancak, Y. et al (2008) 'The Rag GTPases bind raptor and mediate amino acid signaling to mTORC1', Science, Vol 320(5882), p 1496-501.

[300] Ebato, C. et al (2008) 'Autophagy is important in islet homeostasis and compensatory increase of beta cell mass in response to high-fat diet', Cell Metabolism, Vol 8(4), p 325-32.

[301] Wolfson, RL. et al (2016) 'Sestrin2 is a leucine sensor for the mTORC1 pathway', Science, Vol 351(6268), p 43-8.

[302] Albert, FJ. et al (2015) 'USEFULNESS OF B-HYDROXY-B-METHYLBUTYRATE (HMB) SUPPLEMENTATION IN DIFFERENT SPORTS: AN UPDATE AND PRACTICAL IMPLICATIONS', Nutricion Hospitalaria, Vol 32(1), p 20-33.

[303] Jacobs, BL. et al (2013) 'Eccentric contractions increase the phosphorylation of tuberous sclerosis complex-2 (TSC2) and alter the targeting of TSC2 and the mechanistic target of rapamycin to the lysosome', J Physiol, Vol 591(18), p 4611-20.

[304] Yoon, MS et al (2011) 'Phosphatidic acid activates mammalian target of rapamycin complex 1 (mTORC1) kinase by displacing FK506 binding protein 38 (FKBP38) and exerting an allosteric effect', J Biol Chem, Vol 286(34), p 29568-74.

[305] Joy, JM. et al (2014) 'Phosphatidic acid enhances mTOR signaling and resistance exercise induced hypertrophy', Nutr Metab (Lond), Vol 11, p 29.

[306] Tanaka, T. et al (2012) 'Quantification of phosphatidic acid in foodstuffs using a thin-layer-chromatography-imaging technique', J Agric Food Chem, Vol 60(16), p 4156-61.

[307] Ogasawara, R. et al (2013) 'Ursolic acid stimulates mTORC1 signaling after resistance exercise in rat skeletal muscle', Am J Physiol Endocrinol Metab, Vol 305(6), E760-5.

[308] Deldicque, L. et al (2008) 'Effects of resistance exercise with and without creatine supplementation on gene expression and cell signaling in human skeletal muscle', J Appl Physiol 1985, Vol 104(2), p 371-8.

[309] Jorquera, G. et al (2013) 'Testosterone signals through mTOR and androgen receptor to induce muscle hypertrophy', Med Sci Sports Exerc, Vol 45(9), p 1712-20.

[310] Altamirano, F. et al (2009) 'Testosterone induces cardiomyocyte hypertrophy through mammalian target of rapamycin complex 1 pathway', J Endocrinol, Vol 202(2), p 299-307.

[311] Xu, K. et al (2014) 'mTOR signaling in tumorigenesis', BBA, Vol 1846(2), p 638-654.

[312] Lamming, DW. et al (2012) 'Rapamycin-Induced Insulin Resistance Is Mediated by mTORC2 Loss and Uncoupled from Longevity', Science, Vol 335(6076), p 1638-1643.

[313] Liberti, M. and Locasale, J. (2016) 'The Warburg Effect: How Does it Benefit Cancer Cells?', Trends Biochem Sci, Vol 41(3), p 211-218.

[314] Yin, Y. et al (2016) 'mTORC2 promotes type I insulin-like growth factor receptor and insulin receptor activation through the tyrosine kinase activity of mTOR', Cell Research, Vol 26(1), p 46-65.

[315] Chan, S. (2004) 'Targeting the mammalian target of rapamycin (mTOR): a new approach to treating cancer', Br J Cancer, Vol 91(8), p 1420-4.

[316] Zoncu, R. et al (2010) 'mTOR: from growth signal integration to cancer, diabetes and ageing', Molecular Cell Biology, Vol 12, p 21-35.

[317] Chano, T. et al (2007) 'RB1CC1 insufficiency causes neuronal atrophy through mTOR signaling alteration and involved in the pathology of Alzheimer's diseases', Vol 1168(7), p 97-105.

[318] Selkoe, DJ. (2008) 'Soluble oligomers of the amyloid β-protein impair synaptic plasticity and behavior', Behavioural Brain Research, Vol 192(1), p 106-113.

[319] Wilson Powers, R. et al (2006) 'Extension of chronological life span in yeast by decreased TOR pathway signaling', Genes Dev, Vol 20(2), p 174-184.

[320] Julie Wu, J. et al (2013) 'Increased mammalian lifespan and a segmental and tissue-specific slowing of aging following genetic reduction of mTOR expression', Cell Rep, Vol 4(5), p 913-920.

[321] Jia, K. et al (2004) 'The TOR pathway interacts with the insulin signaling pathway to regulate C. elegans larval development, metabolism and life span', Vol 131, p 3897-3906.

[322] Kriete, A. et al (2010) 'Rule-Based Cell Systems Model of Aging using Feedback Loop Motifs Mediated by Stress Responses', PLOS Computational Biology, Vol 6(6), e1000820.

[323] Melnik, B. (2012) 'Dietary intervention in acne: Attenuation of increased mTORC1 signaling promoted by Western diet', Dermatoendocrinol, Vol 4(1), p 20-32.

[324] Fontana, L. et al (2013) 'Dietary protein restriction inhibits tumor growth in human xenograft models', Oncotarget, Vol 4(12), p 2451-61.

[325] Long, X. et al (2005) 'Rheb binding to mammalian target of rapamycin (mTOR) is regulated by amino acid sufficiency', J Biol Chem, Vol 280(25), p 23433-6.

[326] Chen, R. et al (2014) 'The general amino acid control pathway regulates mTOR and autophagy during serum/glutamine starvation', JCB, Vol 206(2), p 173.

[327] Martins, L. et al (2012) 'Hypothalamic mTOR Signaling Mediates the Orexigenic Action of Ghrelin', PLOS One, Vol 7(10), e46923.

[328] McDaniel, S. et al (2011) 'The ketogenic diet inhibits the mammalian target of rapamycin (mTOR) pathway', Epilepsia, Vol 52(3), e7–e11.

[329] Kimball, SR. et al (2004) 'Glucagon represses signaling through the mammalian target of rapamycin in rat liver by activating AMP-activated protein kinase', J Biol Chem, 279(52), p 54103-9.

[330] Watson, K. and Baar, K. (2014) 'mTOR and the health benefits of exercise', Seminars in Cell & Developmental Biology, Vol 36, p 130-139.

[331] Shimizu, H. et al (2010) 'Glucocorticoids increase NPY gene expression in the arcuate nucleus by inhibiting mTOR signaling in rat hypothalamic organotypic cultures', Peptides, Vol 31(1), p 145-9.

[332] Liu, X. et al (2014) 'Discrete mechanisms of mTOR and cell cycle regulation by AMPK agonists independent of AMPK', PNAS, Vol 111(4), E435-E444.

[333] Faivre, S. et al (2006), 'Current development of mTOR inhibitors as anticancer agents', Nature Reviews Drug Discovery volume 5, pages 671–688.

[334] Liu, M. et al (2010) 'Resveratrol inhibits mTOR signaling by promoting the interaction between mTOR and DEPTOR', J Biol Chem, Vol 285(47), p 36387-94.

[335] Beevers, CS. et al (2006) 'Curcumin inhibits the mammalian target of rapamycin-mediated signaling pathways in cancer cells', Int J Cancer, Vol 119(4), p757-64.

[336] Suarez-Arroyo, IJ. et al (2013) 'Anti-tumor effects of Ganoderma lucidum (reishi) in inflammatory breast cancer in in vivo and in vitro models', PLOS One, Vol 8(2), e57431.

[337] Liu, Z. et al (2012) 'Rhodiola rosea extracts and salidroside decrease the growth of bladder cancer cell lines via inhibition of the mTOR pathway and induction of autophagy', Mol Carcinog, Vol 51(3), p 257-67.

[338] Law, PC. et al (2012) 'Astragalus saponins downregulate vascular endothelial growth factor under cobalt chloride-stimulated hypoxia in colon cancer cells', The official journal of the International Society for Complementary Medicine Research (ISCMR)201212:160.

[339] Lee, YK. et al (2010) 'Anthocyanins are novel AMPKα1 stimulators that suppress tumor growth by inhibiting mTOR phosphorylation', Oncol Rep, Vol 24(6), p 1471-7.

[340] Banerjee, N. et al (2013) 'Pomegranate polyphenolics suppressed azoxymethane-induced colorectal aberrant crypt foci and inflammation: possible role of miR-126/VCAM-1 and miR-126/PI3K/AKT/mTOR', Carcinogenesis, Volume 34, Issue 12, 1 December 2013, Pages 2814–2822.

[341] Hong-Brown, LQ. et al (2013) 'Activation of AMPK/TSC2/PLD by alcohol regulates mTORC1 and mTORC2 assembly in C2C12 myocytes', Alcohol Clin Exp Res, Vol 37(11), p 1849-61.

[342] Zhou, R. et al (2011) 'Inhibition of mTOR signaling by oleanolic acid contributes to its anti-tumor activity in osteosarcoma cells', J Orthop Res, Vol 29(6), p 846-52.

[343] Zhang, Z. et al (2014) 'Carnosine Inhibits the Proliferation of Human Gastric Carcinoma Cells by Retarding Akt/mTOR/p70S6K Signaling', J Cancer, Vol 5(5), p 382-9.

[344] Brook, MS. et al (2015) 'Skeletal muscle homeostasis and plasticity in youth and ageing: impact of nutrition and exercise', Acta Physiologica, Vol 216(1), p 15-41.

[345] Brioche, T. et al (2016) 'Muscle wasting and aging: Experimental models, fatty infiltrations, and prevention', Molecular Aspects of Medicine, Vol 50, p 56-87.

[346] Wang, Y. et al (2010) 'Coordinate Control of Host Centrosome Position, Organelle Distribution, and Migratory Response by Toxoplasma gondii via Host mTORC2', JBC, Vol 285, p 15611-15618.

[347] Myamoto, S. et al (2008) 'Akt mediates mitochondrial protection in cardiomyocytes through phosphorylation of mitochondrial hexokinase-II', Cell Death Differ, Vol 15(3), p 521-9.

[348] Laplante, M. and Sabatini, D. (2012) 'mTOR Signaling in Growth Control and Disease', Cell, Vol 149(2), p 274-293.

[349] Betz, C. et al (2013) 'mTOR complex 2-Akt signaling at mitochondria-associated endoplasmic reticulum membranes (MAM) regulates mitochondrial physiology', PNAS, Vol 110(31), p 12526-12534.

[350] Kohn, AD. et al (1996) 'Expression of a constitutively active Akt Ser/Thr kinase in 3T3-L1 adipocytes stimulates glucose uptake and glucose transporter 4 translocation', J Biol Chem, Vol 271(49), p 31372-8.

[351] Hoeffer, CA. and Klann, E. (2010) 'mTOR signaling: At the crossroads of plasticity, memory and disease', Trends in Neuroscience, Vol 33(2), p 67-75.

[352] Kougias, DG. et al (2016) 'Beta-hydroxy-beta-methylbutyrate ameliorates aging effects in the dendritic tree of pyramidal neurons in the medial prefrontal cortex of both male and female rats', Neurobiology of Aging, Vol 40, p 78-85.

[353] Watson, K. and Baar, K. (2014) 'mTOR and the health benefits of exercise', Seminars in Cell & Developmental Biology, Vol 36, p 130-139.

[354] Ricoult, SJ. and Manning, BD. (2013) 'The multifaceted role of mTORC1 in the control of lipid metabolism', EMBO Rep, Vol 14(3), p 242-51.

[355] Sengupta, S. et al (2010) 'mTORC1 controls fasting-induced ketogenesis and its modulation by ageing', Nature, Vol 468(7327), p 1100-4.

[356] Higashi, Y. et al (2013) 'Insulin-Like Growth Factor-1 Regulates Glutathione Peroxidase Expression and Activity in Vascular Endothelial Cells: Implications for Atheroprotective Actions of Insulin-Like Growth Factor-1', Biochim Biophys Acta, Vol 1832(3), p 391-399.

[357] Higashi, Y. et al (2014) 'The Interaction Between IGF-1, Atherosclerosis and Vascular Aging', Front Horm Res, Vol 43, p 107-124.

[358] Ni, F. et al (2013) 'IGF-1 promotes the development and cytotoxic activity of human NK cells', Nat Commun, Vol 4, p 1479.

[359] Puche, JE. et al (2012) 'Human conditions of insulin-like growth factor-I (IGF-I) deficiency', J Transl Med, Vol 10, p 224.

[360] Yang, Y. and Yee, D. (2012) 'Targeting Insulin and Insulin-Like Growth Factor Signaling in Breast Cancer', Journal of Mammary Gland Biology and Neoplasia, Vol 17(3-4), p 251-261.

[361] Kenyon, C. et al (1993) 'A C. elegans mutant that lives twice as long as wild type', Nature volume 366, pages 461–464.

[362] Sattler, F. (2013) 'Growth hormone in the aging male', Clinical Endocrinology & Metabolism, Vol 27(4), p 541-555.

[363] Levine, M. et al (2014) 'Low Protein Intake Is Associated with a Major Reduction in IGF-1, Cancer, and Overall Mortality in the 65 and Younger but Not Older Population', Cell Metabolism, Vol 19(3), p 407-417.

[364] Puche, JE. and Castilla-Cortázar I. (2012) 'Human conditions of insulin-like growth factor-I (IGF-I) deficiency', J Transl Med, Vol 10, p 224.

[365] Burgers, AMG. et al (2011) 'Meta-Analysis and Dose-Response Metaregression: Circulating Insulin-Like Growth Factor I (IGF-I) and Mortality', JCEM, Vol 96(9), p 2912-2920.

[366] Barbier, M. et al (2009) 'Higher circulating levels of IGF-1 are associated with longer leukocyte telomere length in healthy subjects', Mech Ageing Dev, Vol 130(11-12), p 771-6.

[367] Lewis, ME. et al (1993) 'Insulin-like Growth Factor-I: Potential for Treatment of Motor Neuronal Disorders', Experimental Neurology, Vol 124(1), p 73-88.

[368] Lichtenwalner RJ, et al (2001) 'Intracerebroventricular infusion of insulin-like growth factor-I ameliorates the age-related decline in hippocampal neurogenesis', Neuroscience, Vol 107(4), p 603-13.

[369] Aleman, A. et al (1999) 'Insulin-like growth factor-I and cognitive function in healthy older men', J Clin Endocrinol Metab, Vol 84(2), p 471-5.

[370] Carro, E. et al (2002) 'Serum insulin-like growth factor I regulates brain amyloid-beta levels', Nat Med, Vol 8(12), p 1390-7.

[371] Straub, RH. (2014) 'Interaction of the endocrine system with inflammation: a function of energy and volume regulation', Arthritis Research & Therapy, Vol 16(1), p 203.

[372] Kim, T. et al (2015) 'Intense Walking Exercise Affects Serum IGF-1 and IGFBP3', J Lifestyle Med, Vol 5(1), p 21-25.

[373] El-Bahr SM (2013) 'Curcumin regulates gene expression of insulin like growth factor, B-cell CLL/lymphoma 2 and antioxidant enzymes in streptozotocin induced diabetic rats', BMC Complement Altern Med, Vol 13, p 368.

[374] Wang, LM. et al (2012) 'Luteolin inhibits proliferation induced by IGF-1 pathway dependent ERα in human breast cancer MCF-7 cells', Asian Pac J Cancer Prev, Vol 13(4), p 1431-7.

[375] Li, X. et al (2013) 'Epigallocatechin-3-gallate inhibits IGF-I-stimulated lung cancer angiogenesis through downregulation of HIF-1α and VEGF expression', J Nutrigenet Nutrigenomics, Vol 6(3), p 169-78.

[376] Lammintausta R. et al (1976) 'Change in hormones reflecting sympathetic activity in the Finnish sauna', Ann Clin Res, Vol 8(4), p 266-71.

[377] Burd NA et al. (2010) Low-load high volume resistance exercise stimulates muscle protein synthesis more than high-load low volume resistance exercise in young men, PLOS One, Vol 5(8), e12033.

[378] Häkkinen, K. et al (2003) 'Neuromuscular adaptations during concurrent strength and endurance training versus strength training', Eur J Appl Physiol, Vol 89(1), p 42-52.

[379] Schoenfeld BJ et al (2016) 'Effects of Resistance Training Frequency on Measures of Muscle Hypertrophy: A Systematic Review and Meta-Analysis', Sports Med, Vol 46(11), p 1689-1697.

[380] Schoenfeld BJ et al (2015) 'Influence of Resistance Training Frequency on Muscular Adaptations in Well-Trained Men', J Strength Cond Res, Vol 29(7), p 1821-9.

[381] Rafael, Z. et al (2018) 'High Resistance-Training Frequency Enhances Muscle Thickness in Resistance-Trained Men', The Journal of Strength & Conditioning.

[382] Cadore, EL. et al (2008) 'Hormonal responses to resistance exercise in long-term trained and untrained middle-aged men', J Strength Cond Res, Vol 22(5), p 1617-24.

[383] Ochi, E. et al (2018) 'Higher Training Frequency Is Important for Gaining Muscular Strength Under Volume-Matched Training', Frontiers in Physiology 02 July 2018.

[384] Hartman, MJ. et al (2007) 'Comparisons between twice-daily and once-daily training sessions in male weight lifters', Int J Sports Physiol Perform, Vol 2(2), p 159-69.

[385] Heymsfeld, S. et al (1982) Biochemical composition of muscle in normal and semistarved human subjects: Relevance to anthropometric measurements, American Journal of Clinical Nutrition (USA)

[386] Demling, RH. and DeSanti, L. (2000) 'Effect of a hypocaloric diet, increased protein intake and resistance training on lean mass gains and fat mass loss in overweight police officers', Ann Nutr Metab, Vol 44(1), p 21-9.

[387] Nindl, BC. et al (2000), 'Regional body composition changes in women after 6 months of periodized physical training', J Appl Physiol (1985), Vol 88(6), p 2251-9.

[388] Paoli, A. et al (2012) 'Ketogenic diet does not affect strength performance in elite artistic gymnasts', J Int Soc Sports Nutr, Vol 9, p 34.

[389] Garthe, I. et al (2011) 'Effect of two different weight-loss rates on body composition and strength and power-related performance in elite athletes', Int J Sport Nutr Exerc Metab, Vol 21(2), p 97-104.

[390] Pascoe DD, et al (1993), 'Glycogen resynthesis in skeletal muscle following resistive exercise', Med Sci Sports Exerc, Vol 25(3), p 349-54.

[391] Fournier, PA. et al (2004) 'Post-Exercise Muscle Glycogen Repletion in the Extreme: Effect of Food Absence and Active Recovery', J Sports Sci Med, Vol 3(3), p 139-146.

[392] Peters Futre, EM. et al (1987) 'Muscle glycogen repletion during active postexercise recovery', American Journal of Physiology, Vol 253(3), p E305-E311.

[393] Ho, KY. et al (1988) 'Fasting enhances growth hormone secretion and amplifies the complex rhythms of growth hormone secretion in man', J Clin Invest, Vol 81(4), p 968-975.

[394] Röjdmark S, (1989) 'Pituitary-testicular axis in obese men during short-term fasting', Acta Endocrinol (Copenh), Vol 121(5), p 727-32.

[395] Patterson, RE. et al (2015) 'INTERMITTENT FASTING AND HUMAN METABOLIC HEALTH', J Acad Nutr Diet, Vol 115(8), p 1203-1212.

[396] Masiero, E. et al (2009) 'Autophagy is required to maintain muscle mass', Cell Metabolism, Vol 10(6), p 507-15.

[397] Young, CM. et al (1971) 'Effect of body composition and other parameters in obese young men of carbohydrate level of reduction diet', Am J Clin Nutr, Vol 24(3), p 290-6.

[398] Kadowaki, M. et al (1996) 'Acute effect of epinephrine on muscle proteolysis in perfused rat hindquarters', Am J Physiol, Vol 270(6 Pt 1), p E961-7.

[399] Nair KS et al (1988) 'Effect of beta-hydroxybutyrate on whole-body leucine kinetics and fractional mixed skeletal muscle protein synthesis in humans', J Clin Invest, Vol 82(1), p 198-205.

[400] Layman DK et al (2005) 'Dietary protein and exercise have additive effects on body composition during weight loss in adult women', J Nutr, Vol 135(8), p 1903-10.

[401] Layman DK and Walker DA (2006) 'Potential importance of leucine in treatment of obesity and the metabolic syndrome', J Nutr, Vol 136(1 Suppl):319S-23S.

[402] Slater GJ and Jenkins D (2000) 'Beta-hydroxy-beta-methylbutyrate (HMB) supplementation and the promotion of muscle growth and strength', Sports Med, Vol 30(2), p 105-16.

[403] Harber MP et al (2005) 'Effects of dietary carbohydrate restriction with high protein intake on protein metabolism and the somatotropic axis', J Clin Endocrinol Metab, Vol 90(9), p 5175-81.

[404] Thissen, JP et al (1994) 'Nutritional regulation of the insulin-like growth factors', Endocr Rev, Vol 15(1), p 80-101.

[405] Guo, W. et al (2011) 'Sirt1 overexpression in neurons promotes neurite outgrowth and cell survival through inhibition of the mTOR signaling', J Neurosci Res, Vol 89(11), p 1723-36.

[406] Ghosh HS, McBurney M, Robbins PD. (2010) 'SIRT1 negatively regulates the mammalian target of rapamycin', PLoS One, 2010 Feb 15, Vol 5(2), e9199.

[407] Drummond MJ, Fry CS, Glynn EL, Dreyer HC, Dhanani S, Timmerman KL, Volpi E, Rasmussen BB. Rapamycin administration in humans blocks the contraction-induced increase in skeletal muscle protein synthesis. J Physiol. 2009 Apr 1;587(Pt 7):1535-46.

[408] Shimizu N, Yoshikawa N, Ito N, Maruyama T, Suzuki Y, Takeda S, Nakae J, Tagata Y, Nishitani S, Takehana K, Sano M, Fukuda K, Suematsu M, Morimoto C, Tanaka H.

Crosstalk between glucocorticoid receptor and nutritional sensor mTOR in skeletal muscle. Cell Metab. 2011 Feb 2;13(2):170-82.

[409] Muller AF, Janssen JA, Lamberts SW, Bidlingmaier M, Strasburger CJ, Hofland L, van der Lely AJ. Effects of fasting and pegvisomant on the GH-releasing hormone and GH-releasing peptide-6 stimulated growth hormone secretion. Clin Endocrinol (Oxf). 2001 Oct;55(4):461-7.

[410] Millward, DJ. 'The metabolic basis of amino acid requirements', Accessed 2018: http://archive.unu.edu/unupress/food2/UID07E/UID07E05.HTM

[411] Lemon, PW (2000) 'Beyond the zone: protein needs of active individuals', J Am Coll Nutr, Vol 19(5 Suppl), p 513S-521S.

[412] Anthony TG, McDaniel BJ, Knoll P, Bunpo P, Paul GL, McNurlan MA. J Nutr 2007;137:357-362.

[413] Storr, M et al (2003) 'Endogenous CCK depresses contractile activity within the ascending myenteric reflex pathway of rat ileum', Neuropharmacology, Vol 44(4), p 524-32.

[414] Chandra, R and Liddle RA (2007) 'Cholecystokinin', Curr Opin Endocrinol Diabetes Obes, Vol 14(1), p 63-7.

[415] Garaedts, MC. et al (2011) 'Direct induction of CCK and GLP-1 release from murine endocrine cells by intact dietary proteins', Mol Nutr Food Res, Vol 55(3), p 476-84.

[416] Metcalf AM, et al. (1987) 'Simplified assessment of segmental colonic transit', Gastroenterology, 92:40.

[417] Moore, DR. et al (2009) 'Ingested protein dose response of muscle and albumin protein synthesis after resistance exercise in young men', Am J Clin Nutr, Vol 89(1), p 161-8.

[418] Burd, NA. et al (2011) 'Enhanced Amino Acid Sensitivity of Myofibrillar Protein Synthesis Persists for up to 24 h after Resistance Exercise in Young Men', The Journal of Nutrition, Volume 141, Issue 4, 1 April 2011, Pages 568–573.

[419] Arnal, MA. et al (2000) 'Protein feeding pattern does not affect protein retention in young women', J Nutr, Vol 130(7), p 1700-4.

[420] Soeters, MR et al (2009) 'Intermittent fasting does not affect whole-body glucose, lipid, or protein metabolism', Am J Clin Nutr, Vol 90(5), p 1244-51.

[421] Varady, KA. (2011) 'Intermittent versus daily calorie restriction: which diet regimen is more effective for weight loss?', Obes Rev, Vol 12(7), e593-601.

[422] Stote KS et al (2007) 'A controlled trial of reduced meal frequency without caloric restriction in healthy, normal-weight, middle-aged adults', Am J Clin Nutr, Vol 85(4), p 981-8.

[423] Keogh, JB et al (2014) 'Effects of intermittent compared to continuous energy restriction on short-term weight loss and long-term weight loss maintenance', Clin Obes, Vol 4(3), p 150-6.

[424] Umpleby AM and Russell-Jones DL (1996) 'The hormonal control of protein metabolism', Baillieres Clin Endocrinol Metab, Vol 10(4), p 551-70.

[425] Hoffman, JR et al (2010) 'Effect of a proprietary protein supplement on recovery indices following resistance exercise in strength/power athletes', Amino Acids, Vol 38(3), p 771-8.

[426] Schoenfeld BJ, Aragon A. et al (2017) 'Pre- versus post-exercise protein intake has similar effects on muscular adaptations', Peer J, Vol 5, e2825.

[427] Horowitz, JF et al (1997) 'Lipolytic suppression following carbohydrate ingestion limits fat oxidation during exercise', Am J Physiol, 273(4 Pt 1):E768-75.

[428] Van Proeyen K et al (2010) 'Training in the fasted state improves glucose tolerance during fat-rich diet', J Physiol, Vol 588 (Pt 21), p 4289-302.

[429] Gjedsted J et al (2007) 'Effects of a 3-day fast on regional lipid and glucose metabolism in human skeletal muscle and adipose tissue', Acta Physiol (Oxf), Vol 191(3), p 205-16.

[430] Deldique L et al (2010) 'Increased p70s6k phosphorylation during intake of a protein-carbohydrate drink following resistance exercise in the fasted state', Eur J Appl Physiol, Vol 108(4), p 791-800.

[431] Ho, K. Y., Veldhuis, J. D., Johnson, M. L., Furlanetto, R., Evans, W. S., Alberti, K. G., & Thorner, M. O. (1988). Fasting enhances growth hormone secretion and amplifies the complex rhythms of growth hormone secretion in man. The Journal of clinical investigation, 81(4), 968-75.

[432] Trabelski, K. et al (2013) 'Effect of fed- versus fasted state resistance training during Ramadan on body composition and selected metabolic parameters in bodybuilders', J Int Soc Sports Nutr. 2013 Apr 25;10(1):23.

[433] Faye J, Fall A, Badji L, Cisse F, Stephan H, Tine P. Effects of Ramadan fast on weight, performance and glycemia during training for resistance]. Dakar Med. 2005;50(3):146-51.

[434] Bolster DR, Crozier SJ, Kimball SR, Jefferson LS. AMP-activated protein kinase suppresses protein synthesis in rat skeletal muscle through down-regulated mammalian target of rapamycin (mTOR) signaling. J Biol Chem. 2002 Jul 5;277(27):23977-80.

[435] Moscat J, Diaz-Meco MT. Feedback on fat: p62-mTORC1-autophagy connections. Cell. 2011 Nov 11;147(4):724-7. Review.

[436] Yoshizawa F, Kimball SR, Vary TC, Jefferson LS. Effect of dietary protein on translation initiation in rat skeletal muscle and liver. Am J Physiol. 1998 Nov;275(5 Pt 1):E814-20.

[437] Pitkanen HT et al (2003) 'Free amino acid pool and muscle protein balance after resistance exercise,' Med Sci Sports Exerc. 2003 May;35(5):784-92.

[438] Morton RW, Murphy KT, McKellar SR, et al A systematic review, meta-analysis and meta-regression of the effect of protein supplementation on resistance training-induced gains in muscle mass and strength in healthy adults Br J Sports Med 2018;52:376-384.

[439] Marques-Lopes I, Forga L, Martinez JA. Thermogenesis induced by a high-carbohydrate meal in fasted lean and overweight young men: insulin, body fat, and sympathetic nervous system involvement. 2003 Jan;19(1):25-9.

[440] Finn PF, Dice JF. Proteolytic and lipolytic responses to starvation. Nutrition. 2006 Jul-Aug;22(7-8):830-44. Review.

[441] Thompson JR, Wu G. The effect of ketone bodies on nitrogen metabolism in skeletal muscle. Comp Biochem Physiol B. 1991;100: 209–16.

[442] McCarty MF et al (2015) 'Ketosis may promote brain macroautophagy by activating Sirt1 and hypoxia-inducible factor-1', Med Hypotheses. 2015 Nov;85(5):631-9.

[443] Riechman SE (2007) 'Statins and dietary and serum cholesterol are associated with increased lean mass following resistance training', J Gerontol A Biol Sci Med Sci. 2007 Oct;62(10):1164-71.

[444] https://www.fasebj.org/doi/abs/10.1096/fasebj.25.1_supplement.lb563

[445] Hämäläinen EK, et al (1983) 'Decrease of serum total and free testosterone during a low-fat high-fibre diet', J Steroid Biochem. 1983 Mar;18(3):369-70.

[446] Keys A, et al (1963) 'CORONARY HEART DISEASE AMONG MINNESOTA BUSINESS AND PROFESSIONAL MEN FOLLOWED FIFTEEN YEARS', Circulation 1963 Sep;28:381-95.

[447] Ancel Keys (ed), *Seven Countries: A multivariate analysis of death and coronary heart disease*, 1980. Cambridge, Mass.: Harvard University Press.

[448] Steinberg D (1989). "The cholesterol controversy is over. Why did it take so long?". *Circulation*. **80**(4): 1070–1078.

[449] Katz LN, Keys A, Gofman JW (1952). "Atherosclerosis. A Symposium: Introduction"(PDF). *Circulation*. **5**: 98–100.

[450] Reiser Raymond (1973). "Saturated fat in the diet and serum cholesterol concentration: a critical examination of the literature" (PDF). *Am J Clin Nutr*. **26**: 524–555.

[451] Pett, Kahn, Willett, Katz (2017). "Ancel Keys and the Seven Countries Study: An Evidence-based Response to Revisionist Histories" (PDF). True Health Initiative.

[452] Nago N et al (2011) 'Low cholesterol is associated with mortality from stroke, heart disease, and cancer: the Jichi Medical School Cohort Study', J Epidemiol. 2011;21(1):67-74.

[453] Bae JM et al (2012) 'Low Cholesterol is Associated with Mortality from Cardiovascular Diseases: A Dynamic Cohort Study in Korean Adults', J Korean Med Sci. 2012 Jan; 27(1): 58–63.

[454] Parthasarathy S et al (2012) 'Oxidized Low-Density Lipoprotein', Methods Mol Biol. 2010; 610: 403–417.

[455] Vanderlaan PA et al (2009) 'VLDL best predicts aortic root atherosclerosis in LDL receptor deficient mice', J Lipid Res. 2009 Mar; 50(3): 376–385.

[456] National Health and Nutrition Examination Study, 'Intake of Calories and Selected Nutrients for the United States Population 1999-2000', Accessed: https://www.cdc.gov/nchs/data/nhanes/databriefs/calories.pdf

[457] Johnson, TD (2010) 'Mind your cholesterol for a healthy heart', The Nation's Health April 2010, 40 (3) 24.

[458] Dubois C et al (1994) 'Effects of increasing amounts of dietary cholesterol on postprandial lipemia and lipoproteins in human subjects', J Lipid Res. 1994 Nov;35(11):1993-2007.

[459] Lecerf, J., & De Lorgeril, M. (2011). Dietary cholesterol: From physiology to cardiovascular risk. British Journal of Nutrition, 106(1), 6-14.

[460] Westman EC et al (2006) 'Effect of a low-carbohydrate, ketogenic diet program compared to a low-fat diet on fasting lipoprotein subclasses', Int J Cardiol. 2006 Jun 16;110(2):212-6.

[461] Cordain L et al (2004) 'Optimal low-density lipoprotein is 50 to 70 mg/dl: lower is better and physiologically normal', J Am Coll Cardiol. 2004 Jun 2;43(11):2142-6.

[462] Djousse, L. and Gaziano JM (2009) 'Dietary cholesterol and coronary artery disease: a systematic review', Curr Atheroscler Rep. 2009 Nov;11(6):418-22.

[463] Griffin, JD and Lichenstein, AH (2013) 'Dietary Cholesterol and Plasma Lipoprotein Profiles: Randomized-Controlled Trials', Curr Nutr Rep. 2013 Dec; 2(4): 274–282.

[464] Lecerf, JM and de Lorgeril, M (2011) 'Dietary cholesterol: from physiology to cardiovascular risk', British Journal of Nutrition (2011), 106, 6–14.

[465] Caroll MD et al (2012) 'Total and high-density lipoprotein cholesterol in adults: National Health and Nutrition Examination Survey, 2009-2010'. NCHS Data Brief. 2012 Apr;(92):1-8.

[466] Nago N et al (2011) 'Low cholesterol is associated with mortality from stroke, heart disease, and cancer: the Jichi Medical School Cohort Study', J Epidemiol. 2011;21(1):67-74.

[467] Bueno, NB et al (2013) 'Very-low-carbohydrate ketogenic diet v. low-fat diet for long-term weight loss: a meta-analysis of randomised controlled trials', British Journal of Nutrition, Volume 110, Issue 7 14 October 2013 , pp. 1178-1187.

[468] Brinkworth GD et al (2009) 'Long-term effects of a very-low-carbohydrate weight loss diet compared with an isocaloric low-fat diet after 12 mo', Am J Clin Nutr. 2009 Jul;90(1):23-32.

[469] Dashti HM et al (2004) 'Long-term effects of a ketogenic diet in obese patients', Exp Clin Cardiol. 2004 Fall;9(3):200-5.

[470] *Nutrition Week* Mar 22, 1991 21:12:2-3

[471] Barnard N et al (2014) 'Saturated and trans fats and dementia: a systematic review', Neurobiology of Aging, Volume 35, Supplement 2, September 2014, Pages S65-S73.

[472] Pase CS et al (2013) 'Influence of perinatal trans fat on behavioral responses and brain oxidative status of adolescent rats acutely exposed to stress', Neuroscience. 2013 Sep 5;247:242-52.

[473] Harvard T.H. Chan, The Nutrition Source, 'Omega-3 Fatty Acids: An Essential Contribution', Accessed: https://www.hsph.harvard.edu/nutritionsource/what-should-you-eat/fats-and-cholesterol/types-of-fat/omega-3-fats/

[474] The Source for Objective Science-based DHA/EPA Omega-3 Information, DIFFERENTIATION OF ALA (PLANT SOURCES) FROM DHA + EPA (MARINE SOURCES) AS DIETARY OMEGA-3 FATTY ACIDS FOR HUMAN HEALTH,

Accessed: http://www.dhaomega3.org/Overview/Differentiation-of-ALA-plant-sources-from-DHA-+-EPA-marine-sources-as-Dietary-Omega-3-Fatty-Acids-for-Human-Health

[475] Garg, A (1998) 'High-monounsaturated-fat diets for patients with diabetes mellitus: a meta-analysis', Am J Clin Nutr. 1998 Mar;67(3 Suppl):577S-582S.

[476] Finucane, OM et al (2015) 'Monounsaturated fatty acid-enriched high-fat diets impede adipose NLRP3 inflammasome-mediated IL-1β secretion and insulin resistance despite obesity', Diabetes. 2015 Jun;64(6):2116-28.

[477] PM Kris-Etherton, Denise Shaffer Taylor, Shaomei Yu-Poth, Peter Huth, Kristin Moriarty, Valerie Fishell, Rebecca L Hargrove, Guixiang Zhao, Terry D Etherton; Polyunsaturated fatty acids in the food chain in the United States, The American Journal of Clinical Nutrition, Volume 71, Issue 1, 1 January 2000, Pages 179S–188S.

[478] Leaf A and Weber PC (1988) 'Cardiovascular effects of n-3 fatty acids', AGRIS, Accessed: http://agris.fao.org/agris-search/search.do?recordID=US8845581.

[479] Russo GL (2009) 'Dietary n-6 and n-3 polyunsaturated fatty acids: from biochemistry to clinical implications in cardiovascular prevention', Biochem Pharmacol. 2009 Mar 15;77(6):937-46.

[480] Hiza, HAB and Bente L (2007) 'Nutrient Content of the U.S. Food Supply, 1909-2004, A Summary Report', Center for Nutrition Policy and Promotion, Home Economics Research Report No. 57.

[481] Nair U et al (2007) 'Lipid peroxidation-induced DNA damage in cancer-prone inflammatory diseases: a review of published adduct types and levels in humans', Free Radic Biol Med. 2007 Oct 15;43(8):1109-20.

[482] Ghosh S et al (2007) 'Cardiac proinflammatory pathways are altered with different dietary n-6 linoleic to n-3 α-linolenic acid ratios in normal, fat-fed pigs', American Journal of Physiology-Heart and Circulatory Physiology 2007 293:5, H2919-H2927.

[483] US FDA, Trans Fat, Accessed 2018: https://www.fda.gov/food/ucm292278.htm

[484] Menendez JA et al (2005) 'Oleic acid, the main monounsaturated fatty acid of olive oil, suppresses Her-2/neu (erbB-2) expression and synergistically enhances the growth inhibitory effects of trastuzumab (Herceptin) in breast cancer cells with Her-2/neu oncogene amplification', Ann Oncol. 2005 Mar;16(3):359-71.

[485] Siri-Tarino PW et al (2010) 'Meta-analysis of prospective cohort studies evaluating the association of saturated fat with cardiovascular disease', Am J Clin Nutr. 2010 Mar;91(3):535-46.

[486] Wakai, K. et al 'Dietary intakes of fat and total mortality among Japanese populations with a low fat intake: the Japan Collaborative Cohort (JACC) Study', Nutr Metab (Lond). 2014; 11: 12.

[487] Price, Weston, DDS *Nutrition and Physical Degeneration*, 1945, Price Pottenger Nutrition Foundation, Inc., La Mesa, California

[488] Cohen, L A et al, *J Natl Cancer Inst* 1986 77:43.

[489] World Health Organization (2015) 'Q&A on the carcinogenicity of the consumption of red meat and processed meat', Accessed: http://www.who.int/features/qa/cancer-red-meat/en/

[490] Armstrong B and Doll R (1975) 'Environmental factors and cancer incidence and mortality in different countries, with special reference to dietary practices', Int J Cancer. 1975 Apr 15;15(4):617-31.

[491] Koeth RA et al (2013) 'Intestinal microbiota metabolism of L-carnitine, a nutrient in red meat, promotes atherosclerosis', Nat Med. 2013 May;19(5):576-85.

[492] Alexander DD et al (2011) 'Meta-analysis of prospective studies of red meat consumption and colorectal cancer', Eur J Cancer Prev. 2011 Jul;20(4):293-307.

[493] Micha R et al (2010) 'Red and Processed Meat Consumption and Risk of Incident Coronary Heart Disease, Stroke, and Diabetes Mellitus A Systematic Review and Meta-Analysis', Circulation. 2010;121:2271–2283.

[494] Micha, R., Wallace, S. K., & Mozaffarian, D. (2010). Red and processed meat consumption and risk of incident coronary heart disease, stroke, and diabetes mellitus: a systematic review and meta-analysis. Circulation, 121(21), 2271-83.

[495] Davey GK et al (2003) 'EPIC-Oxford: lifestyle characteristics and nutrient intakes in a cohort of 33 883 meat-eaters and 31 546 non meat-eaters in the UK', Public Health Nutr. 2003 May;6(3):259-69.

[496] LaFleur J, Nelson RE, Sauer BC, et al Overestimation of the effects of adherence on outcomes: a case study in healthy user bias and hypertension Heart 2011;97:1862-1869.

[497] Shrank, W.H., Patrick, A.R. & Alan Brookhart, M. J GEN INTERN MED (2011) 26: 546.

[498] Key TJ et al (1996) Dietary habits and mortality in 11,000 vegetarians and health conscious people: results of a 17 year follow up', BMJ. 1996 Sep 28;313(7060):775-9.

[499] Jonathan M. Hodgson, Natalie C. Ward, Valerie Burke, Lawrence J. Beilin, Ian B. Puddey; Increased Lean Red Meat Intake Does Not Elevate Markers of Oxidative Stress

and Inflammation in Humans, The Journal of Nutrition, Volume 137, Issue 2, 1 February 2007, Pages 363–367.

[500] Wang Z et al (2011) 'Gut flora metabolism of phosphatidylcholine promotes cardiovascular disease', Nature. 2011 Apr 7;472(7341):57-63.

[501] Zhang AQ et al (1999) 'Dietary precursors of trimethylamine in man: a pilot study', Food Chem Toxicol. 1999 May;37(5):515-20.

[502] Juraschek BA et al (2013) 'Effect of a High-Protein Diet on Kidney Function in Healthy Adults: Results From the OmniHeart Trial', American Journal of Kidney Diseases, Volume 61, Issue 4, April 2013, Pages 547-554.

[503] Szwarc S (2008) 'Does banning hotdogs and bacon make sense?', JUNKFOOD SCIENCE, Accessed: https://junkfoodscience.blogspot.com/2008/07/does-banning-hotdogs-and-bacon-make.html

[504] McKnight GM et al (1999) 'Dietary nitrate in man : friend or foe?', British journal of nutrition A. 1999, vol. 81, n° 5, pp. 349-358.

[505] Ward, MH et al (2011) 'Ingestion of Nitrate and Nitrite and Risk of Stomach Cancer in the NIH-AARP Diet and Health Study', Epidemiology: January 2011 - Volume 22 - Issue 1 - p S107-S108.

[506] Norman G Hord, Yaoping Tang, Nathan S Bryan; Food sources of nitrates and nitrites: the physiologic context for potential health benefits, The American Journal of Clinical Nutrition, Volume 90, Issue 1, 1 July 2009, Pages 1–10.

[507] Bogdan, C. (2001) 'Nitric oxide and the immune response', Nature Immunology volume 2, pages 907–916 (2001).

[508] Scrimshaw NS (1991) 'Iron deficiency', Sci Am. 1991 Oct;265(4):46-52.

[509] Xu K. et al (2014) 'mTOR signaling in tumorigenesis', BBA, Volume 1846, Issue 2, December 2014, Pages 638-654.

[510] Siri-Tarino PW et al (2010) 'Meta-analysis of prospective cohort studies evaluating the association of saturated fat with cardiovascular disease', Am J Clin Nutr. 2010 Mar;91(3):535-46.

[511] Chowdhury R et al (2014) 'Association of dietary, circulating, and supplement fatty acids with coronary risk: a systematic review and meta-analysis', Ann Intern Med. 2014 Mar 18;160(6):398-406.

[512] Hu, T., Mills, K. T., Yao, L., Demanelis, K., Eloustaz, M., Yancy, W. S., Kelly, T. N., He, J., … Bazzano, L. A. (2012). Effects of low-carbohydrate diets versus low-fat diets on

metabolic risk factors: a meta-analysis of randomized controlled clinical trials. American journal of epidemiology, 176 Suppl 7(Suppl 7), S44-54.

[513] Brillat-Savarin, Jean Anthelme (1970). *The Physiology of Taste*. trans. Anne Drayton. Penguin Books. pp. 208–209.

[514] Banting, W (1864) 'Letter on corpulence : addressed to the public', London, 1869, p. 21.

[515] Taubes, G. (2002) 'What if It's All Been a Big Fat Lie?', The New York Times Magazine, Accessed: https://www.nytimes.com/2002/07/07/magazine/what-if-it-s-all-been-a-big-fat-lie.html

[516] Sir William, O (1920) 'The principles and practice of medicine', New York, London, D. Appleton and company, Accessed: https://archive.org/details/principlesandpr00mccrgoog/page/n6

[517] The Diabetes Control And Complications Trial Research Group, (2001) 'Influence of Intensive Diabetes Treatment on Body Weight and Composition of Adults With Type 1 Diabetes in the Diabetes Control and Complications Trial', Diabetes Care. 2001 Oct; 24(10): 1711–1721.

[518] Henry, RR et al (1993) 'Intensive Conventional Insulin Therapy for Type II Diabetes: Metabolic effects during a 6-mo outpatient trial', Diabetes Care 1993 Jan; 16(1): 21-31.

[519] Jung CH and Kim MS (2013) 'Molecular mechanisms of central leptin resistance in obesity', Arch Pharm Res. 2013 Feb;36(2):201-7.

[520] Knight ZA, Hannan KS, Greenberg ML, Friedman JM (2010) Hyperleptinemia Is Required for the Development of Leptin Resistance. PLoS ONE 5(6): e11376.

[521] Wang, G. (2010) 'Singularity analysis of the AKT signaling pathway reveals connections between cancer and metabolic diseases', Physical Biology, Volume 7, Number 4.

[522] Modan, M., Halkin, H., Almog, S., Lusky, A., Eshkol, A., Shefi, M., Shitrit, A., ... Fuchs, Z. (1985). Hyperinsulinemia. A link between hypertension obesity and glucose intolerance. The Journal of clinical investigation, 75(3), 809-17.

[523] Michael H et al (2008) 'Insulin Resistance and Hyperinsulinemia', Diabetes Care Feb 2008, 31 (Supplement 2) S262-S268.

[524] Rizza, R.A., Mandarino, L.J., Genest, J. et al. Diabetologia (1985) 28: 70.

[525] Del Prato S et al (1994) 'Effect of sustained physiologic hyperinsulinaemia and hyperglycaemia on insulin secretion and insulin sensitivity in man', Diabetologia. 1994 Oct;37(10):1025-35.

[526] Wang, G (2014) 'Raison d'être of insulin resistance: the adjustable threshold hypothesis', Journal of the Royal Society, Vol 11(101).

[527] Unger RH et al (2012) 'Gluttony, sloth and the metabolic syndrome: a roadmap to lipotoxicity', Trends in Endocrinology and Metabolism 21 (2010) 345–352.

[528] Kraegen, EW et al (1991) 'Development of Muscle Insulin Resistance After Liver Insulin Resistance in High-Fat–Fed Rats', Diabetes Nov 1991, 40 (11) 1397-1403.

[529] Clément L et al (2002) 'Dietary trans-10,cis-12 conjugated linoleic acid induces hyperinsulinemia and fatty liver in the mouse', J Lipid Res. 2002 Sep;43(9):1400-9.

[530] Isganaitis E and Lustig R.H. (2005) 'Fast Food, Central Nervous System Insulin Resistance, and Obesity', Arteriosclerosis, Thrombosis, and Vascular Biology. 2005;25:2451–2462.

[531] Storlien LH et al (1991) 'Influence of Dietary Fat Composition on Development of Insulin Resistance in Rats: Relationship to Muscle Triglyceride and ω-3 Fatty Acids in Muscle Phospholipid', Diabetes 1991 Feb; 40(2): 280-289.

[532] Basciano H et al (2005) 'Fructose, insulin resistance, and metabolic dyslipidemia', Nutrition & Metabolism20052:5.

[533] Bergman, R. N., Kim, S. P., Catalano, K. J., Hsu, I. R., Chiu, J. D., Kabir, M., Hucking, K. and Ader, M. (2006), Why Visceral Fat is Bad: Mechanisms of the Metabolic Syndrome. Obesity, 14: 16S-19S.

[534] Accurso A et al (2008) 'Dietary carbohydrate restriction in type 2 diabetes mellitus and metabolic syndrome: time for a critical appraisal', Nutr Metab (Lond). 2008 Apr 8;5:9.

[535] Boden G, Sargrad K, Homko C, Mozzoli M, Stein TP. Effect of a Low-Carbohydrate Diet on Appetite, Blood Glucose Levels, and Insulin Resistance in Obese Patients with Type 2 Diabetes. Ann Intern Med. ;142:403–411.

[536] Volek JS et al (2004) 'Comparison of a very low-carbohydrate and low-fat diet on fasting lipids, LDL subclasses, insulin resistance, and postprandial lipemic responses in overweight women', J Am Coll Nutr. 2004 Apr;23(2):177-84.

[537] https://docs.google.com/spreadsheets/d/1Ucfpvs2CmKFnae9a8zTZS0Zt1g2tdYSIQBFcohfa1w0/edit#gid=547985667

[538] Newman, John C., and Eric Verdin. "Ketone Bodies as Signaling Metabolites." *Trends in endocrinology and metabolism: TEM* 25.1 (2014): 42–52.

[539] Jornayvaz FR et al (2010) 'A high-fat, ketogenic diet causes hepatic insulin resistance in mice, despite increasing energy expenditure and preventing weight gain', American Journal of Physiology-Endocrinology and Metabolism, Vol. 299, No. 5.

[540] Balkau, B., Mhamdi, L., Oppert, J. M., Nolan, J., Golay, A., Porcellati, F., Laakso, M., Ferrannini, E., EGIR-RISC Study Group (2008). Physical activity and insulin sensitivity: the RISC study. Diabetes, 57(10), 2613-8.

[541] Lund, S., Holman, G. D., Schmitz, O., & Pedersen, O. (1995). Contraction stimulates translocation of glucose transporter GLUT4 in skeletal muscle through a mechanism distinct from that of insulin. Proceedings of the National Academy of Sciences of the United States of America, 92(13), 5817-21.

[542] Attvall S et al (1993) 'Smoking induces insulin resistance--a potential link with the insulin resistance syndrome', J Intern Med. 1993 Apr;233(4):327-32.

[543] Shiloah E et al (2003) 'Effect of Acute Psychotic Stress in Nondiabetic Subjects on β-Cell Function and Insulin Sensitivity', Diabetes Care 2003 May; 26(5): 1462-1467.

[544] Piroli GG et al (2007) 'Corticosterone Impairs Insulin-Stimulated Translocation of GLUT4 in the Rat Hippocampus', Neuroendocrinology 2007;85:71–80.

[545] Paul-Labrador M. et al (2006) 'Effects of a randomized controlled trial of transcendental meditation on components of the metabolic syndrome in subjects with coronary heart disease', Arch Intern Med. 2006 Jun 12;166(11):1218-24.

[546] Ken C Chiu, Audrey Chu, Vay Liang W Go, Mohammed F Saad; Hypovitaminosis D is associated with insulin resistance and β cell dysfunction, The American Journal of Clinical Nutrition, Volume 79, Issue 5, 1 May 2004, Pages 820–825.

[547] Huang YJ et al (1997) 'Amelioration of insulin resistance and hypertension in a fructose-fed rat model with fish oil supplementation', Metabolism Clinical and Experimental, November 1997Volume 46, Issue 11, Pages 1252–1258.

[548] Shoelson, S. E., Lee, J., & Goldfine, A. B. (2006). Inflammation and insulin resistance. The Journal of clinical investigation, 116(7), 1793-801.

[549] Alessandra Puddu and Giorgio L. Viviani, " Advanced Glycation Endproducts and Diabetes. Beyond Vascular Complications", Endocrine, Metabolic & Immune Disorders - Drug Targets (2011) 11: 132.

[550] Frazier, TH et al (2011) 'Gut microbiota, intestinal permeability, obesity-induced inflammation, and liver injury', JPEN J Parenter Enteral Nutr. 2011 Sep;35(5 Suppl):14S-20S.

[551] Aeberli I et al (2011) 'Low to moderate sugar-sweetened beverage consumption impairs glucose and lipid metabolism and promotes inflammation in healthy young men: a randomized controlled trial', Am J Clin Nutr. 2011 Aug;94(2):479-85.

[552] Joseph SV et al (2016) 'Fruit Polyphenols: A Review of Anti-inflammatory Effects in Humans', Crit Rev Food Sci Nutr. 2016;56(3):419-44.

[553] Sugimura T et al (2004) 'Heterocyclic amines: Mutagens/carcinogens produced during cooking of meat and fish', Cancer Science, Vol 95(4), p 290-299.

[554] Skog et al (1998) 'Carcinogenic Heterocyclic Amines in Model Systems and Cooked Foods: A Review on Formation, Occurrence and Intake', Food and Chemical Toxicology, Volume 36, Issues 9–10, September–October 1998, Pages 879-896.

[555] Philips DH (1999) 'Polycyclic aromatic hydrocarbons in the diet', Mutation Research/Genetic Toxicology and Environmental Mutagenesis, Volume 443, Issues 1–2, 15 July 1999, Pages 139-147.

[556] Uribarri J et al (2010) 'Advanced Glycation End Products in Foods and a Practical Guide to Their Reduction in the Diet', Journal of the American Dietetic Association, Volume 110, Issue 6, June 2010, Pages 911-916.e12.

[557] Šebeková, K., Krajčovičová-Kudláčková, M., Schinzel, R. et al. Eur J Nutr (2001) 40: 275.

[558] Hipkiss A (1998) 'Carnosine, a protective, anti-ageing peptide?', The International Journal of Biochemistry & Cell Biology, Volume 30, Issue 8, August 1998, Pages 863-868.

[559] Song DH and Wolfe MM (2007) 'Glucose-dependent insulinotropic polypeptide and its role in obesity', Curr Opin Endocrinol Diabetes Obes. 2007 Feb;14(1):46-51.

[560] Cheung IN et al (2016) 'Morning and Evening Blue-Enriched Light Exposure Alters Metabolic Function in Normal Weight Adults', PLoS One. 2016 May 18;11(5):e0155601.

[561] Tuso, P et al (2015) 'A Plant-Based Diet, Atherogenesis, and Coronary Artery Disease Prevention', Perm J. 2015 Winter; 19(1): 62–67.

[562] Dart, Raymond, Adventures With the Missing Link, New York, Viking Press, p. 255, (1969).

[563] Schaller, George B and Gordon Lowther, The Relevance of Carnivore Behavior to the Study of Early Hominids, *Southwest J Anthro*, 25:307-41, 1969.

[564] Aiello, Leslie C., and Peter Wheeler. "The Expensive-Tissue Hypothesis: The Brain and the Digestive System in Human and Primate Evolution." Current Anthropology 36, no. 2 (1995): 199-221.

[565] Braidwood, Robert J., *Prehistoric Men*, 8th edition, Glenville, Ill., Scott, Foresman and Co., pp. 52-113, (1975).

[566] Cohen, Mark Nathan, *The Food Crisis in Pre-History*, New Haven, Yale University Press, p. 15, (1977).

[567] Martin, Paul S., Pleistocene Overkill, *Natural History*, 76:32-38, 1967.

[568] Loren Cordain, Janette Brand Miller, S Boyd Eaton, Neil Mann, Susanne HA Holt, John D Speth; Plant-animal subsistence ratios and macronutrient energy estimations in worldwide hunter-gatherer diets, The American Journal of Clinical Nutrition, Volume 71, Issue 3, 1 March 2000, Pages 682–692

[569] Lindeberg S et al (1994) 'Cardiovascular risk factors in a Melanesian population apparently free from stroke and ischaemic heart disease: the Kitava study', J Intern Med. 1994 Sep;236(3):331-40.

[570] Gundry, S (2017) 'The Plant Paradox: The Hidden Dangers in "Healthy" Foods That Cause Disease and Weight Gain', Harper Wave; 1 edition.

[571] J X Chipponi, J C Bleier, M T Santi, D Rudman; Deficiencies of essential and conditionally essential nutrients, The American Journal of Clinical Nutrition, Volume 35, Issue 5, 1 May 1982, Pages 1112–1116.

[572] Sanders, T. A. (2010), The role of fat in the diet – quantity, quality and sustainability. Nutrition Bulletin, 35: 138-146.

[573] EFSA Panel on Dietetic Products, Nutrition and Allergies (NDA); Scientific Opinion related to the Tolerable Upper Intake Level of eicosapentaenoic acid (EPA), docosahexaenoic acid (DHA) and docosapentaenoic acid (DPA). EFSA Journal 2012;10(7):2815.

[574] Ottoboni F. et al (2005) 'Ascorbic Acid and the Immune System', Journal of Orthomolecular Medicine, Vol. 20, No. 3, 2005.

[575] Martinez-Cordero C et al (2012) 'Testing the Protein Leverage Hypothesis in a free-living human population', Appetite. 2012 Oct;59(2):312-5.

[576] Holt SH et al (1995) 'A satiety index of common foods', Eur J Clin Nutr. 1995 Sep;49(9):675-90.

[577] Pontzer H et al (2012) 'Hunter-Gatherer Energetics and Human Obesity', PLOS One.

[578] Smits SA et al (2017) 'Seasonal cycling in the gut microbiome of the Hadza hunter-gatherers of Tanzania', SCIENCE25 AUG 2017 : 802-806.

[579] Phinney SD et al (1983) 'The human metabolic response to chronic ketosis without caloric restriction: preservation of submaximal exercise capability with reduced carbohydrate oxidation', Metabolism. 1983 Aug;32(8):769-76.

[580] Clemente FJ et al (2014) 'A Selective Sweep on a Deleterious Mutation in CPT1A in Arctic Populations', Am J Hum Genet. 2014 Nov 6;95(5):584-589.

[581] Achten, J. et al (2002) 'Determination of the exercise intensity that elicits maximal fat oxidation', Medicine & Science in Sports & Exercise. 34(1):92-97.

[582] L. S. Sidossis et al (1997) 'Regulation of plasma fatty acid oxidation during low- and high-intensity exercise', American Journal of Physiology-Endocrinology and Metabolism 1997 272:6, E1065-E1070.

[583] E. F. Coyle et al (1997) 'Fatty acid oxidation is directly regulated by carbohydrate metabolism during exercise', American Journal of Physiology-Endocrinology and Metabolism 1997 273:2, E268-E275.

[584] Volek JS. et al (2016) 'Metabolic characteristics of keto-adapted ultra-endurance runners', Metabolism, Vol 65(3), p 100-110.

[585] Phinney, SD. et al (1983) 'The human metabolic response to chronic ketosis without caloric restriction: Preservation of submaximal exercise capability with reduced carbohydrate oxidation', Metabolism, Vol 32(8), p 769-776.

[586] McKenzie E. et al (2005) 'Recovery of Muscle Glycogen Concentrations in Sled Dogs during Prolonged Exercise', Medicine & Science in Sports & Exercise. 37(8):1307-1312.

[587] https://avmajournals.avma.org/doi/abs/10.2460/ajvr.69.8.1097

[588] Walberg JL et al (1988) 'Exercise capacity and nitrogen loss during a high or low carbohydrate diet', Med Sci Sports Exerc. 1988 Feb;20(1):34-43.

[589] Van Hall G et al (2002) 'Human skeletal muscle fatty acid and glycerol metabolism during rest, exercise and recovery', J Physiol. 2002 Sep 15; 543(Pt 3): 1047–1058.

[590] Peters SJ and LeBlanc PJ (2004) 'Metabolic aspects of low carbohydrate diets and exercise', Nutr Metab (Lond). 2004; 1: 7.

[591] Kephart, WC et al (2017) 'The Three-Month Effects of a Ketogenic Diet on Body Composition, Blood Parameters, and Performance Metrics in CrossFit Trainees: A Pilot Study, Sports 2018, 6(1), 1.

[592]Wilson JM and Lowery RP et al (2017) 'The Effects of Ketogenic Dieting on Body Composition, Strength, Power, and Hormonal Profiles in Resistance Training Males', J Strength Cond Res. 2017 Apr 7.

[593] Rauch JT et al (2014) 'The effects of ketogenic dieting on skeletal muscle and fat mass', Journal of the International Society of Sports Nutrition201411 (Suppl 1) :P40.

[594] Paoli, A. et al (2012) 'Ketogenic diet does not affect strength performance in elite artistic gymnasts', J Int Soc Sports Nutr. 2012 Jul 26;9(1):34.

[595] Pietrocola F et al (1987) 'Coffee induces autophagy in vivo', Cell Cycle. 2014;13(12):1987-94.

[596] Zhou J et al (2014) 'Epigallocatechin-3-gallate (EGCG), a green tea polyphenol, stimulates hepatic autophagy and lipid clearance', PLoS One. 2014 Jan 29;9(1):e87161.

[597] Xiao K et al (2013) 'Curcumin induces autophagy via activating the AMPK signaling pathway in lung adenocarcinoma cells', J Pharmacol Sci. 2013;123(2):102-9.

[598] Ouyang DY et al (2013) 'Piperine inhibits the proliferation of human prostate cancer cells via induction of cell cycle arrest and autophagy', Food Chem Toxicol. 2013 Oct;60:424-30.

[599] Hung JY et al (2009) '6-Shogaol, an active constituent of dietary ginger, induces autophagy by inhibiting the AKT/mTOR pathway in human non-small cell lung cancer A549 cells', J Agric Food Chem. 2009 Oct 28;57(20):9809-16.

[600] Lin YT et al (2017) 'Capsaicin Induces Autophagy and Apoptosis in Human Nasopharyngeal Carcinoma Cells by Downregulating the PI3K/AKT/mTOR Pathway', Int J Mol Sci. 2017 Jun 23;18(7). pii: E1343.

[601] Fu Y et al (2014) 'Resveratrol inhibits breast cancer stem-like cells and induces autophagy via suppressing Wnt/β-catenin signaling pathway', PLoS One. 2014 Jul 28;9(7):e102535.

[602] Fan X et al (2015) 'Berberine alleviates ox-LDL induced inflammatory factors by up-regulation of autophagy via AMPK/mTOR signaling pathway', Journal of Translational Medicine, 13:92.

[603] Ci, X., Zhou, J., Lv, H., Yu, Q., Peng, L., & Hua, S. (2017). Betulin exhibits anti-inflammatory activity in LPS-stimulated macrophages and endotoxin-shocked mice through an AMPK/AKT/Nrf2-dependent mechanism. Cell death & disease, 8(5), e2798.

[604] Pietrocola F, et al. (2016) 'Caloric Restriction Mimetics Enhance Anticancer Immunosurveillance', Cancer Cell, 2016 Jul 11;30(1):147-160.

[605] Brosnan JT et al (2007) 'Methionine: A metabolically unique amino acid', Livestock Science, Volume 112, Issues 1–2, October 2007, Pages 2-7.

[606] López-Torres M and Barja G (2008) 'Lowered methionine ingestion as responsible for the decrease in rodent mitochondrial oxidative stress in protein and dietary restriction possible implications for humans', Biochim Biophys Acta. 2008 Nov;1780(11):1337-47.

[607] Sanchez-Roman I. and Barja G (2013) 'Regulation of longevity and oxidative stress by nutritional interventions: Role of methionine restriction', Experimental Gerontology, Volume 48, Issue 10, October 2013, Pages 1030-1042.

[608] Miller, R. A., Buehner, G., Chang, Y., Harper, J. M., Sigler, R. and Smith-Wheelock, M. (2005), Methionine-deficient diet extends mouse lifespan, slows immune and lens aging, alters glucose, T4, IGF-I and insulin levels, and increases hepatocyte MIF levels and stress resistance. Aging Cell, 4: 119-125.

[609] https://www.fasebj.org/doi/abs/10.1096/fasebj.25.1_supplement.528.2

[610] Anthony TG, et al J Nutr 2007;137:357-362.

[611] DiNicolantonio, JJ et al (2018) 'Subclinical magnesium deficiency: a principal driver of cardiovascular disease and a public health crisis', Open Heart 2018;5.

[612] Lai PL et al (2015) 'Neurotrophic properties of the Lion's mane medicinal mushroom, Hericium erinaceus (Higher Basidiomycetes) from Malaysia', Int J Med Mushrooms. 2013;15(6):539-54.

[613] Mori K et al (2009) 'Improving effects of the mushroom Yamabushitake (Hericium erinaceus) on mild cognitive impairment: a double-blind placebo-controlled clinical trial', Phytother Res. 2009 Mar;23(3):367-72.

[614] Gao Y et al (2003) 'Effects of ganopoly (a Ganoderma lucidum polysaccharide extract) on the immune functions in advanced-stage cancer patients', Immunol Invest. 2003 Aug;32(3):201-15.

[615] Bao PP et al (2012) 'Ginseng and Ganoderma lucidum use after breast cancer diagnosis and quality of life: a report from the Shanghai Breast Cancer Survival Study', PLoS One. 2012;7(6):e39343.

[616] Finimundy, T., Dillon, A., Henriques, J. and Ely, M. (2014) A Review on General Nutritional Compounds and Pharmacological Properties of the Lentinula edodes Mushroom. Food and Nutrition Sciences, 5, 1095-1105.

[617] Wu YZ et al (2008) 'Ginkgo biloba extract improves coronary artery circulation in patients with coronary artery disease: contribution of plasma nitric oxide and endothelin-1', Phytother Res. 2008 Jun;22(6):734-9.

[618] Rungapamestry V et al (2007) 'Effect of meal composition and cooking duration on the fate of sulforaphane following consumption of broccoli by healthy human subjects', Br J Nutr. 2007 Apr;97(4):644-52.

[619] Ali SS et al (2004) 'A biologically effective fullerene (C60) derivative with superoxide dismutase mimetic properties', Free Radic Biol Med. 2004 Oct 15;37(8):1191-202.

[620] Galvan YP et al (2017) 'Fullerenes as Anti-Aging Antioxidants', Curr Aging Sci. 2017;10(1):56-67.

[621] Kato S et al (2009) 'Biological safety of liposome-fullerene consisting of hydrogenated lecithin, glycine soja sterols, and fullerene-C60 upon photocytotoxicity and bacterial reverse mutagenicity', Toxicol Ind Health. 2009 Apr;25(3):197-203.

[622] Singh, P., Mishra, S. K., Noel, S., Sharma, S., & Rath, S. K. (2012). Acute exposure of apigenin induces hepatotoxicity in Swiss mice. PloS one, 7(2), e31964.

[623] Stefek M and Karasu C (2011) 'Eye Lens in Aging and Diabetes: Effect of Quercetin', Rejuvenation ResearchVol. 14, No. 5.

[624] Eisenberg T et al (2016) 'Cardioprotection and lifespan extension by the natural polyamine spermidine', Nature Medicine volume 22, pages 1428–1438.

[625] Kumazawa T et al (1995) 'Levels of pyrroloquinoline quinone in various foods', Biochem J. 1995 Apr 15;307 (Pt 2):331-3.

[626] Watanabe A et al (1989) 'Nephrotoxicity of pyrroloquinoline quinone in rats', Hiroshima J Med Sci. 1989 Mar;38(1):49-51.

[627] McCormack D and McFadden D (2013) 'A review of pterostilbene antioxidant activity and disease modification', Oxid Med Cell Longev. 2013;2013:575482.

[628] Xu YF (2016) 'Effect of Polysaccharide from Cordyceps militaris (Ascomycetes) on Physical Fatigue Induced by Forced Swimming', Int J Med Mushrooms. 2016;18(12):1083-1092.

[629] Wilson JM and Lowery RP et al (2013) 'β-Hydroxy-β-methylbutyrate free acid reduces markers of exercise-induced muscle damage and improves recovery in resistance-trained men', Br J Nutr. 2013 Aug 28;110(3):538-44.

[630] Wilkinson DJ et al (2013) 'Effects of leucine and its metabolite β-hydroxy-β-methylbutyrate on human skeletal muscle protein metabolism', J Physiol. 2013 Jun 1;591(11):2911-23.

[631] Madison LL et al (1964) 'TheHypoglycemicActionofKetones', Journal of Clinical Investigation, Vol.43, No.3, 1964.

[632] Biden, T. J., & Taylor, K. W. (1983). Effects of ketone bodies on insulin release and islet-cell metabolism in the rat. The Biochemical journal, 212(2), 371-7.

[633] Finn PF and Dice JF (2005) 'Ketone bodies stimulate chaperone-mediated autophagy', J Biol Chem. 2005 Jul 8;280(27):25864-70.

[634] McCarty MF et al (2015) 'Ketosis may promote brain macroautophagy by activating Sirt1 and hypoxia-inducible factor-1', Med Hypotheses. 2015 Nov;85(5):631-9.

[635] Lee WY et al (2015) 'Phellinus linteus extract induces autophagy and synergizes with 5-fluorouracil to inhibit breast cancer cell growth', Nutr Cancer. 2015;67(2):275-84.

[636] Sinha RA et al (2014) 'Caffeine stimulates hepatic lipid metabolism by the autophagy-lysosomal pathway in mice', Hepatology. 2014 Apr;59(4):1366-80.

[637] H E Carlson, J H Shah; Aspartame and its constituent amino acids: effects on prolactin, cortisol, growth hormone, insulin, and glucose in normal humans, The American Journal of Clinical Nutrition, Volume 49, Issue 3, 1 March 1989, Pages 427–432.

[638] Wolf-Novak LC et al (1990) 'Aspartame ingestion with and without carbohydrate in phenylketonuric and normal subjects: effect on plasma concentrations of amino acids, glucose, and insulin', Metabolism. 1990 Apr;39(4):391-6.

[639] Møller SE (1991) 'Effect of aspartame and protein, administered in phenylalanine-equivalent doses, on plasma neutral amino acids, aspartate, insulin and glucose in man', Pharmacol Toxicol. 1991 May;68(5):408-12.

[640] Just T et al (2008) 'Cephalic phase insulin release in healthy humans after taste stimulation?', Appetite. 2008 Nov;51(3):622-7.

[641] Horwitz DL et al (1988) 'Response to single dose of aspartame or saccharin by NIDDM patients', Diabetes Care. 1988 Mar;11(3):230-4.

[642] Liang Y et al (1987) 'The effect of artificial sweetener on insulin secretion. 1. The effect of acesulfame K on insulin secretion in the rat (studies in vivo)', Horm Metab Res. 1987 Jun;19(6):233-8.

[643] Ma J et al (2009) 'Effect of the artificial sweetener, sucralose, on gastric emptying and incretin hormone release in healthy subjects', Am J Physiol Gastrointest Liver Physiol. 2009 Apr;296(4):G735-9.

[644] Anton, S. D., Martin, C. K., Han, H., Coulon, S., Cefalu, W. T., Geiselman, P., & Williamson, D. A. (2010). Effects of stevia, aspartame, and sucrose on food intake, satiety, and postprandial glucose and insulin levels. Appetite, 55(1), 37-43.

[645] Renwick, A., & Molinary, S. (2010). Sweet-taste receptors, low-energy sweeteners, glucose absorption and insulin release. British Journal of Nutrition, 104(10), 1415-1420.

[646] Mourier A et al (1997) 'Combined effects of caloric restriction and branched-chain amino acid supplementation on body composition and exercise performance in elite wrestlers', Int J Sports Med. 1997 Jan;18(1):47-55.

[647] Nørrelund H et al (2003) 'The decisive role of free fatty acids for protein conservation during fasting in humans with and without growth hormone', J Clin Endocrinol Metab. 2003 Sep;88(9):4371-8.

[648] Zhang Y et al (2011) 'Effects of branched-chain amino acid supplementation on plasma concentrations of free amino acids, insulin, and energy substrates in young men', J Nutr Sci Vitaminol (Tokyo). 2011;57(1):114-7.

[649] Jungas RL et al (1992) 'Quantitative analysis of amino acid oxidation and related gluconeogenesis in humans', Physiol Rev. 1992 Apr;72(2):419-48.

[650] Schliess, F. et al (2006), 'Cell hydration and mTOR-dependent signalling', Acta Physiologica, 187: 223-229.

[651] Röjdmark S et al (1989) 'Pituitary-testicular axis in obese men during short-term fasting', Acta Endocrinol (Copenh). 1989 Nov;121(5):727-32.

[652] Patterson, R. E., Laughlin, G. A., LaCroix, A. Z., Hartman, S. J., Natarajan, L., Senger, C. M., Martínez, M. E., Villaseñor, A., Sears, D. D., Marinac, C. R., ... Gallo, L. C. (2015). Intermittent Fasting and Human Metabolic Health. Journal of the Academy of Nutrition and Dietetics, 115(8), 1203-12.

[653] de Mairan JJO (1729). "Observation Botanique". *Histoire de l'Academie Royale des Sciences*: 35–36.

[654] Bruce, Victor G.; Pittendrigh, Colin S. (1957). "Endogenous Rhythms in Insects and Microorganisms". *The American Naturalist*. **91** (858): 179–195.

[655] Halberg, F. (1959). "Physiologic 24-hour periodicity: general and procedural considerations with reference to the adrenal cycle". *Zeitschrift für Vitamin- Hormone- und Fermentforschung*. **10**: 225–296.

[656] Halberg F, Carandente F, Cornelissen G, Katinas GS (1977). "[Glossary of chronobiology (author's transl)]". *Chronobiologia*. 4 Suppl 1: 1–189.

[657] Hall JC, Rosbash, M, and Young MW. (2017) The Nobel Prize in Physiology or Medicine 2017, Accessed; https://www.nobelprize.org/prizes/medicine/2017/press-release/

[658] Nagoshi E et al (2004) 'Circadian Gene Expression in Individual Fibroblasts', Cell, VOLUME 119, ISSUE 5, P693-705.

[659] Duffy, J. F., & Wright, K. P. (2005). Entrainment of the Human Circadian System by Light. Journal of Biological Rhythms, 20(4), 326–338.

[660] Cromie, William (1999-07-15). "Human Biological Clock Set Back an Hour". *Harvard Gazette*.

[661] Cheung IN et al (2016) 'Morning and Evening Blue-Enriched Light Exposure Alters Metabolic Function in Normal Weight Adults', PLoS One. 2016 May 18;11(5):e0155601.

[662] Garcia-Saenz, A. et al (2018) 'Evaluating the Association between Artificial Light-at-Night Exposure and Breast and Prostate Cancer Risk in Spain (MCC-Spain Study)', Environmental Health Perspectives, 126 (04).

[663] Longo, V. D., & Panda, S. (2016). Fasting, Circadian Rhythms, and Time-Restricted Feeding in Healthy Lifespan. Cell metabolism, 23(6), 1048-1059.

[664] Gill, S., & Panda, S. (2015). A Smartphone App Reveals Erratic Diurnal Eating Patterns in Humans that Can Be Modulated for Health Benefits. Cell metabolism, 22(5), 789-98.

[665] Haturi M et al (2012) 'Time-Restricted Feeding without Reducing Caloric Intake Prevents Metabolic Diseases in Mice Fed a High-Fat Diet', Cell Metabolism, VOLUME 15, ISSUE 6, P848-860.

[666] Scheer FA et al (2013) 'The internal circadian clock increases hunger and appetite in the evening independent of food intake and other behaviors', Obesity (Silver Spring). 2013 Mar;21(3):421-3.

[667] Eve Van Cauter, Kenneth S. Polonsky, André J. Scheen; Roles of Circadian Rhythmicity and Sleep in Human Glucose Regulation, Endocrine Reviews, Volume 18, Issue 5, 1 October 1997, Pages 716–738.

[668] Andreani, T. S., Itoh, T. Q., Yildirim, E., Hwangbo, D. S., & Allada, R. (2015). Genetics of Circadian Rhythms. Sleep medicine clinics, 10(4), 413-21.

[669] Mulder H et al (2009) 'Melatonin receptors in pancreatic islets: good morning to a novel type 2 diabetes gene', Diabetologia. 2009 Jul;52(7):1240-9.

[670] Lindseth, G., Lindseth, P., & Thompson, M. (2013). Nutritional Effects on Sleep. Western Journal of Nursing Research, 35(4), 497–513.

[671] Ritchie, K. et al (2007) 'The neuroprotective effects of caffeine', Neurology August 7, 2007 vol. 69 no. 6 536-545

[672] Coffee consumption and risk of type 2 diabetes mellitus. The Lancet. Volume 360, No. 9344, p1477–1478, 9 November 2002

[673] Caffeine reduces the risk of Alzheimer's disease. https://biosingularity.wordpress.com/2006/09/15/caffeine-reduces-the-rish-of-alzheimers-disease/

[674] Association of Coffee and Caffeine Intake With the Risk of Parkinson Disease. JAMA. 2000;283(20):2674-2679

[675] Adenosine, Adenosine Receptors and the Actions of Caffeine. BCPT. Volume 76, Issue 2

[676] Modified-release hydrocortisone to provide circadian cortisol profiles. J Clin Endocrinol Metab. 2009 May;94(5):1548-54.

[677] Serum caffeine half-lives. Healthy subjects vs. patients having alcoholic hepatic disease. Am J Clin Pathol. 1980 Mar;73(3):390-3.

[678] Caffeine: A User's Guide to Getting Optimally Wired - http://scienceblogs.com/developingintelligence/2008/02/11/optimally-wired-a-caffeine-use/

[679] Low-dose repeated caffeine administration for circadian-phase-dependent performance degradation during extended wakefulness. Sleep. 2004 May 1;27(3):374-81.

[680] Human sleep, sleep loss and behaviour. Implications for the prefrontal cortex and psychiatric disorder. Br J Psychiatry. 1993 Mar;162:413-9.

[681] Influence of caffeine and carbohydrate feedings on endurance performance. Med Sci Sports. 1979 Spring;11(1):6-11.

[682] A high carbohydrate diet negates the metabolic effects of caffeine during exercise. Med Sci Sports Exerc. 1987 Apr;19(2):100-5.

[683] Post-Exercise Caffeine Helps Muscles Refuel - http://www.the-aps.org/mm/hp/audiences/public-press/archive/08/24.html

[684] Cognitive and physiological effects of an "energy drink": an evaluation of the whole drink and of glucose, caffeine and herbal flavouring fractions. Psychopharmacology. November 2004, Volume 176, Issue 3, pp 320–330

[685] Inhibitory effect of grapefruit juice and its bitter principal, naringenin, on CYP1A2 dependent metabolism of caffeine in man. Br J Clin Pharmacol. 1993 Apr;35(4):431-6.

Printed by Amazon Italia Logistica S.r.l.
Torrazza Piemonte (TO), Italy